Handbook of
Lipids in Human Nutrition

Edited by
Gene A. Spiller

CRC Press
Boca Raton   London   New York   Washington, D.C.

Cover design: Chemical structure of β–cyclodextrin. (From Oakenfull, D.G., Pearce, R.J., and Sidhu, G.S., *Aust. J. Dairy Technol.*, *46, 110*, 1991. With permission.)

Library of Congress Cataloging-in-Publication Data

Handbook of lipids in human nutrition / edited by Gene A. Spiller.
    p. cm.
    Includes bibliographical references and index.
    ISBN 0-8493-4248-1 (hard : alk. paper)
    1. Lipids in human nutrition—Handbooks, manuals, etc.
I. Spiller, Gene A.
    [DNLM: 1. Lipids—adverse effects. 2. Lipids—metabolism.
3. Nutrition. 4. Food.   QU 85 H2364 1995]
QP751.H334 1995
612.3'97—dc20
DNLM/DLC
                                  95-17919
                                    CIP

This book contains information obtained from authentic and highly regarded sources. Reprinted material is quoted with permission, and sources are indicated. A wide variety of references are listed. Reasonable efforts have been made to publish reliable data and information, but the author and the publisher cannot assume responsibility for the validity of all materials or for the consequences of their use.

No claim to original U.S. Government works
International Standard Book Number 0-8493-4248-1
Library of Congress Card Number 95-17919
Printed in the United States of America   3 4 5 6 7 8 9 0
Printed on acid-free paper

DEDICATION

To Drs. Denis Burkitt and Hugh Trowell, who have given me a unique perception of the correlation of health and disease with food, and to Drs. John Farquhar and David Jenkins, who always inspire me with their work on the relation of diet to chronic diseases.

ACKNOWLEDGMENT

The author wishes to acknowledge Monica Alton Spiller and Rosemary Schmele for their assistance in various phases of the editing process and in coordinating the final manuscript.

PREFACE

The field of lipids in human nutrition is so vast that to attempt to cover it all in a single handbook would be impossible. With this in mind, I have chosen to limit the choice of topics, rather than to edit a 4000- or 5000-page multivolume handbook. I hope not only to have gathered chapters that will give us the basic facts or techniques needed in lipid research, but to have succeeded in reminding the medical and nutrition community that we must always consider the complex interactions of nutrients to arrive at sound recommendations in disease prevention. The effects of fiber, plant sterols, certain saponins present in foods, and the interaction of antioxidants need to be considered very carefully in the health-disease processes. The reader will find many chapters with challenging ideas. An example is the chapter on plant sterols that describes how lowering the fat content of the diet to an extreme point or using highly refined oils may also decrease the intake of these beneficial compounds.

Has the ultimate study on the long-range effects of food lipids been published? Of course not. It is probably impossible to carry out the ultimate study on the correlation of nutrition to chronic diseases under present conditions: we must accept the pieces of evidence derived from good epidemiology and from controlled human and animal studies. The lifetimes of many of us would be needed to satisfy the purists who hope for the ultimate study, in this or any other field in which we deal with the lifetime of a human being. Meanwhile, in the field of effect of lipids in foods on health, the risk of becoming too *reductionist* and of trying to focus on isolated compounds rather than whole foods can lead to results that can be very confusing not just to the scientific mind, but to the general public. With all this in mind, I have edited this handbook to open the way to more challenging thinking on lipids and health.

Sections 1 and 2 cover chemistry, analytical methodologies, and definitions and effects on health of food lipids as related to blood lipids and carcinogenesis. These topics needed to be gathered in a single volume for quick, up-to-date reference for both researchers and students. Section 3 covers the effects of selected whole foods, and food components other than lipids or their biological byproducts, such as short chain fatty acids, on the health effects of food fatty acids and cholesterol. It was impossible to cover all the possible foods and their effects, so I have chosen some ancient foods such as nuts and olive oil. Section 3 also covers antioxidants, fiber, and a biological product of fiber, short chain fatty acids. The Appendices gather basic data on fatty acid, cholesterol, and plant sterol composition of foods; some data on olive oil, which I have again chosen as a very ancient part of the human diet deserving special attention; and, finally, a chapter on removing cholesterol from animal foods.

THE EDITOR

Gene Alan Spiller, D.Sc., Ph.D., is the director of the Health Research and Studies Center and of the SPHERA Foundation in Los Altos, California. He is the editor of many clinical nutrition books.

Dr. Spiller received his doctorate in chemistry from the University of Milan (Italy), and a Master's degree and a Ph.D. in nutrition from the University of California at Berkeley. He did additional studies at the Stanford University School of Medicine at Stanford, California. He is a member of many professional nutrition societies.

In the 1970s, Dr. Spiller was in charge of the Nutritional Physiology Section of Syntex Research in Palo Alto, California, where he did extensive human and animal research. At the same time he edited many clinical nutrition books. He continued his work in clinical nutrition research and publishing in the 1980s, first as an independent consultant and later as the director of the Health Research and Studies Center and of the SPHERA Foundation in Los Altos, California. Many human clinical studies, reviews, and other publications were the results of this work. Dr. Spiller has carried out clinical studies on the effect of complex whole foods and has focused on fiber, lipids such as monounsaturated fats and foods high in lipids such as nuts, antioxidants, and other beneficial nutrients. In addition, Dr. Spiller has been a lecturer in nutrition in the San Francisco Bay Area at Mills College and Foothill College.

Two of his latest multiauthor books are *The Mediterranean Diets in Health and Disease* (Van Nostrant, 1991) and the *CRC Handbook of Dietary Fiber in Human Nutrition 2nd Edition* (CRC Press, 1993).

CONTRIBUTORS

Fabio Armellini, M.D.
Cattedra di Nutririone Clinica
Ospedale Policlinico
Università di Verona
Verona, Italy

Heather E. T. Bell
Graduate Assistant
Department of Nutrition
Loma Linda University
School of Public Health
Loma Linda, California

Luisa Bissoli
Institute of Internal Medicine
University of Verona
Verona, Italy

Ottavio Bosello, M.D.
Cattedra di Nutririone Clinica
Ospedale Policlinico
Università di Verona
Verona, Italy

M. Campagnola
Istituto di Semeiotica e
  Nefrologia Medica
Università di Verona
Ospedale Policlinico
Verona, Italy

L. Cominacini
Istituto di Semeiotica e
  Nefrologia Medica
Università di Verona
Ospedale Policlinico
Verona, Italy

A. Davoli
Istituto di Semeiotica e
  Nefrologia Medica
Università di Verona
Ospedale Policlinico
Verona, Italy

A. De Santis
Istituto di Semeiotica e
  Nefrologia Medica
Università di Verona
Ospedale Policlinico
Verona, Italy

A. Fratta Pasini
Istituto di Semeiotica e
  Nefrologia Medica
Università di Verona
Ospedale Policlinico
Verona, Italy

John W. Farquhar, M.D.
Director
Stanford Center for Research in
  Disease Prevention
Professor of Medicine
Stanford University School of Medicine
Stanford, California

Gary E. Fraser, M.D., Ch.B., Ph.D.
Professor of Medicine
Departments of Epidemiology and
  Biostatistics
Director, Center for Health Research
Loma Linda University
Loma Linda, California

U. Garbin
Istituto di Semeiotica e
  Nefrologia Medica
Università di Verona
Ospedale Policlinico
Verona, Italy

Maria Hassapidou
Professor
School of Food Technology and
  Human Nutrition
Technological Education Institution (TEI)
Sindos, Thessaloniki, Greece

Debra Geary Hook, M.P.H.
Graduate Assistant
Department of Nutrition
Loma Linda University
School of Public Health
Loma Linda, California

Thomas J. Hudson
Group Leader
Shaklee Corp.
Hayward, California

Alexandra L. Jenkins, B.Sc., R.D.
Department of Nutritional Sciences
Faculty of Medicine
University of Toronto
Toronto, Ontario, Canada

David J. A. Jenkins, M.D., Ph.D.
Professor
Department of Nutritional Sciences
Faculty of Medicine
University of Toronto
Toronto, Ontario, Canada

Antony Kafatos
Professor
Department of Social Medicine
School of Medicine
University of Crete
Greece

Irena B. King, Ph.D.
Fred Hutchinson Cancer Research Center
Seattle, Washington

Apostolos (Paul) Kiritsakis, B.Sc.,
  M.Sc., Ph.D.
Professor
School of Food Technology and Human
  Nutrition
Technological Education Institution (TEI)
Sindos, Thessaloniki, Greece

Elizabeth Teng Leary, Ph.D.
Pacific Biometrics, Inc.
Seattle, Washington

V. Lo Cascio
Istituto di Semeiotica e
  Nefrologia Medica
Università di Verona
Ospedale Policlinico
Verona, Italy

Ruth McPherson, M.D., Ph.D.
University of Ottawa Heart Institute
Ottawa Civic Hospital
Ottawa, Ontario, Canada

David Oakenfull, B.Sc., Ph.D.
Senior Principal Research Scientist
Division of Food Processing
CSIRO Food Research Laboratory
North Ryde, New South Wales, Australia

A.M. Pastorino
Istituto di Semeiotica e
  Nefrologia Medica
Università di Verona
Ospedale Policlinico
Verona, Italy

Cynthia G. Rainey-Macdonald, B.Sc.
Department of Nutritional Sciences
Faculty of Medicine
University of Toronto
Toronto, Ontario, Canada

Bandaru S. Reddy, DVM, Ph.D.
Chief, Division of Nutritional
  Carcinogenesis
American Health Foundation
Valhalla, New York

Joan Sabaté, M.D., Dr.P.H.
Associate Professor
Departments of Nutrition and
  Epidemiology and Biostatistics
School of Public Health
Loma Linda University
Loma Linda, California

Artemis P. Simopoulos, M.D.
President
The Center for Genetics,
  Nutrition and Health
Washington, D.C.

Peter J. Spadafora, M.Sc.
Department of Nutritional Sciences
Faculty of Medicine
University of Toronto
Toronto, Ontario, Canada

Gene A. Spiller, D. Sc., Ph.D.
Director
Health Research and Studies Center, Inc.
SPHERA Foundation
Los Altos, California

Tiziana Todesco, M.D.
Cattedra di Nutrizione Clinica
Ospedale Policlinico
Università di Verona
Verona, Italy

Emanuela Turcato
Cattedra di Nutririone Clinica
Ospedale Policlinico
Verona, Italy

G. Russell Warnick, M.S., M.B.A.
Pacific Biometrics, Inc.
Seattle, Washington

Mauro Zamboni, M.D.
Cattedra di Nutririone Clinica
Ospedale Policlinico
Università di Verona
Verona, Italy

TABLE OF CONTENTS

Section 1. Chemistry, Nomenclature, and Analyses

Section 2. Effect of Food Lipids on Health

TABLES

Section 1: Chemistry, Nomenclature, and Analyses

Chapter 1.1

LIPIDS IN FOODS: CHEMISTRY AND NOMENCLATURE

Irena B. King

INTRODUCTION

Lipids are chemically diverse compounds which can be extracted from animal, plant, and microbial sources with a variety of methods. There is no universal definition of the term lipid. Lipids are usually described, broadly, as those compounds which are insoluble in water and soluble in selected organic solvents such as chloroform, hexane, benzene, diethyl ether, or methanol.[1] Christie[2] recommends a more constricted definition that closely reflects the origins of the term, i.e., "lipids are fatty acids and their derivatives, and substances related biosynthetically or functionally to these compounds". In foods, especially those high in fat, lipids are predominantly triesters of fatty acids with glycerol as a main derivative. The purpose of this discussion is to briefly review the nomenclature, structure, and classification of the major lipids normally found in foods.

NOMENCLATURE

The Commission on Biochemical Nomenclature (CBN) formed by the International Union of Pure and Applied Chemistry (IUPAC) and the International Union of Biochemistry (IUB) has established appropriate nomenclature for lipid terminology.[3] However, in addition to the systematic names, lipid nomenclature often includes a shorthand notation and a wide range of trivial names which are firmly established in the literature of lipid methodology.

The systematic names of fatty acids are based on the CBN nomenclature and indicate the carbon length, position, and configuration of the double bond (Table 1).

CLASSIFICATION

Lipids are most clearly classified into simple, complex, and derived lipids (Table 2). The simple lipids have a fatty acid ester or ether linkage with alcohols. The complex lipids contain other groups in addition to fatty acids with simple derivatives. The derived lipids include compounds obtained by hydrolysis from simple and complex lipids.

Fatty Acids

Although fatty acids are major building blocks of simple and complex lipids, they occur naturally only in trace amounts in this unesterified form and are usually obtained by hydrolysis of the backbone derivatives (glycerols, sterols, or fatty alcohols). There are more than 1000 different kinds of fatty acids identified; fortunately, only a small fraction of these are important from the perspective of food analysis. All fatty acids have a hydrocarbon chain and a carboxyl terminal. Fatty acids differ from each other primarily in chain length and in the number, position, and configuration of the double bonds (Figure 1).

Fatty acids can be either saturated or unsaturated. The saturated fatty acids do not contain any double bonds, while the unsaturated ones have at least one double bond. Fatty acids that have only one double bond are known as monounsaturated; those that have two or more bonds are polyunsaturated. The double bonds can be designated from carboxyl or methyl ends. The presence of the double bond allows for configurational isomerism in cis or trans position. Naturally occurring fatty acids are mostly in the cis configuration.

0-8493-4248-1/96/$0.00+$.50
© 1996 by CRC Press Inc.

TABLE 1
Typical Fatty Acids in Foods

Systematic name	Shorthand notation	Trivial name	Major sources
Saturated			
Tetranoic	C4:0	Butyric	Butter
Hexanoic	6:0	Caproic	Butter
Octanoic	8:0	Caprylic	Coconut
Dodecanoic	12:0	Lauric	Palm kernel, coconut
Tetradecanoic	14:0	Myristic	Palm kernel, coconut
Hexadecanoic	16:0	Palmitic	Palm
Octadecanoic	18:0	Stearic	Most animal fats, cocoa
Eicosanoic	20:0	Arachidic	Peanut
Docosanoic	22:0	Behenic	Seeds
Tetracosanoic	24:0	Lignoceric	Peanut
Monounsaturated			
Cis			
9-Tetradecanoic	14:1n5	Myristoleic	Butter
9-Hexadecanoic	16:1n7	Palmitoleic	Seafood, beef
9-Octadecanoic	18:1n9	Oleic	Olive, canola
11-Octadecanoic	18:1n7	Vaccenic	Seafood
13-Docosenoic	22:1n9	Erucic	Rapeseed
Trans			
9-Octadecanoic	t-18:1n9	Elaidic	Hydrogenated fats
11-Octadecanoic	t-18:1n7	Transvaccenic	Hydrogenated fats, butter
Polyunsaturated			
All *cis*			
9,12-Octadienoic	18:2n6	Linoleic	Sunflower, safflower
6,9,12-Octadecatrienoic	18:3n6	γ-Linolenic	Primrose
8,11,14-Eicosatrienoic	20:3n6	Dihomo-γ-linolenic	Shark liver[a]
5,8,11,14-Eicosatetraenoic	20:4n6	Arachidonic	Eggs, most animal fats
9,12,15-Octatrienoic	18:3n3	Linolenic	Soybean, canola
5,8,11,14,17-Eicosapentaenoic	20:5n3	Timnodonic	Seafood
7,10,13,16,19-Docosapentaenoic	22:5n3	Clupadonic	Seafood
4,7,10,13,16,19-Docosahexaenoic	22:6n3	Cervonic	Seafood

[a] From Perkins.[1]

Acylglycerols

Acylglycerols are esters of fatty acids and glycerol as shown in Figure 2. Although they are often called glycerides or neutral lipids, neither term is recommended by the standard nomenclature rules. One, two, or three fatty acids may be attached to the glycerol backbone, resulting in mono-, di-, or triacylglycerols, respectively. Triacylglycerols are the most abundant in foods. Although the three acyl groups may be identical in nature, the fatty acids are usually mixed. The kind and location of fatty acids on the glycerol molecule influence their biological response.

TABLE 2
Lipid Classification

Simple lipids
    Acylglycerols
    Ether acylglycerols
    Sterols and sterol esters
    Wax esters
Complex lipids
    Glycerophospholipids
    Glyceroglycolipids
    Sphingolipids
    Plasmalogens
Derived lipids
    Fatty acids
    Alcohols (glycerols, sterols, fatty alcohols)

Ether Acylglycerols

These lipids contain fatty acids esterified to two of the hydroxyl groups of glycerol; the remaining hydroxyl group is joined in ether linkage with a long alkyl or alkenyl chain (Figure 3). These lipids are difficult to separate from the triacylglycerols (mild hydrolysis to remove FA and yield glyceryl ethers). Alkyl ethers are not abundant in most foods, but occur in larger amounts in some marine sources.

Sterols and Sterol Esters

Sterols (Figure 4) belong to a large subgroup of steroids, consisting predominantly of cholesterol in animals and sitosterol in plants. Sterol esters contain various long chain fatty acids esterified to the C3-hydroxyl group of the steroid ring system.

Waxes

These lipids (Figure 5) contain esters of fatty acids with long-chain monohydroxylic fatty alcohols or with sterols. Waxes function as protective coatings on leaves, fruits, and skin.

Glycerophospholipids (Phosphoglycerides)

Phosphoglycerides are derivatives of glyceryl phosphoric acid. They have the general structure shown in Figure 6 where X may represent choline, ethanolamine, serine, or inositol.

Glyceroglycolipids (Glycosylacylglycerols)

All plant and bacterial sources contain appreciable amounts of lipids where 1,2-diacyl-sn-glycerols are joined by a glycosidic linkage at position sn-3 to a carbohydrate moiety (Figure 7).

Sphingolipids

Sphingolipids (Figure 8) contain sphingosine or a related derivative as the backbone plus a fatty acid and a polar head group. These lipids consist of ceremides, sphingomyelins, and cerebrosites. Sphingolipids are found in the membranes of some plant and animal cells.

Plasmalogens

Plasmalogens (Figure 9) are complex ether phosphoglycerides. The ether-linked alkyl chain contains a double bond between C1 and C2; it is an alk-1-enyl ether or a vinyl ether, with a double bond in the cis configuration. Ethanolamine is the most common base in the plasmalogen. Small amounts of plasmalogens are found in the membranes of muscle cells, plants, and fungi.

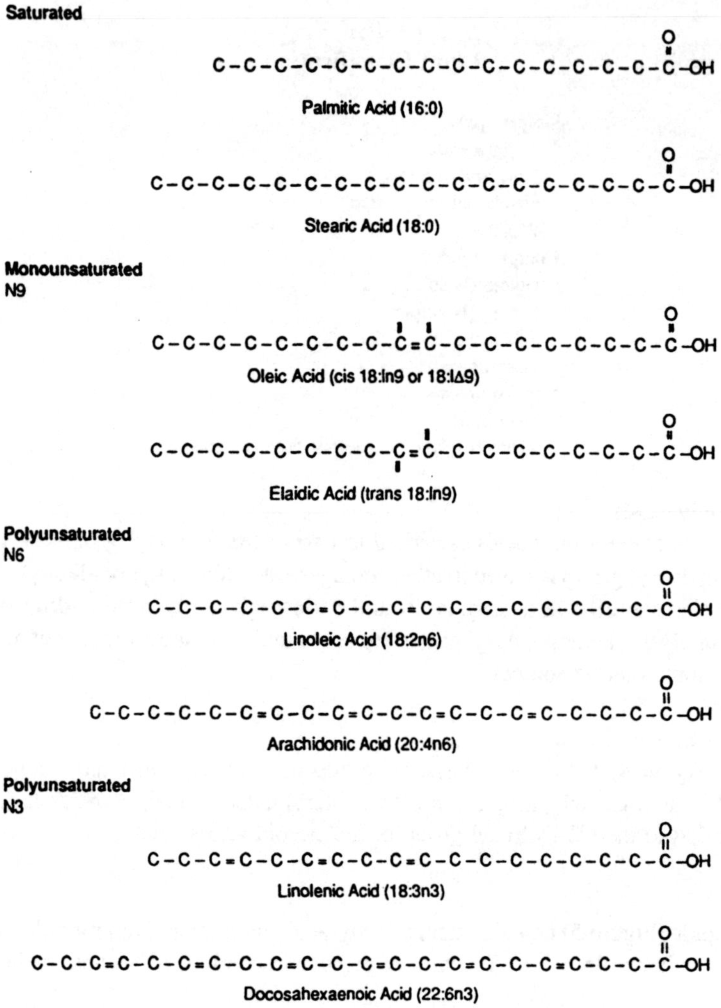

Saturated

$$C-C-C-C-C-C-C-C-C-C-C-C-C-C-C-\overset{\overset{O}{\|}}{C}-OH$$

Palmitic Acid (16:0)

$$C-C-C-C-C-C-C-C-C-C-C-C-C-C-C-C-C-\overset{\overset{O}{\|}}{C}-OH$$

Stearic Acid (18:0)

Monounsaturated
N9

$$C-C-C-C-C-C-C-C-C-C=C-C-C-C-C-C-C-C-\overset{\overset{O}{\|}}{C}-OH$$

Oleic Acid (cis 18:1n9 or 18:1Δ9)

$$C-C-C-C-C-C-C-C-C-C=C-C-C-C-C-C-C-C-\overset{\overset{O}{\|}}{C}-OH$$

Elaidic Acid (trans 18:1n9)

Polyunsaturated
N6

$$C-C-C-C-C-C=C-C-C=C-C-C-C-C-C-C-C-C-\overset{\overset{O}{\|}}{C}-OH$$

Linoleic Acid (18:2n6)

$$C-C-C-C-C-C=C-C-C=C-C-C=C-C-C=C-C-C-C-C-\overset{\overset{O}{\|}}{C}-OH$$

Arachidonic Acid (20:4n6)

Polyunsaturated
N3

$$C-C-C=C-C-C=C-C-C=C-C-C-C-C-C-C-C-C-\overset{\overset{O}{\|}}{C}-OH$$

Linolenic Acid (18:3n3)

$$C-C-C=C-C-C=C-C-C=C-C-C=C-C-C=C-C-C-C-\overset{\overset{O}{\|}}{C}-OH$$

Docosahexaenoic Acid (22:6n3)

FIGURE 1.   Structures of common fatty acids in foods.

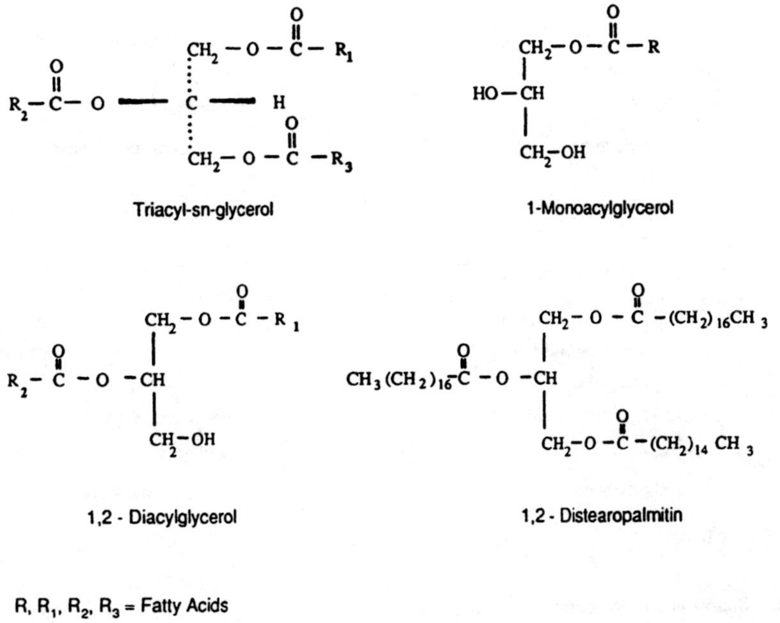

Triacyl-sn-glycerol

1-Monoacylglycerol

1,2 - Diacylglycerol

1,2 - Distearopalmitin

R, R_1, R_2, R_3 = Fatty Acids

FIGURE 2. Acylglycerols.

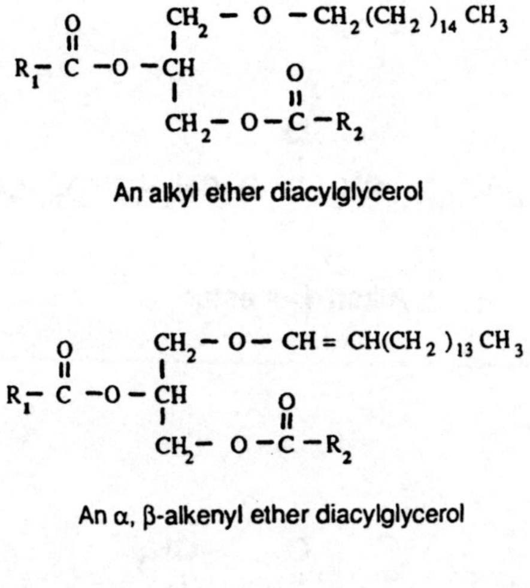

An alkyl ether diacylglycerol

An α, β-alkenyl ether diacylglycerol

FIGURE 3. Ether acylglycerol.

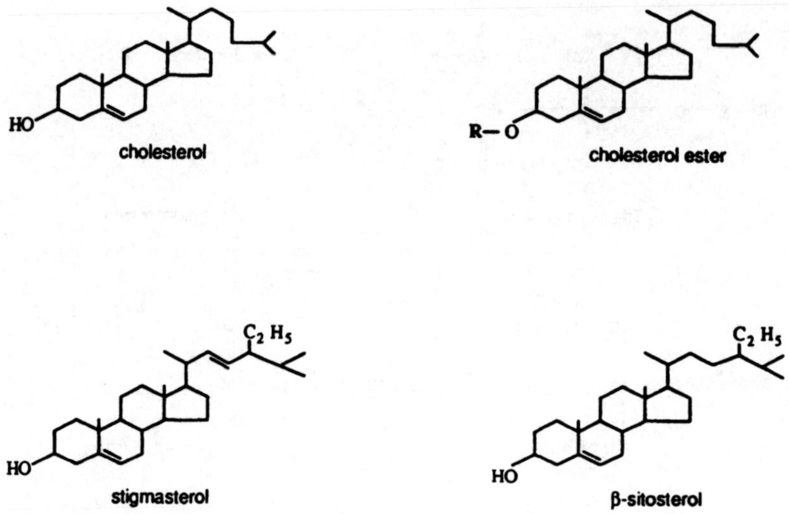

R = Fatty Acid

FIGURE 4. Sterols and sterol esters.

$$CH_3-(CH_2)_{15}-CH_2-\overset{\overset{\large O}{||}}{C}-O-CH_2-(CH_2)_{26}-CH_3$$

Alkan-1-ol ester

$$CH_3-(CH_2)_{15}-CH_2-\overset{\overset{\large O}{||}}{C}-O-\overset{\overset{\large CH_3}{|}}{C}H-(CH_2)_{12}-CH_3$$

Alkan-2-ol ester

FIGURE 5. Wax esters.

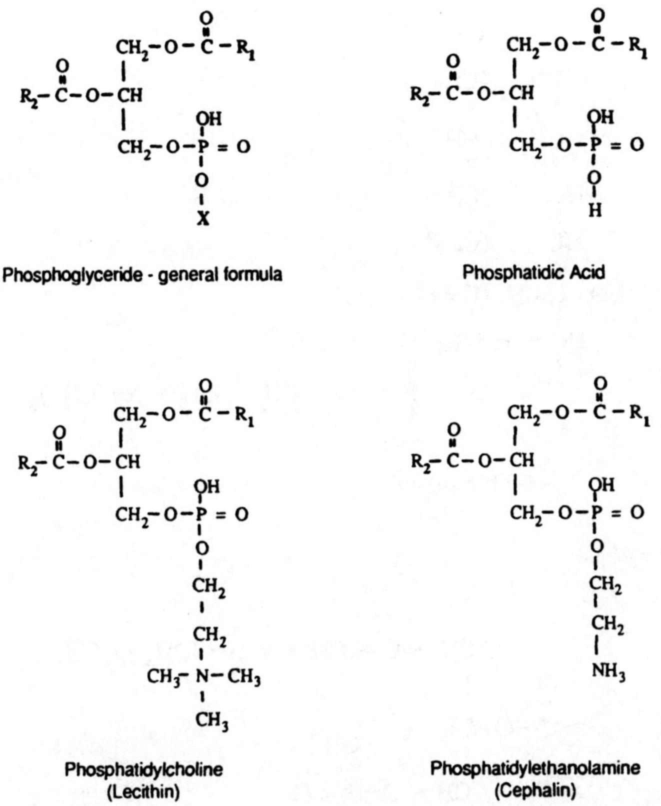

FIGURE 6. Glycerophospholipids.

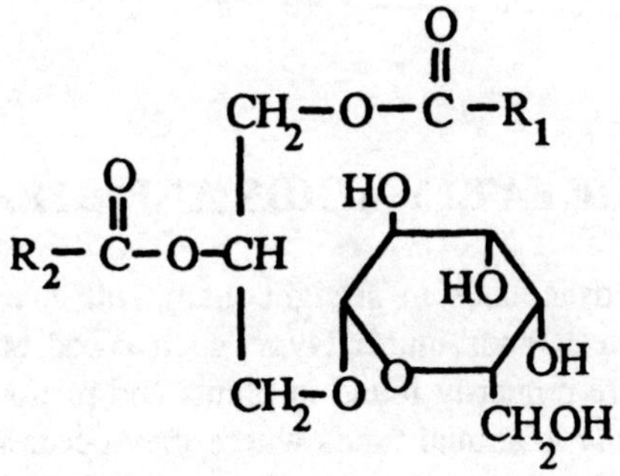

FIGURE 7. Glyceroglycolipid.

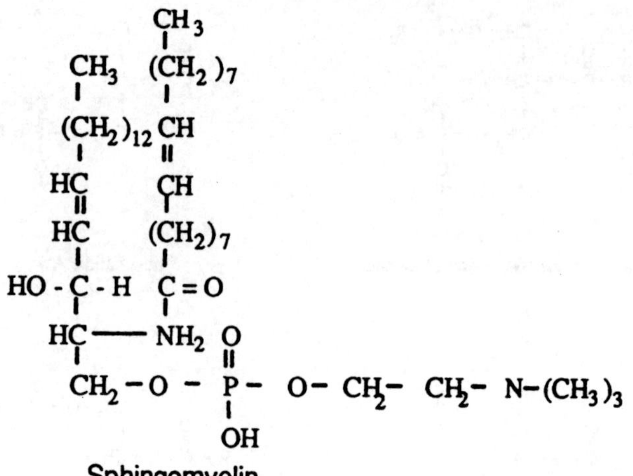

$$CH_3$$
$$|$$
$$CH_3 \quad (CH_2)_7$$
$$| \qquad |$$
$$(CH_2)_{12} \quad CH$$
$$| \qquad \parallel$$
$$HC \qquad CH$$
$$\parallel \qquad |$$
$$HC \qquad (CH_2)_7$$
$$| \qquad |$$
$$HO-C-H \quad C=O$$
$$| \qquad |$$
$$HC \;\text{———}\; NH_2 \quad O$$
$$| \qquad\qquad \parallel$$
$$CH_2-O-P-O-CH_2-CH_2-N-(CH_3)_3$$
$$|$$
$$OH$$

Sphingomyelin

FIGURE 8.   Sphingolipid.

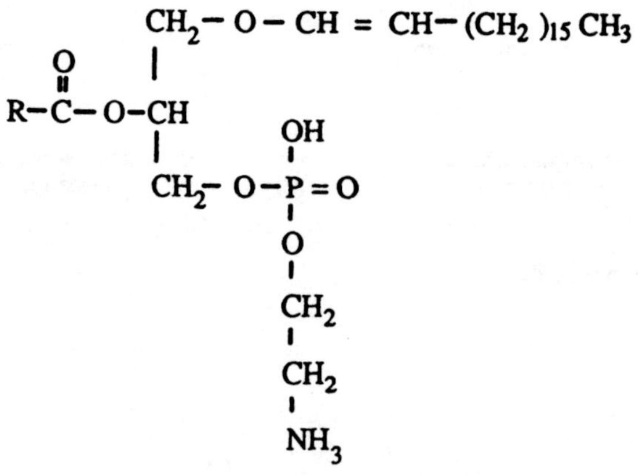

$$CH_2-O-CH=CH-(CH_2)_{15}\,CH_3$$
$$O \qquad |$$
$$\parallel$$
$$R-C-O-CH$$
$$| \qquad\qquad OH$$
$$CH_2-O-P=O$$
$$|$$
$$O$$
$$|$$
$$CH_2$$
$$|$$
$$CH_2$$
$$|$$
$$NH_3$$

FIGURE 9.   Plasmalogen.

OCCURRENCE OF FATTY ACIDS AND STEROLS IN FOODS

The majority of fatty acids in foods are straight chain, with an even carbon number. While some fatty acids are found in all foods, unusual types such as odd, branched, hydroxy, or cyclic fatty acids also exist and are primarily found in plants and microorganisms. A few of these unusual fatty acids are found in animal foods where they occur as a consequence of either ingestion, microbial action, or contamination.[2] The carbon length of fatty acids in foods is highly variable (C2-C80);[4] however, most common fatty acids are C16, C18, C20, and C22. The unsaturated fatty acids usually are in the cis position; however, trans forms exist, especially in naturally occurring microbial sources and commercially processed oils. These isomeric fatty acids are formed during the process of hydrogenation (industrial or ruminant). Most common polyunsaturated fatty acids usually have a methylene interrupted double-bond pattern, although conjugated and other forms exist.[2]

There are great varieties of the naturally occurring phytosterols; however, most foods usually have a simple sterol profile, with cholesterol being the primary sterol from animal sources and sitosterol from plants. Other plant sterols in foods may include campesterol, stigmasterol, brassicasterol, and dehydrocholesterol.[5,6]

REFERENCES

1. **Perkins, E.G.,** Nomenclature and classification of lipids, in *Analyses of Fats, Oils and Lipoproteins,* Perkins, E.G., Ed., American Oil Chemists' Society, Champaign, IL, 1991, chap. 1.
2. **Christie, W.W.,** *Gas Chromatography and Lipids,* Oily Press, Ayr, Scotland, 1989.
3. IUPAC-IUB Commission on Biochemical Nomenclature, The nomenclature of lipids, in *Fundamentals of Lipid Chemistry,* Burton, R.M. and Guena, F., Eds., BI-Science Publications Division, Webster Groves, MO, 1974, Appendix I.
4. **Gunstone, F.D.,** Fatty acids — structural identification, in *Methods in Plant Biochemistry,* Vol. 4, Dey, P.M. and Harborne, J.B., Ser. Eds., Harwood, J.L. and Bowyer, J.R., Eds., Academic Press, New York, 1990, chap. 1.
5. **Harwood, J.L. and Russell, N.J.,** Major lipid types in plants and microorganisms, in *Lipids in Plants and Microbes,* George Allen and Unwin Ltd., London, 1984, chap. 2.
6. **King, I.B., Childs, M.T., Dorsett, C.S., Ostrander, J.G., and Monsen, E.R.,** Shellfish: proximate composition, minerals, fatty acids, and sterols, *J. Am. Diet. Assoc.,* 90(5), 677, 1990.

Chapter 1.2

ANALYTICAL METHODOLOGIES FOR LIPIDS IN FOODS

Irena B. King

INTRODUCTION

The lipid composition of foods plays a central role in food science, food industry and technology, and nutrition and health. A thorough knowledge of lipid composition is essential to our understanding of the relationship between diet and disease, and for establishing national or international nutritional guidelines. Moreover, the fat and oil industries are very important agricultural and economic suppliers that rely on lipid analyses for production maintenance and new product development. Furthermore, lipid analyses are routinely employed to assess changes during food harvesting, processing, and storage, to monitor and control food quality and purity, to evaluate sensory quality of foods, and to comply with food labeling regulations.

Since foods are defined only from the standpoint of their edibility, they represent diverse biological matrices. In addition, there are many physical properties of foods (raw vs. cooked, dry vs. moist, pure vs. mixed diets) which contribute to this diversity. Thus, a detailed lipid analysis of foods can be a difficult and lengthy process that may require several techniques until desired goals are achieved.[1]

The focus of this discussion is to provide the reader with a general overview of lipid methodology applicable to food analysis. The emphasis is on foods as a source of major fatty acids and sterols and not necessarily on the identification of these compounds in numerous lipid classes. There are many factors that influence lipid levels and lipid analyses in foods.[2] Here, only factors influencing lipid analyses will be considered.

COMPOSITED SAMPLE

One of the most common variables in food and lipid analyses is the selection and preparation of a representative sample of the food material for analysis. Food chemists recognize that the procedures used for sample compositing and homogenizing have to meet certain criteria because they can have significant impact on the accuracy and precision of the final results. Many compositing and homogenizing techniques do not necessarily produce an average or homogeneous sample. Potential losses can also occur due to autoxidation, various enzymatic actions, and other destructive reactions.[3] Thus, prior to any lipid analysis in foods, the statistical methods for proper compositing and homogenizing need to be considered and the selected techniques must be validated.

MOISTURE CONTENT

Because lipid analyses are equally applicable to raw and cooked foods and because foods may lose or gain moisture during harvesting, preparation, and storage, the moisture content of foods can be another major variable. Therefore, it is necessary to determine the "normal" weight of the food sample as soon as possible. The weight percent of lipid in the wet food sample should be recorded with the weight percent of lipid relative to the amount of dry matter. It is necessary to determine moisture content separately under standardized conditions.[4]

0-8493-4248-1/96/$0.00+$.50
© 1996 by CRC Press Inc.

GENERAL CONSIDERATIONS IN LIPID EXTRACTION

One of the most important factors to consider in lipid analyses is the type of information required. For example, a detailed knowledge of the composition and structure of all lipid components and their molecular species in a food sample can be a long, tedious, and costly matter and may not be necessary when analyzing a large number of routine samples. The sample size and the type of equipment available are other factors that will influence the selection of the analytic procedures.[5] Today, however, the gas chromatograph equipped with a flame ionization detector is *de rigueur* in any lipid laboratory.

In general, precautions must be taken to protect samples from autoxidation. All solvents should be of high quality and peroxide-free before use. Lipid extracts should not be allowed to dry completely and operations should be carried out under nitrogen with the addition of exogenous antioxidants, such as BHT (butylated hydroxy toluene). The temperature during the extraction procedure should not exceed 50°C (under nitrogen). Furthermore, the old procedures in which the sample was extracted in a Soxhlet extractor with refluxing solvent for many hours should be avoided. All plastics should be avoided. Special attention should be paid to cleanliness of the glassware, although routine acid soaks are usually unnecessary.

In lipid analyses of foods there has been a traditional tendency to use particular methods for different foods. Although some of these preferences are due to tradition, other methods have been developed to improve the efficiency of the lipid extraction for specific foodstuffs. Foods are derived from materials of biological origin and the analyses of lipids in foods cover a wide variety of matrices. To some extent the type of food matrix will influence the selection of extraction procedures (for example, milk, cereal, or liver require different modifications). In leafy plant materials, significant enzymatic degradation of lipids can occur during the extraction.[6] Therefore, pretreatment with hot water or alcohol (isopropanol) may be required to inactivate many of the phospholipases to preserve lipid classes. There is an inexhaustible amount of literature on the various modifications in lipid methodology; consequently, any beginning (or not so beginning) lipid analyst should start with a thorough literature search. The reference methods published by the Association of Official Analytical Chemists[4] are a good place to start. Other excellent publications on lipid methodologies are Christie,[5,7,8] Perkins,[9] and Kates.[10]

LIPID EXTRACTION

The purpose of lipid extraction is to quantitatively recover all lipids from a given source in an undegraded state and uncontaminated with nonlipid components. Although various extraction methods are available, the most reliable are those that use a combination of polar and nonpolar solvents.[11-13] There are several types of associations in which lipids participate in a given matrix: namely, hydrophobic associations, electrostatic forces, hydrogen bonding, and covalent bonding.[10] Usually, in foods, most of the lipids are associated with proteins via noncovalent bonding and the first step of an extraction procedure is to destroy these associations. A polar solvent such as methanol accomplishes this goal. Then a nonpolar solvent such as chloroform "dissolves" the lipids.

TOTAL LIPID

The most direct and reliable method for determination of total lipids in foods is the gravimetric procedure. However, it is critical to purify all the interfering non-lipid substances that will extract in polar solvents along with the lipids.[14] After lipid extraction and purification, a measured aliquot of the extract is transferred to a tared disposable aluminum weighing dish. The dishes containing the extracts are placed in a vacuum desiccator, then flushed thoroughly with nitrogen. The solvent is evaporated until the dishes reach a constant weight. After weighing, the dried residue may be dissolved in a small volume of chloroform for further analyses.[15] It is a usual laboratory practice to discard the samples after the gravimetric determination because of the lipid exposure to air. However, depending on the sample source and the precautions taken, often these samples are not compromised. The main limitation of the gravimetric procedure is determined by the amount of the lipid to be weighed. Weighing samples less than 5 mg is challenging and is not recommended unless very sensitive microbalance is available in a proper setup.

FOLCH METHOD

This method, adapted for a small sample size, is used routinely in the author's laboratory. It is a useful, multipurpose extraction technique for biological specimens, mixed diets, and all types of animal tissues with recoveries greater than 96%. Sample size may range from 0.5 to 2.5 g; however, the final ratio of the extracting solvents to the sample is always maintained at least 19:1. The extraction procedure requires (depending on sample weight) 20, 50, or 250 ml Pyrex® glass or Teflon® centrifuge tubes or bottles and the following solutions:

    A 2:1 (v:v) chloroform:methanol
    B Saline 9 g NaCl/l
    C Upper phase solvent (3:48:47, v:v:v) chloroform:methanol:saline

1. Weigh out appropriate amount of homogenized food sample into the centrifuge tube with a Teflon®-lined cap. Work quickly to avoid moisture loss.
2. Add 9.5 ml of solution A (based on 0.5 g sample). Vortex three times for 30 s over a period of 30 min. For some samples, such as muscle tissue, it is more effective to add the methanol aliquot first in order to denature proteins. Then after vortexing, add the appropriate amount of chloroform.
3. Centrifuge at $800 \times g$ at 4°C for 10 to 20 min.
4. Transfer the supernatant into a clean glass centrifuge tube. If the protein does not precipitate to the bottom of the tube, add a few drops of methanol until it does.
5. Extract the precipitate two more times with 2 ml of solution A. Shake, centrifuge, and then add the supernatant from each extract to the tube in Step 4.
6. Add 2.8 ml of solution B to the pooled extracts and shake vigorously. Spin at $800 \times g$ for 10 min.
7. Aspirate and discard the upper phase.
8. Add 6 ml of solution C, mix vigorously, then spin at $800 \times g$ for 10 min.
9. Aspirate and discard the upper phase. Repeat Steps 8 and 9.

TRANSESTERIFICATION OF LIPIDS

There are several transesterification procedures that are satisfactory such as boron trifluoride-methanol, diazomethane, or hydrochloric acid-methanol. However, each method has its unique disadvantages which need to be evaluated for a given sample.[16,17]

The following transesterification by the acetyl chloride method[18] is used in the author's laboratory for various total fatty acid determinations.

1. Prepare and aliquot internal standard such as nonadecanoate or trinonadecanoin (see note below).
2. Weigh out 50 mg oil or total lipid or 200 mg tissue homogenate in 20 ml Pyrex® glass tubes with Teflon®-lined caps.

3. Add 5 ml acetyl chloride reagent (40:10:5, v:v:v, methanol:hexane:acetyl chloride). When preparing this reagent, drop in acetyl chloride slowly while mixing constantly. Use immediately.
4. Heat at 100°C for 1 h with frequent mixing. Check for leaks.
5. After cooling, add 10 ml 6% potassium carbonate.
6. Extract three times with 3 ml hexane.
7. Evaporate hexane under nitrogen and reconstitute in 500 µl hexane.
8. Inject 1 µl into GC.

Note: To prepare internal standard, weigh out a minimum of 10 mg of each lipid and dissolve in 100 ml of chloroform or another solvent such as isooctane in a volumetric flask with a Teflon®-lined screw cap. Estimate desired concentrations to aliquot 1-ml volumes. For many analyses 1 µg/µl is usually satisfactory. The choice of internal standard is critical. For more discussions see References 16 and 19.

GAS CHROMATOGRAPHY OF FATTY ACIDS

It is not possible here, even briefly, to describe all the major concepts and developments in gas chromatography and fatty acids analyses. Others have written comprehensively on this subject.[20-22] The emphasis is on the rudimentary application of gas chromatography to fatty acids analyses.

The fatty acids are commonly determined by gas chromatography (GC) in the form of fatty acid methyl esters (FAME). Depending on the objective of the analysis, FAME may be expressly obtained from many foodstuffs by direct transesterification or after total lipid extraction and/or lipid class isolation and then by transesterification as described previously (Figure 1). Sometimes it is necessary to do a preliminary separation of FAME prior to GC analysis using silver ion chromatography. This technique is especially useful in the identification of trans fatty acids.[23-25]

The determination of methyl esters by GC involves separation, identification, and quantitation. A GC separation is performed on FAME samples usually dissolved in hexane,

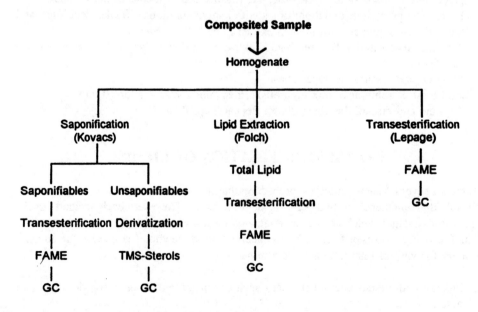

FIGURE 1.   Flow sheet for total lipid, fatty acids, and sterols analyses.

isooctane, or dichloromethane. The methyl esters are separated on a gas chromatograph such as Hewlett-Packard (Avondale, PA) or Varian (Walnut Creek, CA) equipped with a flame ionization detector (FID). An automatic sampler and a computerized software that operate the instrument and process data are not only modern features, but also contribute to the accuracy and precision of the analysis.[26,27]

In the author's laboratory, FAME are usually separated in a split mode (split ratio 1:50) on a 30 m × 0.25 mm ID wall-coated open-tubular fused silica column (DB23) with 0.25-μm coating (J & W, Folsom CA). The carrier gas is helium at 60 PSI (at the tank) equipped with an inline oxygen trap and an alarm sensor; makeup gas is nitrogen at 30 ml/min (60 PSI at the tank). At the detector end hydrogen is at 35 ml/min (30 PSI) and breathing air is at 420 ml/min (20 PSI). Column linear velocity is set at 33 cm/s (oven temperature of 200 °C). The injector and detector port temperatures (T) are at 250 and 275°C, respectively. The oven temperature is programmed using 3 ramp settings: T1 = 165, hold 1 (H1) = 9 min, rate 1 (R1) = 6, T2 = 188, H2 = 15, R2 = 4, T3 = 248. Analysis time to 24:1n9 is 33 min.

IDENTIFICATION AND QUANTITATION

The elution order and the identification of the FAME peaks on the chromatogram are commonly based on the relative retention times (RRT) and the equivalent chain lengths (ECL) which are reproducible for a given stationary phase and temperature.[28] These procedures can be used for a verification of known fatty acid profiles, but they do not permit complete identification of the true unknowns. Additional column phase separations, spectroscopic procedures, or silver ion chromatography should be employed for full identification.[29]

Qualitative identification and quantitative precision are evaluated with the use of model mixtures of known FAME such as GC-87 or NIH methyl ester standards from Nu-Chek Prep, Elysian, MN. It is important to select standards that are similar in composition to the experimental samples. In many cases such standards are not available and it is necessary to rely on other well-characterized products.[30] For quantitative results, a known amount of internal standard is added to the sample at an early stage of analysis and procedures are carried out through the GC data acquisition. The weight of each identified fatty acid (WFA) is determined by comparison of individual areas of fatty acids to the area of internal standard as shown by the following formula:

$$WFA = (AFA:AIS) \times WIS \times RFFA$$

WFA — weight of fatty acid (as FAME)
AFA — area of fatty acid
AIS — area of internal standard
WIS — weight of internal standard
RFFA — theoretical response factor

Theoretical response factors are small correction factors that are applied to compensate for the lack of influence of the carbonyl carbon on the FID response.[21] These theoretical factors must be determined and optimized prior to running any real samples and they should be confirmed when chromatographic conditions change.[26,27] Proper dilution, weight, and moisture are other factors that also need to be taken into account for calculation of the final results.

ANALYSIS OF STEROLS

Sterols are analyzed using direct saponification and hexane extraction of a 1- to 1.5-g sample from composited homogenate as described by the method of Kovacs et al.[31]

The extracted sterols are converted to the trimethylsilyl derivatives and analyzed by GC as described by Berg et al.[32] using a fused silica capillary column (60 m × 0.25 mm i.d.) coated with SE-30 (Supelco, Inc., Bellefonte, PA). The following temperature program can be used: oven temperature programmed at 270°C for 15 min, followed by a 0.4°C increase/minute to 285°C. Column pressure is set at 25 PSI with helium as a carrier gas. Injector and detector temperature are both set at 300°C. Other procedures which do not require derivatization may be used to separate most common plant and animal sterols.[4]

Sterols are identified by comparison with retention times of plant sterol standards purchased from Supelco which should be confirmed by mass spectrometry. Sterols are quantified using 5-α-cholestane as an internal standard.

QUALITY CONTROL

Today, with modern instrumentation and almost instant data production, it is especially important not to overlook the quality control aspect in lipid analyses. The accuracy and precision of lipid results is influenced by many factors which include extraction, instrument, and human errors. In order to obtain high accuracy and precision it is necessary to optimize each of the steps in the lengthy procedures of the lipid analysis. Quality control protocols must be clearly established and recovery studies documented. Reference and commercial standard samples should be run concurrently and statistical methods must be used to evaluate the obtained quality control results.[33]

REFERENCES

1. **Christie, W.W. and Noble, R.C.,** Recent developments in lipid analysis, in *Food Constituents and Food Residues. Their Chromatographic Determinations,* Lawrence, J. F., Ed., Marcel Dekker, New York, 1984, chap. 1.
2. **Kinsella, J.E., Posati, L., Weihrauch, J., and Anderson, B.,** Lipids in foods: problems and procedures in collecting data, *CRC Crit. Rev. Food Technol.,* p. 299, January 1975.
3. **Stewart, K.K.,** Compositing and homogenizing, *J. Food Comp. Anal.,* 5, 99, 1992.
4. AOAC, *Official Methods of Analysis of the Association of Official Analytical Chemists,* 15th ed., Helrich, K., Ed., Association of Official Analytical Chemists, Arlington, VA, 1990.
5. **Christie, W.W.,** *Lipid Analysis. Isolation, Identification and Structural Analysis of Lipids,* 2nd ed., Pergamon Press, Oxford, 1982.
6. **Harwood, J.L. and Russell, N.J.,** Major lipid types in plants and microorganisms, in *Lipids in Plants and Microbes,* George Allen and Unwin Ltd., London, 1984, chap. 2.
7. **Christie, W.W.,** Extraction and hydrolysis of lipids and some reactions of their fatty acid components, in *Handbook of Chromatography, Lipids,* Vol. 1, Mangold, H. K., Ed., CRC Press, Boca Raton, FL, 1984, chap. 2.
8. **Christie, W.W.,** Solid-phase extraction columns in the analysis of lipids, in *Advances in Lipid Methodology — One,* Christie, W. W., Ed., Oily Press, Ayr, Scotland, 1992, chap. 1.
9. **Perkins, E.G.,** Nomenclature and classification of lipids, in *Analyses of Fats, Oils and Lipoproteins,* Perkins, E. G., Ed., American Oil Chemists' Society, Champaign, IL, 1991, chap. 1.
10. **Kates, M.,** Techniques of lipidology. Isolation, analysis and identification of lipids, in *Laboratory Techniques in Biochemistry and Molecular Biology,* Vol. 3, Work, T. S. and Work, E., Eds., North-Holland, Amsterdam, 1973.
11. **Folch, J., Lees, M., and Sloane, G.H.,** A simple method for isolation and purification of total lipids from animal tissues, *J. Biol. Chem.,* 226, 497, 1957.
12. **Bligh, E.G. and Dyer, W.J.,** A rapid method of total lipid extraction and purification, *J. Biochem. Physiol.,* 37, 911, 1959.
13. **Radin, N.S.,** Extraction of tissue lipids with a solvent of low toxicity, *Methods Enzymol.,* 72, 5, 1981.
14. **Nelson, G.,** Isolation and purification of lipids from biological matrices, in *Analyses of Fats, Oils and Lipoproteins,* Perkins, E. G., Ed., American Oil Chemists' Society, Champaign, IL, 1991, chap. 2.

15. **Dryer, L.R.,** The lipids, in *Fundamentals of Clinical Chemistry,* Tietz, N. W., Ed., W. B. Saunders, Philadelphia, 1970, chap. 7.

16. **Kuksis, A.,** Quantitative and positional analysis of fatty acids, in *Lipid Research Methodology,* Story, J. A., Ed., Alan R. Liss, New York, 1984, chap. 2.

17. **Wood, R.,** Sample preparation, derivatization and analysis, in *Analyses of Fats, Oils and Lipoproteins,* Perkins, E. G., Ed., American Oil Chemists' Society, Champaign, IL, 1991, chap. 15.

18. **Lepage, G. and Roy, C.C.,** Direct transesterification of all classes of lipids in a one-step reaction, *J. Lipid Res.,* 27, 114, 1986.

19. **Shantha, N.C. and Ackman, R.G.,** Nervonic acid versus tricosanoic acid as internal standards in quantitative GC analyses of fish oil longer-chain n-3 polyunsaturated fatty acid methyl esters, *J. Chromatogr.,* 553, 1, 1990.

20. **Christie, W.W.,** *Gas Chromatography and Lipids. A Practical Guide,* Oily Press, Ayr, Scotland, 1989.

21. **Ackman, R.G.,** The analysis of fatty acids and related materials by gas-liquid chromatography, in *Progress in the Chemistry of Fats and Other Lipids,* Vol. 12, Holman, R. T., Ed., Pergamon Press, Elmsford, NY, 1972, 165.

22. **Kuksis, A.,** Newer developments in determination of structure of glycerides and phosphoglycerides, in *Progress in the Chemistry of Fats and Other Lipids,* Vol. 12, Holman, R. T., Ed., Pergamon Press, Elmsford, NY, 1972, 1.

23. **Sampugna, J., Pallansch, L.A., Enig, M.G., and Keeney, M.,** Rapid analysis of trans fatty acids on SP-2340 glass capillary columns, *J. Chromatogr.,* 249, 245, 1982.

24. **Firestone, D. and Sheppard, A.,** Determination of trans fatty acids, in *Advances in Lipid Methodology — One,* Christie, W. W., Ed., Oily Press, Ayr, Scotland, 1992, chap. 8.

25. **Nikolova-Damyanova, B.,** Silver ion chromatography and lipids, in *Advances in Lipid Methodology — One,* Christie, W. W., Ed., Oily Press, Ayr, Scotland, 1992, chap. 6.

26. **Craske, J.D.,** Separation of instrumental and chemical errors in the analysis of oils by gas chromatography a collaborative evaluation, *J. Am. Oil Chem. Soc.,* 70(4), 325, 1993.

27. **Craske, J.D. and Bannon, C.D.,** Gas liquid chromatography analysis of the fatty acid composition of fats and oils: a total system for high accuracy, *J. Am. Oil Chem. Soc.,* 64(10), 1413, 1987.

28. **Ackman, R.G.,** Application of gas-liquid chromatography to lipid separation and analysis: qualitative and quantitative analysis, in *Analyses of Fats, Oils and Lipoproteins,* Perkins, E. G., Ed., American Oil Chemists' Society, Champaign, IL, 1991, chap. 16.

29. **Gunstone, F.D.,** Fatty acids — structural identification, in *Methods in Plant Biochemistry,* Vol. 4, Dey, P. M. and Harborne, J. B., Ser. Eds., Harwood, J. L. and Bowyer, J. R., Eds., Academic Press, New York, 1990, chap. 1.

30. **Ackmam, R.G.,** Omega-3 PUFA in marine oil products, *J. Am. Oil Chem. Soc.,* 64(4), 1987.

31. **Kovacs, M.I.P., Anderson, W.E., and Ackman, R.G.,** A simple method for the determination of cholesterol and some plant sterols in fishery-based food products, *J. Food Sci.,* 44, 1299, 1979.

32. **Berg, C.J., Jr., Krzynowek, J., Alatalo, P., and Wiggin, K.,** Sterol and fatty acid composition of the clam, *Codakia orbicularis* with chemoautotropic symbionts, *Lipids,* 20, 116, 1985.

33. **Naito, H.K. and David, J.A.,** Laboratory considerations: determination of cholesterol, triglyceride, phospholipid, and other lipids in blood and tissues, in *Lipid Research Methodology,* Story, J. A., Ed., Alan R. Liss, New York, 1984, chap. 1.

Chapter 1.3

ANALYTICAL PROCEDURES FOR MEASUREMENT OF THE LIPIDS AND LIPOPROTEINS IN CARDIOVASCULAR RISK ASSESSMENT

G. Russell Warnick and Elizabeth Teng Leary

The term lipid designates the class of compounds which are soluble in organic solvents and insoluble, or nearly so, in water. Serum lipids are carried in the aqueous circulation in the form of lipoproteins, in combination with certain proteins designated apolipoproteins. The primary focus of clinical and analytical interest in the human serum lipids derives from their predictive association with coronary heart disease. The relationship between serum cholesterol and risk of coronary heart disease has long been known.[1,2] The well-established atherogenic effect of cholesterol is primarily associated with the low density lipoprotein (LDL) fraction which contributes to deposition of cholesterol in the arterial wall. The high density (HDL) fraction is considered to be anti-atherogenic or protective, presumably through the process of reverse cholesterol transport, that is, mobilization of cholesterol out of the body for excretion via the liver. The other major lipoprotein classes, very low density (VLDL) and chylomicrons, transport the triglycerides, the body's major fuel source originating from the diet and from the liver, respectively. The other major lipid class of moderate diagnostic interest is the phospholipids.

The major lipoproteins and their physical and chemical characteristics are illustrated in Figure 1 and Tables 1 and 2.[3,4] The largest lipoprotein particles, the chylomicrons, composed primarily of triglycerides, transport dietary fat into the circulation to meet the energy needs of cells. Triglycerides are hydrolyzed by lipoprotein lipase in the capillary endothelial beds and transported into cells. Triglycerides in excess of current energy needs are stored in fat cells.

The other class of triglyceride-rich particles, designated VLDL, are synthesized in the liver and transport endogenous triglycerides back into the circulation. As triglycerides are removed, the transition stage is designated intermediate density lipoprotein (IDL), a subset of the LDL class. The end product after catabolism of the bulk of triglycerides is designated LDL. Removal of the triglycerides leaves cholesterol as the major lipid. Esterified cholesterol, being hydrophobic, resides in the core, and unesterified cholesterol is on the surface with the relatively polar phospholipids. Apolipoprotein B and other minor apolipoproteins (Table 3) are on the lipoprotein surface providing structure as well as receptor recognition. Apolipoproteins have amphipathic properties; certain regions attract to the lipid environment of the lipoproteins while other regions are hydrophilic, giving the lipoproteins solubility in the plasma. LDL is positively associated with risk of coronary heart disease, contributing to plaque formation by mediating cholesterol deposition in the arterial wall.

The Lp(a) class [lipoprotein (a)] is a variant of LDL with the additional (a) protein linked by a disulfide bridge.[5,6] This variant is genetically determined and highly atherogenic. The particle appears to inhibit clot lysis by virtue of similarity in structure to plasminogen as well as appearing to promote cholesterol uptake in the arterial wall.

HDL designates a smaller and heterogeneous class of lipoproteins, rich in cholesterol and protein.[3] The major protein of HDL is apoAI and this protein is sometimes assayed directly in serum as an approximation of the HDL level. The HDL class has been inversely associated with coronary heart disease risk in epidemiological studies, presumably by mobilizing cholesterol out of the arterial plaques. Cholesterol taken up by HDL is esterified and can be transferred to triglyceride-rich lipoproteins which are ultimately removed by the liver. HDL

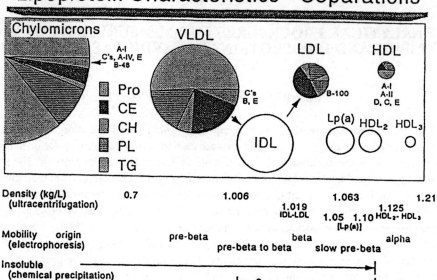

FIGURE 1.   Lipoprotein characteristics — separations.

TABLE 1
Physical Properties of Human Plasma Lipoprotein Families

	Electrophoretic definition	Particle size (nm)	Molecular weight	Density (g/ml)
Chylomicrons	Remains at origin[a]	75-1200	~4000,000,000	0.93
VLDL	Pre-β lipo- proteins	30-80	10-80,000,000	0.93-1.006
IDL	Slow pre-β lipoproteins[b]	25-35	5-10,000,000	1.006-1.019
LDL	β-Lipoproteins	18-25	2,300,000	1.019-1.063
HDL₂	α-Lipoproteins	9-12	360,000	1.063-1.125
HDL₃	α-Lipoproteins	5-9	175,000	1.125-1.210

From Smith, L.C., Massey, J.B., Sparrow, J.T., Gotto, Jr., A.M., and Pownall, H.J., in *Supramolecular Structure and Function*, G. Pifat and J.N. Herak, Eds., Plenum, New York, 1983, 210. With permission.

[a] On paper.
[b] On geon pevikon or agarose.

particles may also be bound and removed directly by the liver. Major subclasses of HDL are designated HDL2 and HDL3 based on density.[7] HDL subclasses may also be categorized based on the apolipoproteins residing on their surface.[8,9] For example, some particles contain only the major protein, apoAI, while other particles contain apoAI with apoAII or other minor proteins. These subclasses appear to have different functional roles and may have different associations with risk of coronary disease.

The lipoproteins are traditionally defined in terms of their hydrated densities, indicated in Figure 1 and Table 1, which provide a useful basis for their separation by ultracentrifugation. The lipoproteins also have characteristic mobilities on electrophoresis as shown which corre-

TABLE 2

Chemical Composition of Normal Human Plasma Lipoproteins

	Surface components			Core lipids	
	Cholesterol[a]	Phospholipids (mol%)	Apolipoprotein	Triglycerides (mol%)	Cholesteryl esters
Chylomicrons	35	63	2	95	5
VLDL	43	55	2	76	24
IDL	38	60	2	78	22
LDL	42	58	0.2	19	81
HDL$_2$	22	75	2	18	82
HDL$_3$	23	72	5	16	84

From Smith, L.C., Massey, J.B., Sparrow, J.T., Gotto, Jr., A.M., and Pownall, H.J., in Supramolecular Structure and Function, G. Pifat and J.N. Herak, Eds., Plenum, New York, 1983, 213. With permission.

[a]   May be distributed between the surface and core and, in the case of large chylomicrons, more cholesterol may be in the core than on the surface.

TABLE 3

Characteristics of Plasma Apolipoproteins in Normal Fasting Humans

	Plasma concentration		Distribution in lipoproteins (mol%)[b]					
	(mg/dl)	(mol%)[a]	HDL	LDL	IDL	VLDL	Major tissue source	Molecular weight
ApoA-I	130	43	100				Liver and intestine	28,016
ApoA-II	40	22	100				Liver and intestine	14,414
ApoA-IV							Liver and intestine	44,465
ApoB-48	} 80	5		90	8	2	Intestine	264,000
ApoB-100							Liver	550,000
ApoC-I	6	9	97		1	2	Liver	6,630
ApoC-II	3	3	60		10	30	Liver	8,900
ApoC-III	12	13	60	10	10	20	Liver	8,800
ApoD	10	5	100					22,000
ApoE-II								
ApoE-III	} 5	2	50	10	20	20	Liver	34,145
ApoE-IV								

From Gotto, Jr., A.M., Pownall, H.J., and Havel, R.J., *Methods in Enzymology*, 128, 3, 1986. With permission.

[a] Based on total plasma concentration.
[b] For each apolipoprotein.

spond approximately, but not exactly, to the density classifications. The various lipoprotein particles complex to different degrees, with polyanions and divalent cations providing another means of separation.

Recognition of the contribution of the lipids to the high prevalence of coronary heart disease in the developed countries has driven intense research into the epidemiology, mechanisms, pathology, and methodology over recent decades (reviewed in Reference 1). Research studies have established a foundation for development of national guidelines for detection, intervention, and treatment of CHD.[1,10] The U.S. National Cholesterol Education Program (NCEP) was organized in 1986 under the auspices of the National Institutes of Health (NIH) at the urging of an NIH Consensus Conference.[11] The NCEP has become the dominant authority in this arena, forging consensus among researchers, professional societies, advocacy groups, and government agencies. Reports have been issued for intervention measures in the general population,[12] for cholesterol screening programs,[13] and for identification and treatment of high cholesterol in children[14] and adults.[1,2,10] In addition, recommendations have been provided for measurement of cholesterol.[15,16] Guidelines have been completed and will be published shortly for measurement of triglycerides, HDL, and LDL cholesterol.[17-19] Each set of guidelines has been developed by panels of experts after months of review and deliberation and represents a mainstream view among the divergent opinions in these complex areas.

Updated guidelines for detection, evaluation, and treatment of high blood cholesterol in adults (ATP 2) were recently published.[1,2] The recommended workup is illustrated in Figures 2 to 4 and Table 4. The guidelines implicate LDL cholesterol as the major established risk factor for coronary heart disease and the primary basis for treatment decisions. Therapy by fat-

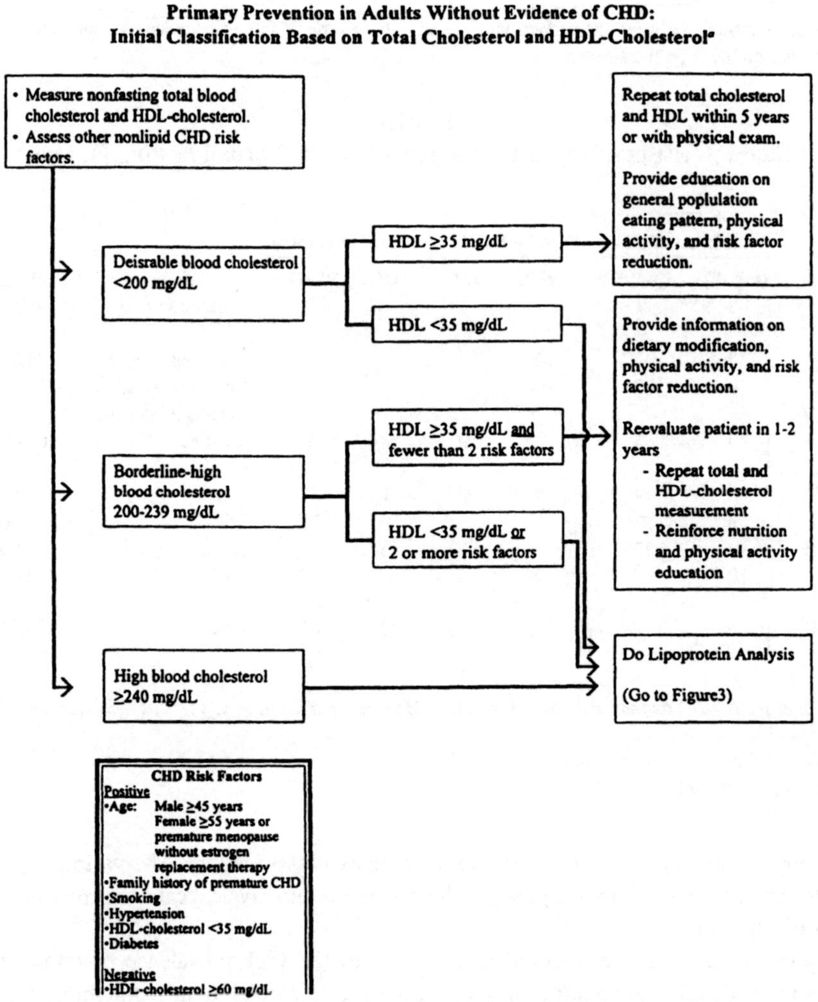

Primary Prevention in Adults Without Evidence of CHD:
Initial Classification Based on Total Cholesterol and HDL-Cholesterol[a]

FIGURE 2. Primary prevention in adults without evidence of CHD: initial classification based on total cholesterol and HDL-cholesterol. (From NCEP. Summary of the second report of the National Cholesterol Education Program [NCEP] expert panel on detection, evaluation, and treatment of high blood cholesterol in adults [Adult Treatment Panel II]. With permission.)

and cholesterol-restricted diet is considered the first line of defense. Drug therapy is reserved for patients refractory to dietary treatment or with symptoms of coronary disease. HDL cholesterol is involved in treatment decisions; values below 35 mg/dl are considered a risk factor, while high levels, above 60, are considered a negative risk factor, subtracting one from the number of risk factors.

Initially total and HDL cholesterol are measured and measurements can be made on a non-fasting specimen. The classification scheme is outlined in Figure 2.[1,2] Patients with high cholesterol at or more than 240 mg/dl require evaluation of the lipoproteins. Patients with borderline elevated cholesterol from 200 to 239 mg/dl and with two or more risk factors or with HDL cholesterol less than or equal to 35 mg/dl also require analysis of the lipoproteins to determine LDL cholesterol. The risk factors considered in the classification scheme are outlined in Figure 2.

The lipoprotein evaluation leads to the classification scheme outlined in Figure 3.[1,2] For patients without symptoms of coronary disease, LDL cholesterol values at 160 mg/dl or higher require treatment. Borderline cases with LDL cholesterol between 130 and 159 and two or more risk factors are also candidates for treatment. Since the lipid and lipoprotein values are subject to natural biological fluctuations, it is recommended to confirm the values by repeat measurement before making treatment decisions. Dietary treatment is recommended as the first step with the goal to bring LDL cholesterol values below the triggering cut point.

Primary Prevention in Adults Without Evidence of CHD:
Subsequent Classification Based on LDL-Cholesterol

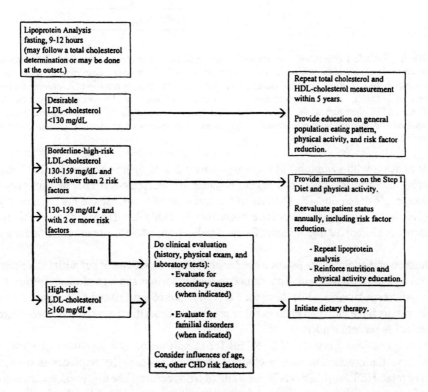

FIGURE 3. Primary prevention in adults without evidence of CHD: subsequent classification based on LDL-cholesterol. (b) On the basis of the average of two determinations. If the first two LDL-cholesterol tests differ by more than 30 mg/dl, a third test should be obtained within 1 to 8 weeks and the average value of three tests used. (From NCEP. Summary of the second report of the National Cholesterol Education Program [NCEP] expert panel on detection, evaluation, and treatment of high blood cholesterol in adults [Adult Treatment Panel II]. With permission.)

Secondary Prevention in Adults with Evidence of CHD:
Classification Based on LDL-Cholesterol[a]

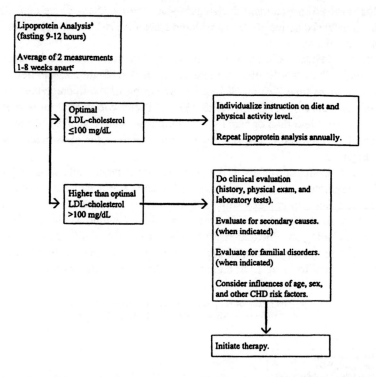

FIGURE 4. Secondary prevention in adults with evidence of CHD: classification based on LDL cholesterol. (b) Lipoprotein analysis should be performed when the patient is not in the recovery phase from an acute coronary or other medical event that would lower his usual LDL-cholesterol level. (c) If the first two LDL-cholesterol tests differ by more than 30 mg/dl, a third test should be obtained within 1 to 8 weeks and the average value of the three tests used. (From NCEP. Summary of the second report of the National Cholesterol Education Program [NCEP] expert panel on detection, evaluation, and treatment of high blood cholesterol in adults [Adult Treatment Panel II]. With permission.)

For patients with symptoms of coronary heart disease, such as angina, the LDL cut point triggering therapy is lower, 100 mg/dl, because of the increased risk in this group. The decision levels based on LDL cholesterol for initiation of treatment and the treatment goal to bring risk to an appropriate level are summarized in Table 4.[1,2] The initiation levels for drug treatment are shifted 30 mg/dl higher for each indication, while the treatment goals remain the same.

Children and adolescents follow a similar classification scheme,[14] but with lower cut points for total and LDL cholesterol. The cut points for borderline and high cholesterol are 170 and 200 mg/dl, respectively, while those for LDL cholesterol are 110 and 130 mg/dl. In children whose parents have elevated cholesterol or premature heart disease, selective measurement of cholesterol is recommended.

These guidelines from the U.S. NCEP for classification and treatment of children and adults drive the practice in routine clinical laboratories where the emphasis is on measurements of total, HDL, and LDL cholesterol and of triglycerides. The triglycerides are associated with atherogenic lipoproteins, but the association of triglycerides with risk of coronary heart disease in epidemiological studies is ambiguous. A recent NIH-sponsored Consensus Conference[20] concluded that evidence to date is insufficient to justify aggressive intervention and recommended treatment by the dietary regimens suggested for total and HDL cholesterol. The NCEP has recommended cut points for triglycerides of 250, 400, and 1000 mg/dl.[10] Below 250

TABLE 4
Treatment Decisions Based on LDL-Cholesterol

	Initiation level (mg/dl)	LDL goal (mg/dl)
Dietary Therapy		
Without CHD and with fewer than two risk factors	≥160	<160
Without CHD and with two or more risk factors	≥130	<130
With CHD	>100	≤100
Drug Treatment		
Without CHD and with fewer than two risk factors	≥190	<160
Without CHD and with two or more risk factors	≥160	<130
With CHD	≥130	≤100

From NCEP. Summary of the second report of the National Cholesterol Education Program (NCEP) expert panel on detection, evaluation, and treatment of high blood cholesterol in adults (Adult Treatment Panel II). With permission.

TABLE 5
Common Techniques Used in the
Separation of Lipoprotein Fractions

Method	Ref.
Ultracentrifugation	22
Sequential density	23
Density gradient	24-27
Fonal	28
Electrophoresis	31,32
Agarose	33-37
Polyacrylamide	40
Chemical precipitation	43,44
(e.g., polyanions with divalent cations)	
Immunochemical interactions	55,56
Chromatography	
Immunoaffinity	57
Polyanion affinity	58,59
Molecular sieving	60
Nuclear magnetic resonance	62,63

is desirable, up to 400 is considered borderline, and above 500 treatment is justified. More than 1000 is considered severe and pancreatitis is likely. Measurements of the apolipoproteins, other lipoprotein fractions, and other lipids are of interest primarily in specialty laboratories and for research studies.

A variety of methods exist for the separation and quantitation of the serum lipoproteins (Table 5; recent review in Reference 21). Lipoproteins may be separated based on their physical properties such as density, charge properties, and apolipoprotein content. The wide range in density observed among the lipoprotein classes is a function of the relative lipid and protein content and enables fractionation by ultracentrifugation. Electrophoretic separations are enabled by differences in charge and size among the particles. Chemical precipitation methods depend on size and differences in the apolipoprotein content. Antibodies specific to certain apolipoproteins can be used to bind and separate lipoprotein classes. Chromatographic methods take advantage of size differences as in molecular sieving methods or composition

TABLE 6
Common Chemical Precipitation
Methods Employed in Separation
of HDL Reagent

	Ref.
Heparin with Mn^{2+}	45-48
Dextran sulfate with Mg^{2+}	51
Sodium phosphotungstate with Mg^{2+}	49, 50
Polyethyleneglycol	52, 53

in affinity methods using, for example, heparin sepharose. Quantitation by nuclear magnetic resonance depends on differences in the proton domain of the different lipoprotein classes.

A variety of ultracentrifugation methods have been used.[22] The earliest methods involved interpreting the flotation patterns of analytical ultracentrifugation. This technically demanding approach was largely superseded by preparative ultracentrifugation, using sequential density adjustments to fractionate major and minor classes.[23] Density gradient methods,[24-28] either non-equilibrium techniques in which separations are based on the rate of flotation or equilibrium techniques in which separations run to completion, permit fractionation of several or all classes in a single run. Methods use different types of rotors: swinging bucket, fixed angle, vertical, and zonal. Recent methods have indicated a trend toward smaller-scale separations in small rotors in table-top ultracentrifuges.[29,30] Ultracentrifugation, even though tedious, expensive, and technically demanding, remains a workhorse approach for both analytical and preparative purposes. Ultracentrifugation has been considered a reference method for lipoprotein quantitation, appropriate because the lipoproteins are defined in terms of hydrated density. The currently accepted reference method involving a combination of ultracentrifugation and chemical precipitation will be described subsequently.

Electrophoretic methods (reviewed in References 31 and 32) allow separation and quantitation of major lipoprotein classes and provide a visual display useful in detecting unusual or variant patterns. Agarose has been the most common media for separation of whole lipoproteins, providing a clear background and convenience.[33-37] Electrophoretic methods have, in general, been considered useful for qualitative analysis, but less than desirable for lipoprotein quantitation because of poor precision and large systematic biases compared to other methods.[38] Evaluations of a commercial automated electrophoretic system demonstrate, however, that electrophoresis can be precise and accurate. Recent studies in the authors' laboratory show separation of the major lipoprotein classes including the Lp(a) and good agreement with the reference method.[39]

Electrophoresis in polyacrylamide gels is used for separation of lipoprotein classes,[40] subclasses, and of the apolipoproteins. Of particular recent interest have been methods which fractionate LDL subclasses.[41,42] Evidence suggests that the heavier, lipid-depleted, but smaller fractions of LDL are more atherogenic than the larger, lighter subclasses.

Separation methods involving chemical precipitation (Table 6), usually with polyanions, such as heparin, and divalent cations, such as manganese, because of their convenience have become common in routine practice.[43,44] Methods have been described for separating all the major lipoprotein classes but have been most successful and common for separating HDL from the atherogenic lipoproteins. ApoB is rich in positively charged amino acids which form complexes with the polyanions; the divalent cations neutralize the charged groups on the lipoproteins which afford solubility. The larger the lipoprotein the greater the tendency for insolubility and precipitation, making possible selective precipitation of the lipoprotein classes and even separation of some subclasses with reasonable specificity.

The earliest common method used heparin in combination with manganese to precipitate the apoB-containing lipoproteins.[45-47] Because manganese produced interference with enzymic assays, alternative reagents were developed.[48] Sodium phosphotungstate[49,50] with magnesium became common in routine use but, because of its sensitivity to reaction conditions and greater variability, is being replaced by dextran sulfate with magnesium.[51] Dextran sulfate is a synthetic heparin prepared by treating dextran with sulfuric acid. Different molecular weights are available: 50 kDa is becoming most common in routine HDL quantitations.

Polyethylene glycol can be used to fractionate lipoproteins, but at 100-fold higher concentrations than the polyanions, requiring larger dilutions and highly viscous reagents which are difficult to pipette precisely.[52-54]

Separation of lipoproteins by immunochemical means using antibodies specific to epitopes on the apolipoproteins has potential for both research and routine use.[55,56] Antibodies can be immobilized on a solid support such as a column matrix or latex beads to facilitate separations. The apoB-containing lipoproteins as a group can be bound by antibodies to apoB. Selectivity within the apoB-containing lipoproteins, for example, removing VLDL while retaining LDL, can be obtained by using antibodies to the minor apolipoproteins. HDL can be selectively removed using antibodies to apoAI, the major protein of HDL. The use of latex-immobilized antibodies is the basis for a recently introduced commercial method for direct quantitation of LDL cholesterol.

Chromatographic methods can involve immobilized antibodies for affinity separations of lipoproteins.[57] Affinity separations have also been accomplished using dextran sulfate or heparin bound to sepharose.[58,59] Separations based on size using column materials such as agarose have also been reported.[60] Field flow fractionation is a new technique using differential flow characteristics, such as the difference in flow rate between the center and edge of a channel to fractionate lipoproteins.[61]

Nuclear magnetic resonance (NMR) spectroscopy has been used for research investigations and has potential for routine quantitation of lipoproteins.[62,63] NMR takes advantage of differing shifts for protons in either methyl or methylene groups in the lipoproteins to deduce concentrations by computer analysis of the signals from the resonance envelope. Recent studies suggest that lipoprotein subclasses can be determined in addition to major lipoprotein classes. The equipment for NMR analysis is expensive and specialized skills are required, making this method possibly appropriate for high volume routine applications but not for general use.

Lipoproteins have traditionally been quantitated in terms of their cholesterol content.[64] Early methods used strong acids, such as sulfuric, with other chemicals such as acetic anhydride or ferric chloride which produced a color with cholesterol. Since the reactions were relatively non-specific, partial or full extraction by organic solvent was required. The accepted reference method for cholesterol uses hexane extraction after hydrolysis with alcoholic KOH and Liebermann Burchard color reagent, composed of sulfuric and acetic acids and acetic anhydride.[65,66] The method is tedious but gives good agreement with the "gold standard" method developed and applied at the U.S. National Institute for Standards and Technology, the so-called definitive method employing isotope dilution-mass spectrometry.[67]

Virtually all routine measurements and most research analyses have shifted to enzymic reagents. A major revolution in laboratory medicine over the last two decades has been the development of analytical methods using linked enzyme catalyzed reactions to quantitate directly in serum analytes of clinical interest. The specificity of the enzymes allows reasonably accurate quantitations without the necessity for extraction or other pre-treatment. Enzymes that react specifically with the analytes of interest are coupled in sequence with enzyme reactions, giving colored reaction products that can be measured using built-in spectrophotometers. Three, four, or more linked enzymic reactions in sequence are not unusual. Enzymic reagents coupled with microprocessor-controlled robotic instruments have made the practice of clinical chemistry highly automated and efficient. There is now an amazing variety of highly sophisticated analyzers using different combinations of mechanical devices to measure sample and reagent, mix, incubate, and monitor the reactions. Systems have evolved from continuous flow instruments, through batch analyzers, to today's highly automated discrete analyzers which can perform one or many tests as requested for each specimen on microliters of specimen. Test requests are processed and results transferred directly on-line with a host computer. Cholesterol and, often, triglycerides are included in common test panels and

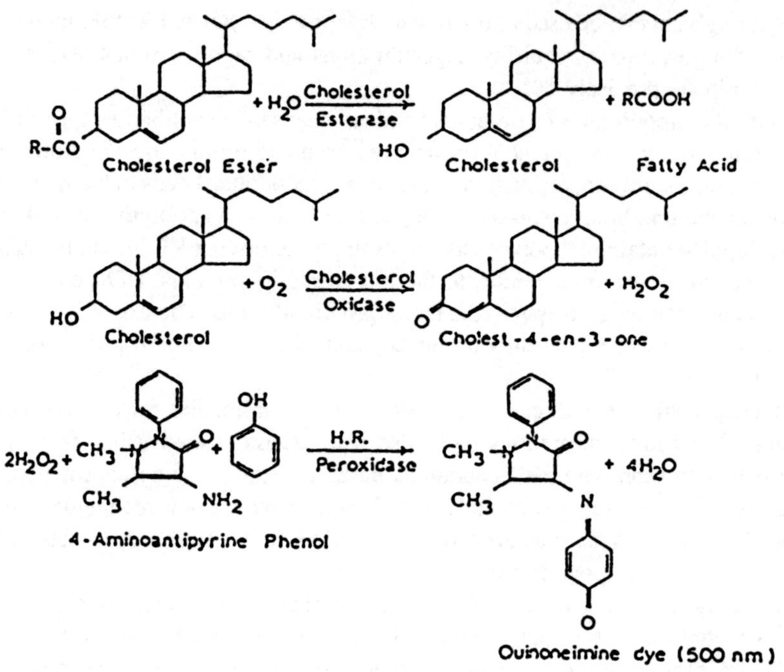

FIGURE 5. Enzymic reaction sequence for cholesterol analysis. (From Warnick, G.R., "Enzymatic Methods for Quantification of Lipoprotein Lipids", in *Methods in Enzymology 129*, Albers, J.J. and Segrest, J.P., Eds., Academic Press, Orlando, FL, 1986, 101. With permission.)

cholesterol, benefiting from this automation revolution, has become the most common routine test performed in clinical laboratories.

Although other enzymic reaction sequences have been described, only one sequence illustrated in Figure 5 is in common use for cholesterol assay.[68,69] Cholesteryl ester hydrolase cleaves the fatty acid residue from the cholesteryl ester producing unesterified or free cholesterol. Cholesterol is reacted by cholesterol oxidase enzyme to produce hydrogen peroxide, which feeds into a common color sequence using horseradish peroxidase to couple two colorless chemicals into a colored compound, monitored at about 500 nm. While reasonably robust and specific, this reaction sequence has minor problems. Some hydrolase enzymes may not fully cleave the esters, underestimating cholesterol.[70] Reducing substances such as vitamin C and bilirubin can interfere with the color reaction.[71] Care must be taken in the choice of reagent and in the application and especially calibration to assure reliable results.

Several reaction sequences are available for the triglyceride assay, illustrated in Figure 6. The reactions all use lipase enzymes to cleave fatty acids from the glycerol backbone. Glycerol, in turn, participates in any one of several sequences. One of the more common earlier reactions, ending in an ultraviolet-absorbing product, used glycerol kinase and pyruvate kinase, culminating in the conversion of NADH to NAD+ with an associated decrease in absorbency, monitored at 340 nm.[72] This reaction is relatively susceptible to interference and side reactions. The UV end point is also less convenient for modern analyzers, so this and other UV sequences are gradually being replaced by the second sequence which feeds into the same peroxidase color reaction described for cholesterol.[73] The intermediate steps involve glycerol kinase and glycerol phosphate oxidase.

All enzymic triglyceride reaction sequences react with endogenous free glycerol, a universal and significant source of interference.[74,75] In most specimens, the endogenous free glycerol contributes to a 10- to 20-mg/dl overestimation of triglycerides. A few specimens, about 20%, will have glycerol with levels increased in certain conditions like diabetes and liver disease.

Ultraviolet Method

$$\text{Triglycerides} \xrightarrow{\text{Lipase}} \text{Glycerol + Free Fatty Acids}$$

$$\text{Glycerol + ATP} \xrightarrow{\text{Glycerol Kinase}} \text{Glycerol-1-Phosphate + ADP}$$

$$\text{ADP + Phosphoenolpyruvate} \xrightarrow{\text{Pyruvate Kinase}} \text{ATP + Pyruvate}$$

$$\text{Pyruvate + NADH + H}^+ \xrightarrow{\text{LDH}} \text{Lactate + NAD}^+$$

Colorimetric Method

$$\text{Triglyceride} \xrightarrow{\text{Lipase}} \text{Glycerol + Free Fatty Acids}$$

$$\text{Glycerol + ATP} \xrightarrow{\text{Glycerol Kinase}} \text{Glycerol-1-PO}_4 + \text{ADP}$$

$$\text{Glycerol-1-PO}_4 + \text{O}_2 \xrightarrow{\text{Glycerol Phosphate Oxidase}} \text{Dihydroxyacetone PO}_4 + \text{H}_2\text{O}_2$$

$$\text{H}_2\text{O}_2 + \text{Phenol + 4 Aminoantipyrine} \xrightarrow{\text{Peroxidase}} \text{Quinoneimine dye + 2H}_2\text{O}$$

FIGURE 6. Enzymic reaction sequences for analysis of triglycerides. (From Warnick, G.R., "Enzymatic Methods for Quantification of Lipoprotein Lipids", in *Methods in Enzymology 129*, Albers, J.J. and Segrest, J.P., Eds., Academic Press, Orlando, FL, 1986, 101. With permission.)

Few routine laboratories, but most research laboratories, incorporate some variation on the method to correct for endogenous free glycerol. The most common is to perform a second parallel measurement without the lipase enzyme to quantitate only the free glycerol blank which is subtracted from the total measurement of the complete reaction sequence to determine a net or blank-corrected glycerol.[76] Another approach, designated the single cuvette blank, is to begin with the lipase-free reagent and, after a brief incubation, take a blank reading which includes only the free glycerol. The lipase is then added as a second separate reagent and, after additional incubation, a total reading is taken which is corrected for the blank reading and this represents net triglycerides.[77] A commercial variation on this is to react the free glycerol to a colorless product and then add a key ingredient of the color reaction with the lipase.

The accepted reference method for triglyceride assay, designated the chromotropic acid method, involves hydrolysis, extraction, and color reaction with chromotropicacid.[78] The assay is tedious, not well characterized, and not applied except at the Centers for Disease Control (CDC). Measurement of phospholipids is rare in routine practice, although some have suggested quantifying HDL in terms of phospholipids as the major lipid class rather than cholesterol.[79] Phospholipids are often measured in research, for example, in studies of dietary influences. A common enzymic sequence, outlined in Figure 7, measures the choline containing phospholipids, lecithin, lysolecithin, and sphingomyelin, comprising about 95%+ of total

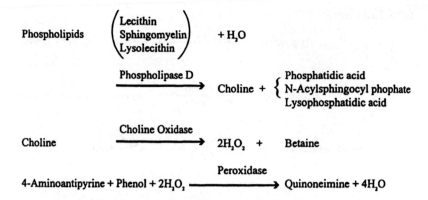

FIGURE 7. Enzymic reaction sequence for analysis of choline containing phospholipids. (From Warnick, G.R., "Enzymatic Methods for Quantification of Lipoprotein Lipids", in *Methods in Enzymology 129*, Albers, J.J. and Segrest, J.P., Eds., Academic Press, Orlando, FL, 1986, 101. With permission.)

phospholipids in serum.[80,81] Commercial kit methods using this sequence are available. Prior to the availability of enzymic reagents, a common approach involved acid digestion and analysis of the freed phosphorus.[82]

A method for lipoprotein quantitation, designated beta-quantification-beta — referring to the electrophoretic term for LDL — common in research and clinical practice involves a combination of ultracentrifugation and chemical precipitation.[38,47] Ultracentrifugation is used to float VLDL and any chylomicrons for separation. The fractions are recovered by slicing the tube between the fractions and pipetting. Ultracentrifugation is preferred for the VLDL separation because other methods such as precipitation are not highly specific for VLDL and subject to interference from chylomicrons. Ultracentrifugation is a robust but tedious method that works well, provided technique is meticulous. Incomplete recovery of the fractionated lipoproteins is a potential and unappreciated problem that can have a substantial effect on the result.

Chemical precipitation is used to separate HDL from either the whole serum or the infranate obtained from ultracentrifugation. Precipitation is efficient, convenient, and relatively robust for the HDL separation step. Cholesterol and sometimes other lipids are quantitated in serum and in the fractions by enzymic or other assay methods. LDL is calculated as the difference between cholesterol in the infranate and in the HDL fraction. VLDL cholesterol is usually calculated as the difference between that in whole serum and in the infranate fraction. Calculation of VLDL cholesterol by difference is considered more reliable than direct measurement in the top fraction because quantitative recovery of the lipoproteins in the top is difficult. Recovery and analysis of the top are nevertheless recommended; comparison of the sum of the top and bottom fractions to the total serum value gives a useful recovery check on technique. Losses in recovery of the bottom can have a substantial effect on reported values for LDL and VLDL.

A shortcut technique which bypasses ultracentrifugation is the so-called "Friedewald calculation" or derived beta-quantification.[38,83] HDL is separated by precipitation and its cholesterol assayed; cholesterol and triglycerides are measured in the serum. VLDL cholesterol is estimated as triglycerides divided by five, an approximation which works reasonably well in most normolipemic specimens. The presence of elevated triglycerides — 400 mg/dl is the accepted limit — chylomicrons, and beta-VLDL characteristic of the rare type III hyperlipoproteinemia precludes this estimation. The estimated VLDL cholesterol and measured HDL cholesterol are subtracted from total cholesterol to estimate or derive LDL cholesterol. This method has been used almost universally in estimating LDL cholesterol in routine clinical practice. Investigations in lipid specialty laboratories have suggested the

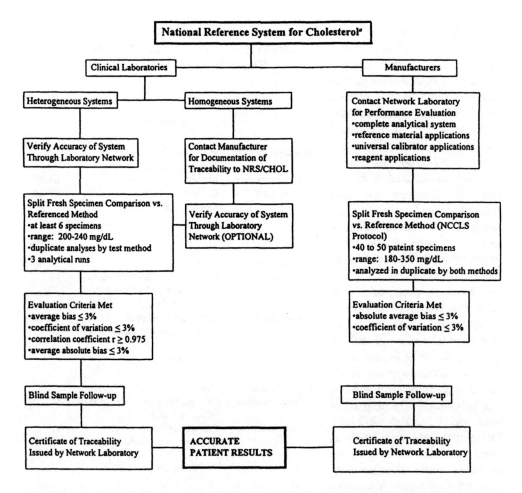

FIGURE 8. National reference system for cholesterol. (From Myers, G.L. and Hassemer, D.J., "Standardization of Lipid and Lipoprotein Measurements", in *Methods for Clinical Laboratory Measurement of Lipid and Lipoprotein Risk Factors,* Rifai, N. and Warnick, G.R., Eds., AACC Press, Washington, D.C., 1991, 101. With permission.)

method is adequate provided the underlying measurements are made with good accuracy and precision.[84,85] There is considerable concern about the reliability in routine laboratories. An expert panel of the NCEP concluded that the level of analytical performance required to derive LDL cholesterol to meet clinical needs was beyond the capability of most routine laboratories and recommended development of alternative methods, preferably ones that separate LDL for direct quantitation.[19]

Absolute accuracy and standardization of results are especially important in the analysis of the lipids and lipoproteins. Since these analytes are risk factors, not diagnostic, decision points cannot be established by an individual laboratory or manufacturer as is done for other diagnostic analytes. Decision points are set by expert panels from review of population distributions and coronary disease estimates established in large, long-term, epidemiological studies. Standardization of the participating research laboratories made results comparable among laboratories and over time. For other research laboratories to generate comparable results and for routine clinical laboratories to obtain reliable classification of patients using the national reference cut points, the methods must be standardized to the same accuracy base of the national studies.

For cholesterol the reference system is quite advanced and complete (Figure 8).[16,86] The definitive method at the National Institute for Standards and Technology provides the accuracy target.[67] A reference method at the CDC, which is calibrated by an approved primary reference standard to the definitive method, provides a transferable, practical reference link to the research laboratories.[66] A network of standardized laboratories, the Cholesterol Reference Method Laboratory Network, established in the U.S. and a few European and Asian countries, provides a link to manufacturers and clinical laboratories.[86] The network was established primarily to provide direct comparisons on fresh serum specimens, necessary for reliable accuracy transfer because of analyte-matrix interactions on processed reference materials. In the early stages of implementing a standardization program directed to manufacturers and routine laboratories, lyophilized or freeze-dried materials were used. These materials were made in large quantities, assayed by the definitive and/or reference methods, and distributed for accuracy transfer. Subsequently, biases were observed in enzymic assays on specimens from patients. Even though such materials are convenient, stable, and amenable to shipment at ambient temperatures, the process of lyophilization altered their properties in enzymic assays, which are sensitive to the nature of the analyte and the specimen matrix, such that results were not representative of results on patient specimens.[87] In order to achieve reliable feedback on accuracy and transfer of the accuracy base, direct comparisons with the reference method on actual patient specimens were determined to be necessary. The Cholesterol Network Program not only performs reliable accuracy comparisons but has developed a formal certification program whereby laboratories and manufacturers can document tracability to the National Reference System for Cholesterol.

This network program will eventually be expanded to include triglycerides, HDL, and LDL cholesterol. At the present time, standardization of these analytes is much less complete than for cholesterol. There are accepted reference methods but no definitive methods (Table 7). The CDC has conducted a Lipoprotein Standardization Program in conjunction with the National Heart Lung and Blood Institute covering total and HDL cholesterol and triglycerides available

TABLE 7

Lipid/Lipoprotein Analysis Methods with the Hierarchy of the Reference System

		Ref.
Cholesterol		
Definitive method	Isotope dilution Mass spectroscopy	67
Reference method	Modified Abell-Kendall	65, 66
Routine method	Enzymic sequence Esterase, oxidase, peroxidase	68, 69
HDL		
Definitive method	None	
Reference method	CDC method (ultracentrifugation, precipitation, Abell-Kendall cholesterol assay)	86
Routine methods	Chemical precipitation with enzymic cholesterol assay	18, 44
LDL		
Definitive method	None	
Reference method	Beta-quantification (CDC method) (same as HDL)	86
Routine methods	Beta-quantification	47
	Friedewald estimation	83
Triglyceride		
Definitive method	None	
Referral method	Chromotropic acid	78
Routine methods	UV 340	72
	GPO	73
Phospholipid		
Definitive method	None	
Reference method	None	
Routine methods	Nonenzymic	82
	Enzymic	80, 81

TABLE 8
Analytical Performance Limits
Recommendations for 1993

	Precision CV (%)	Bias (%)	Total error (%)
Cholesterol	3	±3	±8.9
HDL			
≥42 mg/dl	6	±10	±21.8
<42 mg/dl	SD ≤2.5		
LDL	4	±4	±11.8
Triglycerides			
>220 mg/dl	5	±5	±14.8

to NIH-funded research laboratories. A major deficiency, especially considering that LDL cholesterol is the important decision analyte in clinical practice, is that there is no formal standardization program for LDL cholesterol.

Expert laboratory panels of the NCEP[16-19] have established requisite analytical performance goals based on clinical needs for routine measurements (Table 8). For total cholesterol analysis the performance goal for total error is 8.9%. That is, the bias and imprecision should be such that each individual cholesterol measurement falls within ±8.9% of the reference method value. Actually, a statistical nuance is that since the goals are based on 95% certainty, 95 of 100 measurements should fall within the limits. Imprecision describes the random variability and is specified in terms of CV, coefficient of variation, or relative standard deviation, the standard deviation divided by the mean. CV is used since variation is often proportional to the level. One can assay a specimen many times and calculate the mean to determine the usual value or the central tendency. The standard deviation indicates the imprecision or variation around the mean. The standard deviation divided by the mean is the coefficient of variation. Bias describes the systematic error or overall accuracy; the difference between the mean and the true value gives the bias at that level. Bias is primarily a function of the method's calibration and may be different at different levels. Of greatest interest is bias at the decision points. The bias and CV targets listed in Table 8 are representative of performance which will meet the NCEP total error goals.

Similar performance goals are given for triglycerides and HDL and LDL cholesterol (Table 8). Accuracy in the triglyceride measurement is not as critical because the cut points are widespaced and because the physiological variation is so large, CV about 25 to 30%, that analytical variation is relatively insignificant. Accuracy in LDL and HDL cholesterol measurement is important and a substantial challenge with existing methods. Improvements will need to be made in routine laboratory measurements to provide results which reliably classify patients and monitor their treatment.

Acceptable analytical performance derives from following accepted principles of laboratory quality assurance. Aspects of quality assurance include selection of reliable methods, reagents and equipment, use of appropriate applications, following detailed protocols, provision for well-trained and motivated staff, and vigorous follow-up on performance problems. An essential element is analysis of quality control specimens. Specimens should preferably closely emulate actual patient specimens. For the present, quality control pools that are prepared from freshly collected patient serum, pooled, aliquoted into securely sealed vials, and quickly frozen are most suitable. Pools of fresh-frozen serum are essential for monitoring lipoprotein separation and analysis and preferable for monitoring cholesterol and other lipid measurements. Commercial lyophilized pools undergo matrix alterations which change their analysis characteristics, so results often do not represent results on patients. At least two pools should be analyzed, preferably with levels at or near decision points.

This review provides background on the research and clinical contexts that motivate interest in separation of the lipoproteins and measurement of the serum lipids. Major separation methods and the lipid quantitation methods are reviewed. The importance of accuracy and standardization is emphasized. The central role of the lipoproteins in metabolism and their association with major disease conditions such as coronary disease will maintain interest in their measurement for years to come.

REFERENCES

1. National Cholesterol Education Program, Second report of the expert panel on detection, evaluation, and treatment of high blood cholesterol in adults (Adult Treatment Panel II), NCEP 1993.
2. National Cholesterol Education Program, Summary of the second report of the National Cholesterol Education Program (NCEP) expert panel on detection, evaluation, and treatment of high blood cholesterol in adults (Adult Treatment Panel II), *J.A.M.A.*, 260(23), 3015, 1993.
3. **Gotto, A.M., Jr., Pownall, H.J., and Havel, R.J.,** Introduction to the plasma lipoproteins, in *Methods in Enzymology*, Segrest, J. P. and Albers, J. J., Eds., Academic Press, Orlando, FL, 1986, 128:3.
4. **Smith, L.C., Massey, J.B., Sparrow, J.T., Gotto, A.M., Jr., and Pownall, H.J.,** *Supramolecular Structure and Function*, Pifat, G. and Herak, J. H. N., Eds., Plenum Press, New York, 1983, 210.
4a. **Smith, L.C., Massey, J.B., Sparrow, J.T., Gotto, A.M., Jr., and Pownall, H.J.,** *Supramolecular Structure and Function*, Pifat, G. and Herak, J.H.N., Eds., Plenum Press, New York, 1983, 213.
5. **Loscalzo, J.,** Lipoprotein (a): a unique risk factor for atherothrombotic disease, *Arteriosclerosis*, 10(5), 672, 1990.
6. **Reese, A.,** Lipoprotein (a): a possible link between lipoprotein metabolism and thrombosis, *Br. Heart J.*, 65, 2, 1991.
7. **Stalmpfer, M.J., Sacks, F.M., Salvini, S., Willett, W.C., and Hennekens, C.H.,** A prospective study of cholesterol, apolipoproteins, and the risk of myocardial infarction, *N. Engl. J. Med.*, 325, 6, 1991.
8. **Fruchart, J.C.,** Lipoprotein heterogeneity and its effect on apolipoprotein assays, *Scand. J. Clin. Lab. Invest.*, 50 (Suppl. 198), 51, 1990.
9. **Fruchart, J.D. and Ailhaud, G.,** Apolipoprotein A-containing lipoprotein particles: physiological role, quantification, and clinical significance, *Clin. Chem.*, 38(6), 793, 1992.
10. National Cholesterol Education Program, Report of the National Cholesterol Education Program expert panel on detection, evaluation and treatment of high blood cholesterol in adults, *Arch. Intern. Med.*, 148, 36, 1988.
11. Consensus Development Conference, Lowering blood cholesterol to prevent heart disease, *J.A.M.A.*, 253, 2080, 1985.
12. National Cholesterol Education Program, Population Strategies for Blood Cholesterol Reduction (Executive Summary), NIH Publ. No. 90-3047, 1990.
13. U.S. Department of Health and Human Services, Recommendations Regarding Public Screening for Measuring Blood Cholesterol, NIH Publ. No. 89-3045, 1989.
14. National Cholesterol Education Program, Report of the Expert Panel on Blood Cholesterol Levels in Children and Adolescents, NIH Publ. No. 91-2732, 1991.
15. National Cholesterol Education Program Laboratory Standardization Panel, Current status of blood cholesterol measurements in clinical laboratories in the United States, *Clin. Chem.*, 34, 193, 1988.
16. Laboratory Standardization Panel of the National Cholesterol Education Program, Recommendations for Improving Cholesterol Measurement, NIH Publ. No. 90-2964, 1990.
17. **Stein, E.A. and Meyers, G. L.,** Recommendations for triglyceride measurement, National Cholesterol Education Program Lipoprotein Measurement Working Group, November 1991, in press.
18. **Warnick, G.R. and Wood, P. D.,** Recommendations for measurement of high density lipoprotein cholesterol, National Cholesterol Education Program Lipoprotein Measurement Working Group, September 1992, in press.
19. **Bachorik, P.S. and Ross, J. W.,** Recommendations for measurement of low density lipoprotein cholesterol, National Cholesterol Education Program Lipoprotein Measurement Working Group, November 1991, in press.
20. Triglyceride, high density lipoprotein, and coronary heart disease, *NIH Consens. Dev. Conf. Consens Statement 1992 Feb 26-28*, 10(2), 1, 1992.
21. **Warnick, G.R. and Dominiczak, M.H.,** Separation and quantitation of lipoproteins, *Curr. Opinion Lipidol.*, 1(6), 493, 1990.
22. **Hatch, F.T. and Lees, R.S.,** Practical methods for plasma lipoprotein analysis, *Adv. Lipid Res.*, 6, 1, 1968.
23. **Havel, R.J., Eder, H.A., and Bragdon, J.H.,** The distribution and chemical composition of ultracentrifugally separated lipoproteins in human serum, *J. Clin. Invest.*, 34, 1345, 1955.
24. **Redgrave, T.G., Roberts, D.C.K., and West, C.E.,** Separation of plasma lipoproteins by density-gradient ultracentrifugation, *Anal. Biochem.*, 65, 42, 1975.
25. **Chapman, M.J., Goldstein, S., Lagrange, D., and Laplaud, P.M.,** A density gradient ultracentrifugal procedure for the isolation of the major lipoprotein classes from human serum, *J. Lipid Res.*, 22, 339, 1981.
26. **Rosseneu, M., Van Diervliet, J.P., Bury, J., and Vinaimont, N.,** Isolation and characterization of lipoprotein profiles in newborns by density gradient ultracentrifugation, *Pediatric Res.*, 17(10), 788, 1983.
27. **Kukarni, K.R., Garber, D.W., Schmidt, C.F., Marcovina, S.M., Ho, M.H., Wilhite, B.J., Beaudrie, K.R., and Segrest, J.P.,** Analysis for cholesterol in all lipoprotein classes by single vertical ultracentrifugation of fingerstick blood and controlled-dispersion flow analysis, *Clin. Chem.*, 38(9), 1898, 1992.

28. **Patsch, J.R. and Patsch, W.,** Zonal ultracentrifugation, in *Methods of Enzymology,* Albers, J. J. and Segrest, J. P., Eds., Academic Press, Orlando, FL, 1986, 129:3.
29. **Wu, L.L., Warnick, G.R., Wu, J.T., Williams, R.R., and Lalouel, J.M.,** A rapid micro-scale procedure for determination of the total lipid profile, *Clin. Chem.,* 35(7), 1486, 1989.
30. **Brousseau, T., Clavey, V., Bard, J.M., and Fruchart, J.C.,** Sequential ultracentrifugation micromethod for separation of serum lipoproteins and assays of lipids, apolipoproteins, and lipoprotein particles, *Clin. Chem.,* 39(6), 960, 1993.
31. **Lewis, L.A. and Opplt, J.J.,** *CRC Handbook of Electrophoresis (Volume I),* CRC Press, Boca Raton, FL, 1980.
32. **Lewis, L.A. and Opplt, J.J.,** *CRC Handbook of Electrophoresis (Volume II),* CRC Press, Boca Raton, FL, 1980.
33. **Noble, R.P.,** Electrophoretic separation of plasma lipoproteins in agarose gel, *J. Lipid Res.,* 9, 693, 1968.
34. **Lindgren, F.T., Silvers, J., Jutagir, R., et al.,** A comparison of simplified methods for lipoprotein quantitation using the analytic ultracentrifuge as a standard, *Lipids,* 12, 278, 1977.
35. **Conlon, D., Blankstein, L.A., and Pasakarnis, P.A.,** Quantitative determination of high-density lipoprotein cholesterol by agarose gel electrophoresis updated, *Clin. Chem.,* 24, 227, 1979.
36. **Papadopoulos, N.M.,** Hyperlipoproteinemia phenotype determination by agarose gel electrophoresis updated, *Clin. Chem.,* 24, 227, 1978.
37. **Warnick, G.R., Nguyen, T., Bergelin, R.O., Wahl, P.W., and Albers, J.J.,** Lipoprotein quantification: an electrophoretic method compared with the Lipid Research Clinics method, *Clin. Chem.,* 28, 116, 1982.
38. **Rifai, N., Warnick, G.R., McNamara, J.R., Belcher, J.D., Grinstead, G.F., and Frantz, Jr., I.D.,** Measurement of low-density-lipoprotein cholesterol in serum: a status report, *Clin. Chem.,* 38(1), 150, 1992.
39. **Warnick, G.R., Leary, E.T., and Goetsch, J.,** Electrophoretic quantification of LDL-cholesterol using the Helena REP, *Clin. Chem.,* 39(6)(Abstr. 0011), 1122, 1993.
40. **Muniz, N.,** Measurement of plasma lipoproteins by electrophoresis on polyacrylamide gel, *Clin. Chem.,* 23, 1826, 1977.
41. **Krauss, R.M., Lindgren, F.T., and Ray, R.M.,** Interrelationships among subgroups of serum lipoproteins in normal human subjects, *Clin. Chem. Acta,* 104, 275, 1980.
42. **McNamara, J.R., Campos, H., Ordovas, J.M., Peterson, J., Wilson, P.W.F., and Schaefer, E.,** Effect of gender, age, and lipid status on low density lipoprotein subfraction distribution, *Atheriosclerosis,* 7(5), 483, 1987.
43. **Burstein, M. and Legmann, P.,** Lipoprotein precipitation, in *Monographs on Atherosclerosis,* Vol. 2, Clarkson, T. B., Kritchevsky, D., and Pollak, O. J., Eds., S. Karger, New York, 1982, 1.
44. **Levin, S.J.,** High-density lipoprotein cholesterol: review of methods, American Society of Clinical Pathologists, Check sample, Core Chemistry, No. PTS 89-2(PTS-36):5(2), 1989.
45. **Burstein, M. and Samaille, J.,** Sur undosage rapide du cholestérol lie aux α- et aux β-lipoproteins du serum, *Clin. Chim. Acta,* 5, 609, 1960.
46. **Fredrickson, D.S., Levy, R.I., and Lindgren, F.T.,** A comparison of heritable abnormal lipoprotein patterns as defined by two different techniques, *J. Clin. Invest.,* 47, 2446, 1968.
47. **U.S. Department of Health and Human Services,** Lipid Research Clinics Program. Manual of Laboratory Operations, 2nd ed., Hainline, Jr., A., Karon, J., and Lippel, K., Eds., National Institutes of Health, 1983.
48. **Steele, B.W., Koehler, D.F., Azar, M.M., Blaszkowski, T.P., Kuba, K., and Dempsey, M.E.,** Enzymatic determinations of cholesterol in high-density-lipoprotein fractions prepared by a precipitation technique, *Clin. Chem.,* 22, 98, 1976.
49. **Burstein, M. and Scholnick, H.R.,** Lipoprotein-polyanion-metal interactions, *Adv. Lipid Res.,* 11, 68, 1973.
50. **Lopes-Virella, M.F., Stone, P., Ellis, S., and Colwell, J.A.,** Cholesterol determination in high-density lipoproteins separated by three different methods, *Clin. Chem.,* 23, 882, 1977.
51. **Warnick, G.R., Cheung, M.C., and Albers, J.J.,** Comparison of current methods for high-density lipoprotein cholesterol quantitation, *Clin. Chem.,* 25, 596, 1979.
52. **Viikari, J.,** Precipitation of plasma lipoproteins by PEG-6000 and its evaluation with electrophoresis and ultracentrifugation, *Scand. J. Clin. Lab. Invest.,* 36, 265, 1976.
53. **Allen, J.K., Hensley, W.J., Nicholls, A.V., and Whitfield, J.B.,** An enzymic and centrifugal method for estimating high-density lipoprotein cholesterol, *Clin. Chem.,* 25, 325, 1979.
54. **Demacker, P.N.M., Vos-Janssen, H.E., Hijmans, A.G.M., Van't Laar, A., and Jansen, A.P.,** Measurement of high-density lipoprotein cholesterol in serum: comparison of six isolation methods combined with enzymic cholesterol analysis, *Clin. Chem.,* 1980a(26), 1780, 1980.
55. **Schumaker, V.N., Robinson, M.T., Curtiss, L.K., Butler, R., and Sparkes, R.S.,** Anti-apoprotein B monoclonal antibodies detect human low density lipoprotein polymorphism, *J. Biol. Chem.,* 259, 6423, 1984.
56. **Kerscher, L., Schiefer, S., Draeger, B., Maier, J., and Ziegenhorn, J.,** Precipitation methods for the determination of LDL-cholesterol, *Clin. Biochem.,* 18, 118, 1985.
57. **James, R.W., Proudfoot, A., and Pometta, D.,** Immunoaffinity fractionation of high-density lipoprotein subclasses 2 and 3 using anti-apolipoprotein A-I and A-II immunosorbent gels, *Biochim. Biophys. Acta,* 1002, 292, 1989.

58. **Weisgraber, K.H. and Mahley, R.W.,** Subfractionation of human high density lipoproteins by heparin-Sepharose affinity chromatography, *J. Lipid Res.,* 21, 316, 1980.
59. **Evans, A.J., Huff, M.W., and Wolfe, B.M.,** Accumulation of an apoE-poor subfraction of very low density lipoprotein in hypertriglyceridemic men, *J. Lipid Res.,* 30, 1691, 1989.
60. **Rudel, L.L., Marzetta, C.A., and Johnson, F.L.,** Separation and analysis of lipoproteins by gel filtration, in *Methods in Enzymology 129,* Albers, J. J. and Segrest, J. P., Eds., Academic Press, Orlando, FL, 1986, 45.
61. **Giddings, J.C.,** Hyperlayer field-flow fractionation, *American Laboratory,* p. 20D, November 1992.
62. **Otvos, J.D., Jeyarajah, E.J., and Bennett, D.W.,** Quantification of plasma lipoproteins by proton nuclear magnetic resonance spectroscopy, *Clin. Chem.,* 37(3), 377, 1991.
63. **Otvos, J.D., Jeyarajah, E.J., Bennett, D.W., and Krauss, R.M.,** Development of a proton nuclear magnetic resonance spectroscopic method for determining plasma lipoprotein concentrations and subspecies distributions from a single, rapid measurement, *Clin. Chem.,* 38(9), 1632, 1992.
64. **Zak, B.,** Cholesterol methodologies: a review, *Clin. Chem.,* 23(7), 1201, 1977.
65. **Abell, L.L., Levy, B.B., Brody, B.B., and Kendall, F.C.,** A simplified method for the estimation of total cholesterol in serum and demonstration of its specificity, *J. Biol. Chem.,* 195, 357, 1952.
66. **Duncan, I.W., Mather, A., and Cooper, G.R.,** The Procedure for the Proposed Cholesterol Reference Method, Division of Environmental Health Laboratory Sciences, CEH, Centers for Disease Control, Atlanta, GA, 1982.
67. **Cohen, A., Hertz, H.S., Mandel, J., et al.,** Total serum cholesterol by isotope dilution/mass spectrometry: a candidate definitive method, *Clin. Chem.,* 26, 854, 1980.
68. **Richmond, W.,** Preparation and properties of a cholesterol oxidase from *Nocardia* sp. and its application to the enzymatic assay of total cholesterol in serum, *Clin. Chem.,* 19, 1350, 1973.
69. **Allain, C.C., Poon, L.S., Chan, C.S.G., et al.,** Enzymatic determination of total serum cholesterol, *Clin. Chem.,* 20, 470, 1974.
70. **Wiebe, D.A. and Bernert, J.T., Jr.,** Influence of incomplete cholesteryl ester hydrolysis on enzymic measurements of cholesterol, *Clin. Chem.,* 30(3), 352, 1984.
71. **McGowan, M.W., Artiss, J.D., and Zak, B.,** Spectrophotometric study on minimizing bilirubin interference in an enzyme reagent mediated cholesterol reaction, *Microchem. J.,* 27, 564, 1982.
72. **Bucolo, G., Yabut, J., and Chang, T.Y.,** Mechanized enzymatic determination of triglycerides in serum, *Clin. Chem.,* 21(3), 420, 1975.
73. **McGowan, M., Artiss, J., Strandbergh, D.R., and Zak, B.,** A peroxidase-coupled method for the colorimetric determination of serum triglycerides, *Clin. Chem.,* 29(3), 538, 1983.
74. **Stinshoff, K., Weisshaar, D., Staehler, F., Hesse, D., Gruber, W., and Steier, E.,** Relation between concentrations of free glycerol and triglycerides in human sera, *Clin. Chem.,* 23(6), 1029, 1977.
75. **Cole, T.G.,** Glycerol blanking in triglyceride assays: is it necessary?, *Clin. Chem.,* 36(7), 1267, 1990.
76. **Warnick, G.,** Enzymatic methods for quantification of lipoprotein lipids, in *Methods in Enzymology 129,* Albers, J. J. and Segrest, J. P., Eds., Academic Press, Orlando, FL, 1986, 101.
77. **Sullivan, D.R., Kruijswijk, Z., West, C.E., Kohlmeier, M., and Katan, M.B.,** Determination of serum triglycerides by an accurate enzymatic method not affected by free glycerol, *Clin. Chem.,* 31(7), 1227, 1985.
78. **Lofland, Jr., H. B.,** A semiautomated procedure for the determination of triglycerides in serum, *Anal. Biochem.,* 9, 393, 1964.
79. **Yamaguchi, Y.,** Determination of high-density lipoprotein phospholipids in serum, *Clin. Chem.,* 26(9), 1275, 1980.
80. **Takayama, M., Itoh, S., Nagasaki, T., and Tanimizu, I.,** A new enzymatic method for determination of serum choline-containing phospholipids, *Clin. Chim. Acta,* 79, 93, 1977.
81. **McGowan, M.W., Artiss, J.D., and Zak, B.,** A procedure for the determination of high-density lipoprotein choline-containing phospholipids, *J. Clin. Chem. Clin. Biochem.,* 20, 807, 1982.
82. **Bartlett, G.R.,** Phosphorus assay in column chromatography, *J. Biol. Chem.,* 234(3), 466, 1959.
83. **Friedewald, W.T., Levy, R.I., and Fredrickson, D.S.,** Estimation of the concentration of low-density lipoprotein cholesterol in plasma, without use of the preparative ultracentrifuge, *Clin. Chem.,* 18(6), 499, 1972.
84. **Warnick, G.R., Knopp, R.H., Fitzpatrick, V., and Branson, L.,** Estimating low-density lipoprotein cholesterol by the Freidewald equation is adequate for classifying patients on the basis of nationally recommended cutpoints, *Clin. Chem.,* 36(1), 15, 1990.
85. **McNamara, J.R., Cohn, J.S., Wilson, P.W.F., and Schaefer, E.J.,** Calculated values for low-density lipoprotein cholesterol in the assessment of lipid abnormalities and coronary disease risk, *Clin. Chem.,* 36(1), 36, 1990.
86. **Myers, G.L. and Hassemer, D.J.,** Standardization of lipid and lipoprotein measurements, in *Methods for Clinical Laboratory Measurement of Lipid and Lipoprotein Risk Factors,* Rifai, N. and Warnick, G. R., Eds., AACC Press, Washington, D.C., 1991, 101.
87. **Eckfeldt, J.H. and Copeland, K.R.,** Accuracy verification and identification of matrix effects, *Arch. Pathol. Lab. Med.,* 117, 381, 1993.

Section 2: Effect of Food Lipids on Health

Chapter 2.1

EFFECTS OF DIETARY FATTY ACIDS AND CHOLESTEROL ON CARDIOVASCULAR DISEASE RISK FACTORS IN MAN

Ruth McPherson and Gene A. Spiller

Cardiovascular disease (CVD) constitutes one of the most important health problems in North America and accounts not only for the most deaths, but also for the greatest proportion of lost future earnings, the largest number of hospital days, and the greatest number of pharmaceutical prescriptions per year of any major disease.[1] One of the major risk factors for CVD (and the most prevalent single risk factor) is hypercholesterolemia. This risk factor is clearly related to diet and life-style factors and is certainly amenable to intervention. The National Cholesterol Education Program has once again emphasized the need for more basic science research and public health education to facilitate the necessary changes in nutritional behavior.[2]

Diet is the most important environmental variable affecting the plasma lipoprotein spectrum and ischemic heart disease (IHD) risk. Specific nutritional factors altering human lipoprotein metabolism include the content and composition of dietary fat, particularly carbon chain length, degree of saturation, position of the first double bond, and cis or trans configuration. Other significant dietary elements include dietary cholesterol, dietary fibers, alcohol, energy balance, and obesity.

This chapter addresses an important topic: the effects of fatty acids (saturated fatty acids [SFA], monounsaturated fatty acids [MUFA], and polyunsaturated fatty acids [PUFA]) and of dietary cholesterol on plasma lipoproteins. Also addressed is a question believed to be of significance in the industrial processing of fats and oils, that is, the relative effect of trans fatty acids (TFA) (mainly elaidic acid, a partially hydrogenated product of oleic acid) and stearic acid (produced by more complete hydrogenation of oleic acid) on plasma lipoproteins. The relative effects of these two fatty acids are of importance, since moderate amounts of TFA are present in many foods such as margarines and elimination of TFAs would be possible if (the apparently lipid-neutral SFA) stearic acid was substituted for TFA as a source of hardened fat.

SATURATED FATTY ACIDS

There is agreement that certain, but not all, SFAs raise plasma levels of low density lipoprotein (LDL) cholesterol.

MEDIUM CHAIN TRIGLYCERIDES (MCT) C4:0-C10:0

Medium chain fatty acids, although present in a small number of food sources, do not appear to raise LDL cholesterol levels. MCTs behave much like complex carbohydrates in terms of effects on LDL cholesterol levels.[3]

LAURIC (C12:0), MYRISTIC (C14:0), AND PALMITIC ACIDS (C16:0)

The early studies of Keys[4] and Hegsted[5] demonstrated that myristic (C14:0) and palmitic (C16:0) acids were more hypercholesterolemic than stearic acid (C18:0). Based on a series of experiments on the cholesterol raising effects of butterfat and other natural fats and oils, Hegsted[5] concluded that myristic acid is more hypercholesterolemic than palmitic acid. Mensink has recently reviewed the published results of controlled dietary trials and also concluded that myristic acid is more hypercholesterolemic than palmitic acid.[6] In agreement with the studies of Hegsted and colleagues and the recent studies of Denke and Grundy,[7]

Mensink has also reviewed experimental data that demonstrates that lauric acid is less hypercholesterolemic than myristic or palmitic acids.[8]

MECHANISMS FOR THE HYPERCHOLESTEROLEMIC EFFECTS OF C:12-C:16 FATTY ACIDS

Kinetic studies suggest that the LDL-C raising effect of the C12 to C16 saturated fatty acids is mediated by suppression of LDL-receptor clearance of LDL.[9-13] The mechanism of this effect is unclear but may be related to changes in a putative regulatory pool of hepatic unesterified cholesterol.[12] SFAs have little effect on very low density lipoprotein (VLDL) triglyceride concentration and there are few data to suggest that VLDL apo B synthesis is altered. Reports are conflicting regarding the effect of SFA on hepatic mRNA for the LDL receptor;[14,15] the mRNA for apoB does not appear to be altered in nonhuman primates.[15]

EFFECTS OF STEARIC ACID (C18:0) ON LDL-CHOLESTEROL CONCENTRATIONS

The early studies of Ahrens et al.[16] demonstrated that cocoa butter, which is rich in C18:0, does not raise cholesterol levels. Recently, Bonanome and Grundy[17] demonstrated that substitution of stearic acid for palmitic acid in a liquid formula diet was as effective as oleic acid in reducing LDL cholesterol in 11 subjects. The mechanism by which stearic acid lowers LDL is not clear, but may be related to the fact that this fatty acid is rapidly desaturated to oleic acid[18] and does not accumulate in plasma triglycerides in subjects fed a C18:0-enriched diet.[17] Although stearic acid is found in significant quantities in beef and pork fat, it is a prominent component in few fats other than cocoa butter. However, stearic acid could provide a useful substitute for other hard fats in foods such as margarine. The hypocholesterolemic effect of stearic acid deserves confirmation in a larger number of subjects, consuming solid food diets.

UNSATURATED FATTY ACIDS

The major PUFA found in human diets (linoleic acid [C18:2,n-6]) has a net LDL cholesterol lowering effect which is approximately half of the LDL raising effect of SFAs.[4,5] For many years, linoleate was considered the only prevalent fatty acid with significant hypocholesterolemic effect. The early work by Keys[4] and Hegsted[5] suggested that MUFAs (mainly oleic acid [C18:1]) were relatively lipid neutral. However, more recent studies by Mattson and Grundy,[19] albeit in patients ingesting a liquid formula diet, demonstrated that oleic acid was as effective as linoleic acid in lowering LDL-C. Later studies by this group, again using liquid formula diets, suggested that a diet enriched in MUFA was also as effective as a very low fat (LF) diet in lowering LDL-C and prevented the fall in HDL-C associated with a LF diet.[20] Mensink and Katan, in a less rigorously controlled study, reported that MUFAs were as effective as PUFA in lowering LDL-C; in this study no change was seen in HDL-C in either of these diets as compared to a SFA-rich diet.[21] Valsta et al.[22] demonstrated that a diet containing oleic acid-rich rapeseed oil lowered LDL-C to a similar extent as a high linoleate sunflower oil, but reduced HDL_2 and VLDL-C to a lesser extent. Finally, Ginsberg and colleagues[23] reported that replacement of carbohydrate by MUFA in an American Heart Association (AHA) Step I diet did not jeopardize the LDL-C lowering effect of the Step I diet. Nonetheless, there are also several reports[4-6,24] demonstrating that PUFAs are somewhat more effective LDL-C lowering fatty acids as compared to MUFAs.

MUFAs have the dietary advantage of being less susceptible to oxidation than are PUFAs.[25] Endogenous antioxidants may be limiting in human diets relative to linoleic acid intake.[26] Rabbits fed diets rich in linoleic acid for 10 weeks were shown to have LDL that was more susceptible to oxidation than that from oleate-fed animals.[27] Three recent human studies have also reported that LDL is more easily oxidized when the diet is rich in PUFA as compared to

MUFA,[28-30] but the level of dietary linoleic acid in these studies (up to 30% of calories) exceeded plausible dietary intakes. The extent and significance of oxidative modification of lipoprotein fatty acids in diets containing variable levels of polyunsaturated fatty acids and antioxidants deserve further study.[25]

PUFA may lower total HDL-C or HDL_2-C concentrations when incorporated into diets, but only at high[19,22,31,32] and not at moderate[33] levels; MUFAs appear to maintain HDL levels[19,22,23,31] almost comparable to those when SFAs are the prominent dietary fatty acids. The mechanism of the effect of different unsaturated fatty acids on HDL concentration and composition and the significance of such effects in terms of reverse cholesterol transport are not clear.

TRANS ISOMERS OF C18:1 VS. STEARIC ACID

Trans isomers of natural cis-oleic acid may be present in significant amounts in hydrogenated vegetable oils. Anderson et al.[34] demonstrated that incorporation of elaidic acid (C18:2,n-9,trans) into a typical American diet caused significant increases in serum cholesterol, whereas McOsker et al.[35] found little effect on blood lipid when this trans fatty acid was incorporated into a cholesterol-poor diet. More recently, Mensink and Katan[36] have demonstrated that replacement of dietary oleic acid with trans fatty acids at a level of 10% of calories (about threefold usual American intakes) had LDL-C raising effects similar to SFA and, more significantly, also lowered HDL cholesterol. Thus, trans fatty acids appear to have effects on plasma lipoproteins that are at least as detrimental as those of SFA. If these findings are confirmed in a larger group of subjects, it may be necessary to find substitutes, such as stearic acid, for TFA in human diets. Thus more complete hydrogenation of oleic acid to stearic acid could well be a useful strategy to eliminate trans fatty acids from the American diet without increasing the amount of LDL-raising fatty acids.

EFFECTS OF DIETARY CHOLESTEROL ON SERUM LIPOPROTEINS

The effects of dietary cholesterol effects on plasma lipoproteins have been an area of controversy in human nutrition. Responses appear to vary considerably for different individuals, raising the public health question as to the degree of reduction in cholesterol intake required for the population at large. As yet, rapid methods are not available to identify those individuals most susceptible to the effects of dietary cholesterol on these variables.

HETEROGENEITY OF RESPONSE

Even under conditions of rigorous dietary control, virtually all studies have demonstrated a marked heterogeneity of response to dietary cholesterol. In most studies, approximately 10% of individuals appear to be hyperresponders, while a similar proportion exhibit little change in plasma cholesterol in response to dietary cholesterol. In a review of the Minnesota experiments, Jacobs and colleagues[37] reported that 64% of individuals tested showed plasma cholesterol responses within 30% of the predicted value of Keys.[4] The factors influencing individual responsiveness are not known. In the first well-controlled metabolic studies carried out by Keys,[4] the magnitude of the plasma cholesterol change was directly related to baseline cholesterol levels. However, Mistry et al.[38] and McNamara et al.[39] found no effect of initial cholesterol concentration on response to dietary cholesterol. In a small number of subjects, Packard and Shepherd[40] found that baseline HDL-cholesterol levels did not predict response. It has been suggested that individuals may adapt to an increased cholesterol intake by a reduction in cholesterol absorption. This response appears to occur at very high cholesterol intakes and a modest difference in cholesterol absorption was noted between hyper- and hyporesponders to an 800-mg/d load of dietary cholesterol.[39] It has been proposed that apolipoprotein E polymorphism may influence the lipoprotein response to dietary cholesterol. In population studies, the apoprotein E phenotypes E4/4 and E3/4 have been reported to be associated with higher cholesterol levels when the indigenous diet is high in fat and cholesterol as in Germany, Canada, and Finland.[41] In population groups consuming a low cholesterol diet, the effect of E phenotype is less striking, suggesting that the E phenotype may alter metabolic responses to dietary cholesterol.[41,42] Kesaniemi et al.[43] have reported that cholesterol absorption efficiency is reduced in subjects heterozygous or homozygous for the E2 allele and increased in subjects carrying an E4 allele. However, a recent well-controlled study[44] demonstrated that the LDL-C response to dietary cholesterol did not vary according to the apo E genotype. However, the increase in HDL-C with cholesterol feeding varied significantly according to the apo E genotype (E3/2: no change; E3/3: +4%; E4/3: +9%).

Effects of Dietary Cholesterol on Hepatic Lipoprotein Metabolism

The liver may respond to an excess of cholesterol by increasing biliary secretion of cholesterol.[45] Bile acid secretion does not appear to be altered in humans.[46] *De novo* cholesterol synthesis may also be reduced by cell regulation of HMG-coA-reductase activity. The latter response has been demonstrated in peripheral cells in humans in response to cholesterol feeding.[39,47] These compensatory mechanisms may not be sufficient to prevent net cholesterol accumulation. As cell cholesterol content increases, down regulation of the B/E receptor has been demonstrated in the rabbit,[48] thus reducing catabolism of LDL and IDL. Receptor-dependent chylomicron remnant metabolism may be less susceptible to alteration in hepatocyte cholesterol content.[49]

Hepatic cholesterol may also be repackaged into lipoproteins in the VLDL spectrum. Cholesterol enrichment of VLDL has been demonstrated,[50] although it is not clear that VLDL-apoB secretion increases.[51] Intermediate density lipoprotein (IDL) (1.006-1.019) cholesterol concentrations may increase in response to dietary cholesterol.[38,51] IDL cholesterol transport was reported to increase markedly in seven of eight subjects consuming 1700 mg/d of cholesterol.[51] Most IDL are converted to particles of higher density or taken up directly by the liver. This process is in part dependent on the size and lipid composition of these particles. Hepatic lipase is important in the hydrolysis of triglyceride and phospholipid, primarily in HDL but also in the VLDL remnant, and may play a role in the hepatic uptake of remnant cholesterol and in the conversion of IDL to LDL.[52] An atherogenic cholesterol-rich lipoprotein in the VLDL density range, but with beta mobility (VLDL), appears in response to cholesterol feeding in many animal species.[53] This lipoprotein, which is characteristic of type III dyslipoproteinemia, also appears with cholesterol feeding in some individuals.[38,54] It is not yet clear whether the rise in LDL pool size which occurs in many individuals in response to cholesterol feeding is due to an increase in apolipoprotein B secretion or to a down regulation of the B/E receptor and reduced LDL catabolism. There is disagreement on the effect of dietary cholesterol on the secretion rate of apoprotein B in the LDL density range. Packard and colleagues demonstrated a 23% increase in LDL synthesis in seven normal subjects when dietary cholesterol was increased from 180 to 1470 mg/d. This change was associated with a 40-mg/dl rise in LDL-cholesterol and a 10% decrease in the fractional catabolic rate (FCR) of LDL.[40] Ginsberg and colleagues, on the other hand, reported that cholesterol feeding did not alter LDL apoB secretion.[55]

Cholesterol feeding may be associated with changes in LDL composition as well as particle number. The effect of dietary cholesterol on the composition of LDL subclasses is not clear. Cholesterol enrichment of LDL has been reported by three groups of investigators.[38,56-58] Schonfeld and colleagues[59] found greater heterogeneity of LDL particle size in response to dietary cholesterol. Diet-induced atherosclerosis in nonhuman primates has been associated with an increase in the size and cholesterol content of LDL.[60]

SUMMARY OF RELATIVE EFFECTS OF FATTY ACIDS AND DIETARY CHOLESTEROL ON SERUM CHOLESTEROL

SUMMARY EQUATIONS OF KEYS AND COLLEAGUES

Keys et al.,[61] in the late 1950s, developed a formula to predict the change in serum cholesterol concentration in response to dietary fatty acids. Based on their own data suggesting that MUFAs were lipid-neutral, MUFAs were not included in the equation.

$$\delta \text{ Chol (mg/dl)} = 2.74 \text{ S} - 1.31 \text{ P}$$

where S = change in the percentage of calories derived from SFA and P = change in percentage of calories derived from PUFA. The equation was later modified to include the effects of dietary cholesterol:[4]

$$\delta \text{ Chol (mg/dl)} = 1.35 \text{ (2S - P)} + 1.5 \text{ Z}$$

where Z = difference between the square root of the initial intake of cholesterol and the experimental intake in cholesterol as milligrams cholesterol/1000 kcal.

SUMMARY EQUATION OF HEGSTED AND COLLEAGUES

Hegsted et al.[5] first developed an equation in 1965[5] after studying the effects of dietary fatty acids and cholesterol on serum cholesterol levels. This differed from that of Keys mainly in terms of the effect of dietary cholesterol.

$$\delta \text{ Chol} = 2.16 \text{ S} - 1.65 \text{ P} + 6.77 \text{ C} - 0.53$$

where C = each 100 mg change in dietary cholesterol/day, S = change in % calories derived from SFA, and P = changes in % calories derived from PUFA.

In 1986[62] Hegsted revised his formula to simplify the last part of his equation as follows:

$$\delta \text{ Chol (mg/dl)} = 2.16 \text{ S} - 1.65 \text{ P} + 0.097 \text{ C}$$

The major difference between the conclusions of Hegsted and Keys is that Hegsted found a linear correlation between dietary cholesterol from 0 to 400 mg/1000 kcal, whereas Keys[4] reported a nonlinear relationship based on the square root of cholesterol intake.

EQUATION OF MENSINK AND KATAN

Mensink and Katan have recently reported the results of a meta-analysis of 27 major dietary studies investigating the effects of specific dietary fatty acids on serum lipoproteins (Table 1).[6,14,18-20,27,30,63-83] They report multiple regression equations for mean changes in serum lipids and lipoproteins when carbohydrate in the diet is replaced isocalorically by SFAs, MUFAs, or PUFAs. The equations for HDL demonstrated that all three classes of fatty acids would raise HDL relative to carbohydrate, with SFA being more effective than MUFA, which in turn was more effective than PUFA. However, the small decrease expected in HDL-C when SFA is replaced by PUFA was minor and not considered to be biologically significant at maximal anticipated levels of intake. Their regression equation for LDL-C demonstrated that SFA markedly elevates LDL-C relative to carbohydrate; PUFAs lower LDL-C relative to carbohydrate, whereas MUFAs have a small but not statistically significant LDL-C lowering effect relative to carbohydrate. Of note in this revised equation, only C12:0 to C16:0 fatty acids are considered cholesterol-raising.

$$\delta \text{ Chol (mg/dl)} = 1.2 \text{ X } (1.8 \text{ } \delta \text{ S}' - 0.1 \text{ } \delta \text{ M} - 0.5 \text{ } \delta \text{ P})$$

TABLE 1
Major Studies of Effects of Dietary Fatty Acids on Plasma Lipids in Man

Study	Year	Ref.	No.	Length (d)	Design	Diet	SFA/MFA/PFA[a]
Grande et al.	1972	63	38	28	Crossover	A	2.2/1.5/0.6
						B	3.2/6.5/2.5
						C	5.0/16.1/6.3
						D	8.3/6.7/12.7
Anderson et al.	1976	64	12	14	Crossover	A	19.6/8.64/5.2
						B	4.8/5.1/22.7
Brussard et al.	1980	65	60	35	Parallel	A	8.0/10.0/3.0
						B	10.0/8.0/11.0
						C	11.0/8.0/11.0
						D	18.0/16.0/3.0
Lewis et al.	1981	30	12	35	Crossover	A	8.7/8.5/8.7
						B	12.7/12.9/12.8
Brussard et al.	1982	66	35	91	Parallel	A	7.0/8.0/4.0
						B	9.0/10.0/11.0
Laine et al.	1982	67	24	20	Crossover	A	16.3/14.1/3.3
						B	7.4/12.4/14.4
						C	7.7/12.7/15.3
Becker et al.	1983	68	12	28	Crossover	A	2.7/29.2/6.5
						B	4.0/15.1/17.5
Harris et al.	1983	69	7	28	Crossover	A	14.4/16.4/7.2
						B	6.4/10.8/21.6
Wolf and Grundy	1983	70	11	30	Crossover	A	18.8/10.0/9.6
						B	9.3/9.9/9.6
						C	14.1/7.2/7.2
Mattson and Grundy	1985	71	12	28	Crossover	A	19.1/15.4/3.9
						B	3.3/28.2/6.9
						C	4.3/5.6/28.1
Reiser et al.	1985	72	19	35	Crossover	A	23.1/6.6/3.0
						B	14.3/15.9/3.1
						C	6.5/7.7/18.9
Grundy	1986	73	7	28	Crossover	A	24.0/7.7/6.7
						B	6.4/6.4/6.4
						C	3.8/26.9/7.7
Grundy et al.	1986	74	9	60	Crossover	A	9.6/12.5/16.3
						B	9.6/9.6/9.6
Mensink and Katan	1987	75	48	35	Parallel	A	6.7/9.3/5.2
						B	9.8/24.0/5.1
Bonanome and Grundy	1988	14	11	21	Crossover	A	19.6/14.9/3.7
						B	3.1/30.6/4.7
Grundy et al.	1988	76	10	42	Crossover	A	6.7/25.9/5.8
						B	6.7/6.7/5.8
Katan et al.	1988	77	47	21	Crossover	A	10.5/12.0/21.0
						B	23.0/14.5/5.2
McDonald et al.	1989	78	8	18	Crossover	A	5.1/20.2/10.3
						B	6.8/7.4/21.6
Mensink and Katan	1989	18	58	35	Parallel	A	12.9/15.1/7.9
						B	12.6/10.8/12.7
Ginsberg et al.	1990	20	24	70	Parallel	A	9.0/10.6/10.0
						B	8.8/17.2/10.1
Mensink and Katan	1990	79	59	21	Crossover	A	9.5/24.1/4.6
						B	19.4/14.7/3.4
Wardlaw and Snook	1990	80	20	35	Crossover	A	6.7/26.9/5.8
						B	7.7/13.4/18.2
Berry et al.	1991	27	18	84	Crossover	A	8.7/17.1/6.3
						B	8.3/6.8/17.4

TABLE 1 (continued)
Major Studies of Effects of Dietary Fatty Acids on Plasma Lipids in Man

Study	Year	Ref.	No.	Length (d)	Design	Diet	SFA/MFA/PFA[a]
Chan et al.	1991	81	8	18	Crossover	A	6.5/18.7/7.4
						B	5.3/18.3/8.5
						C	7.1/8.4/16.8
						D	6.4/9.9/16.1
Wardlaw et al.	1991	82	16	56	Parallel	A	7.2/22.1/10.7
						B	7.4/8.1/22.2
Valsta et al.	1992	19	59	17	Crossover	A	12.4/16.2/7.6
						B	12.7/10.2/13.3
Martin et al.	1990	83	38	23	Crossover	A	10.1/9.9/10.1

[a] SFA: saturated fatty acids; MFA: monounsaturated fatty acids; PFA: polyunsaturated fatty acids. From Mensink, R.P. and Katan, M.B., *Arterioscler. Thromb.*, 12, 911, 1992. Used with permission.

where $S' = $ C12:0 (lauric) + C14:0 (myristic) + C16:0 (palmitic) saturated fatty acids only, M = monounsaturated fatty acids, and P = polyunsaturated fatty acids.

GRANDE AND HEGSTED EQUATION FOR DIETARY CHOLESTEROL

These investigators[62,63] used data from 39 reports to derive the following equation:

$$\delta \text{ Chol (mg/dl)} = 1.5 \, (C_2^{1/2} - C_1^{1/2})$$

where δ Chol = change in serum cholesterol in mg/dl, C_1 = intake of dietary cholesterol per 1000 kcal on initial diet, and C_2 = intake of dietary cholesterol per 1000 kcal on experimental diet.

Of interest, the results of this meta-analysis are similar to the earlier conclusions of the groups of Keys[4] and Hegsted[5,62] and differ from the more recent reports of Ginsberg et al.[23] and Grundy and colleagues.[74]

SUMMARY

Dietary fatty acids and cholesterol are the most important food components altering serum lipoproteins. The C12:0 to C16:0 SFA raise LDL cholesterol, while shorter chain SFA and C18:0 do not alter LDL-C. PUFA lowers LDL-C slightly more effectively than MUFA, although MUFA also has a net LDL-C lowering activity. It is generally agreed that the C12 to C16 SFA should be reduced to 7% or less of total energy intake. Nonetheless, the optimal proportion of calories as carbohydrate, PUFA, and MUFA in the context of normal solid food diets is not yet established. Dietary cholesterol increases LDL-C in most, but not all, individuals and the incremental effect appears to be greatest at lower levels of cholesterol intake.

REFERENCES

1. Canadian Centre for Health Information, Causes of Death, *Health Reports*, Statistics Canada, Catalogue 82-003, 1990, 2(1), 3.
2. National Cholesterol Education Program. Adult Treatment Program II, *J.A.M.A.*, 269, 3015, 1993.
3. **Hashim, S.A., Arteaga, A., and Van Itallie, T.B.,** *Lancet*, 1, 1105, 1960.

4. Keys, A., Anderson, J.T., and Grande, F., *Metabolism*, 14, 776, 1965.

5. Hegsted, D.M., McGandy, R.B., Myers, M.L., and Stare, F.J., *Am. J. Clin. Nutr.*, 17, 281, 1965.

6. Mensink, R.P. and Katan, M.B., *Arterioscler. Thromb.*, 12, 911, 1992.

7. Denke, M.A. and Grundy, S.M., *Am. J. Clin. Nutr.*, 56, 895, 1992.

8. Mensink, R.P., *Am. J. Clin. Nutr.*, Suppl., 711S, 1993.

9. Shepherd, J.C., Packard, C.J., Grundy, S.M., Yeshurun, D., Gotto, A.M., and Taunton, O.D., *J. Lipid. Res.*, 21, 91, 1980.

10. Spady, D.K. and Dietschy, J.M., *Proc. Natl. Acad. Sci. U.S.A.*, 82, 4526, 1985.

11. Nicolosi, R.J., Stucchi, A.F., Kowala, M.C., Hennessy, L.K., Hegsted, D.M., and Schaefer, E.J., *Arteriosclerosis*, 10, 119, 1990.

12. Woolett, L.A., Spady, D.K., and Dietschy, J.M., *J. Clin. Invest.*, 89, 1133, 1992.

13. Grundy, S.M. and Denke, M.A., *J. Lipid Res.*, 31, 1149, 1990.

14. Fox, J.C., McGill, H.C., Carey, K.D., and Getz, G.S., *J. Biol. Chem.*, 262, 9014, 1987.

15. Sorci-Thomas, M., Wilson, M.D., Johnson, F.L., Williams, D.L., and Rudel, L.L., *J. Biol. Chem.*, 264, 9039, 1989.

16. Ahrens, E.H., Hirsch, J., Insull, W., Tsaltas, T.T., Blomstrand, R., and Peterson, M.L., *Lancet*, 1, 943, 1957.

17. Bonanome, A. and Grundy, S.M., *N. Engl. J. Med.*, 318, 1244, 1988.

18. Elovson, J., *Biochim. Biophys. Acta*, 106, 480, 1965.

19. Mattson, F.H. and Grundy, S.M., *J. Lipid Res.*, 26, 194, 1985.

20. Grundy, S.M., *N. Engl. J. Med.*, 314, 745, 1986.

21. Mensink, R.P. and Katan, M.B., *N. Engl. J. Med.*, 321, 436, 1989.

22. Valsta, L.M., Jauhiainen, M., Aro, A., Katan, M.B., and Mutanen, M., *J. Lipid Res.*, 12, 50, 1992.

23. Ginsberg, H.N., Barr, S.L., Gilbert, A., Karmally, W., Decklebaum, R., et al., *N. Engl. J. Med.*, 322, 574, 1990.

24. Chang, N.W. and Huang, P.C., *J. Lipid Res.*, 31, 2141, 1990.

25. Witzum, J.L. and Steinberg, D., *J. Clin. Invest.*, 88, 1785, 1991.

26. Esterbauer, H., Striegl, G., Puhl, H., and Rotheneder, M., *Free Radical Res. Commun.*, 6, 67, 1990.

27. Parthasarathy, S., Khoo, J.C., Miller, E., Barnett, J., Witztum, J.L., and Steinberg, D., *Proc. Natl. Acad. Sci. U.S.A.*, 87, 3894, 1990.

28. Bonanome, A., Pagnan, A., Biffanti, S., Opportuno, A., Sorgato, F., Dorella, M., Maiorino, M., and Ursini, F., *Arterioscler. Thromb.*, 12, 529, 1992.

29. Reaven, P.S., Parthasarathy, S., Grasse, E., et al., *Am. J. Clin. Nutr.*, 54, 701, 1991.

30. Berry, E., Eisenberg, S., Haratz, D., et al., *Am. J. Clin. Nutr.*, 53, 899, 1991.

31. Grundy, S.M., Nix, D., Whelan, M., and Franklin, L., *J.A.M.A.*, 256, 2351, 1986.

32. Schaefer, E.F., Levy, R.I., Ernst, N.D., Van Sant, F.D., and Brewer, H.B., *Am. J. Clin. Nutr.*, 34, 1758, 1981.

33. Lewis, B., Hammett, F., Katan, M., McPherson, K.R., et al., *Lancet* 2, 1310, 1981.

34. Anderson, J.T., Grande, F., and Keys, A., *J. Nutr.*, 75, 388, 1961.

35. McOsker, D.E., Mattson, F.H., Sweringer, H.B., and Kligman, A.M., *J.A.M.A.*, 180, 380, 1962.

36. Mensink, R.P. and Katan, M.B., *N. Engl. J. Med.*, 323, 439, 1990.

37. Jacobs, D.R., Anderson, J.T., Hannan, P., Keys, A., and Blackburn, H., *Arteriosclerosis*, 3, 349, 1983.

38. Mistry, P., Miller, N.E., Laker, M., Hazzard, W.R., and Lewis, B., *J. Clin. Invest.*, 67, 493, 1981.

39. McNamara, D.J., Kolb, R., Parker, T.S., Batwin, H., Samuel, P., Brown, C.D., ??

40. Packard, C.J., McKinney, L., Carr, K., and Shepherd, J., *J. Clin. Invest.*, 72, 45, 1983.

41. Utermann, G., *Am. Heart J.*, 113, 433, 1987.

42. Manttari, M., Koskinen, P., Ehnholm, C., Huttunen, J.K., and Manninen, V., *Metabolism*, 40, 217, 1991.

43. Kesaniemi, Y.A., Ehnholm, C., and Miettinen, T.A., *J. Clin. Invest.*, 80, 578, 1987.

44. Martin, L.J., Connelly, P.W., Nancoo, D., Wood, N., Zhang, Z.J., Maguire, G., Tall, A.R., and McPherson, R., *J. Lipid Res.*, 34, 437, 1993.

45. Quintao, E., Grundy, S.M., and Ahrens, E.H., *J. Lipid Res.*, 12, 233, 1971.

46. Grundy, S.M., *Annu. Rev. Nutr.*, 3, 71, 1983.

47. Harwood, H.L., Bridge, D.M., and Stacpoole, P.W., *J. Clin. Invest.*, 79, 1125, 1987.

48. Kovanen, P.T., Brown, M.S., Basu, S.K., Bilheimer, D.W., and Goldstein, J.L., *Proc. Natl. Acad. Sci. U.S.A.*, 78, 1396, 1981.

49. Havel, R.J., *Annu. Rev. Physiol.*, 48, 119, 1986.

50. Oh, S.Y. and Miller, L.T., *Am. J. Clin. Nutr.*, 42, 421, 1985.

51. Nestel, P.J. and Billington, T., *Metabolism*, 32, 320, 1983.

52. Patch, J.F., Prasad, S., Gotto, A.M., and Patsch, W., *J. Clin. Invest.*, 80, 341, 1987.

53. Fainaru, M., Mahley, R.W., Hamilton, R.L., and Innerarity, T.L., *J. Lipid Res.*, 23, 702, 1982.

54. Mahley, R.W., Innerarity, T., Bersot, T.P., Lipson, A., and Margolis, S., *Lancet*, 2, 807, 1978.

55. Ginsberg, H., Le, N.A., Mays, C., Gibson, J., and Brown, W.V., *Arteriosclerosis,* 1, 463, 1981.
56. Sacks, F.M., Handysides, G.H., Marais, G.E., Rosner, B., and Kass, E.H., *Arch. Intern. Med.,* 146, 1573, 1986.
57. Beynen, A.C. and Katan, M.B., *Atherosclerosis,* 54, 157, 1985.
58. Zanni, E.E., Zannis, V.I., Blum, C.B., Herbert, P.N., and Breslow, J.L., *J. Lipid Res.,* 28, 518, 1987.
59. Schonfeld, G., Patsch, W., Rudel, L.L., Nelson, C., Epstein, M., and Olson, R.E., *J. Clin. Invest.,* 69, 1072, 1982.
60. Johnson, F.L., St. Clair, R., and Rudel, L.L., *J. Clin. Invest.,* 72, 221, 1983.
61. Keys, A., Anderson, J.T., and Grande, F., *Lancet,* 2, 959, 1957.
62. Hegsted, D.M., *Am. J. Clin. Nutr.,* 44, 299, 1986.
63. Grande, F., Anderson, J.T., and Keys, A., *Am. J. Clin. Nutr.,* 25, 53, 1972.
64. Anderson, J.T., Grande, F., and Keys, A., *Am. J. Clin. Nutr.,* 29, 1784, 1976.
65. Brussard, J.H., Dallinga-Thie, G., Groot, P.H.E., and Katan, M.B., *Atherosclerosis,* 35, 515, 1980.
66. Brussard, J.H., Katan, M.B., Groot, P.H.E., Havekas, L.M., and Hautvast, J.G., *Atherosclerosis,* 42, 205, 1982.
67. Laine, D.C., Snodgrass, C.M., Dawson, E.A., Ener, M.A., Kuba, K., and Frantz, I.D., *Am. J. Clin. Nutr.,* 35, 683, 1982.
68. Becker, N., Illingworth, D.R., Alaupovic, P., Connor, W.E., and Sundberg, E.E., *Am. J. Clin. Nutr.,* 37, 355, 1983.
69. Harris, W.S., Connor, W.E., and McMurray, M.P., *Metabolism,* 32, 179, 1983.
70. Wolf, R.N. and Grundy, S.M., *J. Nutr.,* 113, 1521, 1983.
71. Mattson, F.H. and Grundy, S.M., *J. Lipid Res.,* 26, 195, 1985.
72. Reiser, R., Probstfield, J.L., Silvers, A., Scott, L.W., et al., *Am. J. Clin. Nutr.,* 42, 190, 1985.
73. Grundy, S.M., *N. Engl. J. Med.,* 314, 745, 1986.
74. Grundy, S.M., Nix, D., Whelan, M.F., and Franklin, L., *J.A.M.A.,* 256, 2351, 1986.
75. Mensink, R.P. and Katan, M.B., *Lancet,* 1, 122, 1987.
76. Grundy, S.M., Florentin, L., Nix, D., and Whelan, M.F., *Am. J. Clin. Nutr.,* 47, 965, 1988.
77. Vries, J.H.M., *J. Lipid Res.,* 29, 883, 1988.
78. McDonald, B.E., Gerrard, J.M., Bruce, V.M., and Corner, E.J., *Am. J. Clin. Nutr.,* 50, 1382, 1989.
79. Mensink, R.P. and Katan, M.B., *N. Engl. J. Med.,* 323, 439, 1990.
80. Wardlaw, G.M. and Snook, J.T., *Am. J. Clin. Nutr.,* 51, 815, 1990.
81. Chan, J.K., Bruce, V.M., and McDonald, B.E., *Am. J. Clin. Nutr.,* 54, 104, 1991.
82. Wardlaw, G.M., Snook, J.T., Lin, M.C., Puangco, M.A., and Kwon, J.S., *Am. J. Clin. Nutr.,* 12, 50, 1991.
83. Martin, H., Wahrburg, U., Sandkamp, M., Schulte, H., and Assmann, G., *Infusionstherapie,* 17(Suppl. 11), 32, 1990.

Chapter 2.2

OMEGA-3 FATTY ACIDS

PART I: METABOLIC EFFECTS OF OMEGA-3 FATTY ACIDS AND ESSENTIALITY

Artemis P. Simopoulos

INTRODUCTION

Research on the role of omega-3 fatty acids in growth and development and in health and disease has expanded over the past 15 years. In fact, today we know more about the metabolism and functions of these fatty acids than any other. Advances in methodology, particularly the availability of deuterated materials, have contributed to our understanding of processes involved in the desaturation and elongation of fatty acids and their metabolites. Whereas in the 1950s the emphasis was on vegetable oils and their hypocholesterolemic effects, today the important functions of omega-3 fatty acids, particularly eicosapentanoic and docosahexanoic acid from fish or fish oil, in atherogenesis, inflammation, thrombus formation, gene expression, and cell-to-cell communication have taken central stage and have led to intervention studies and clinical trials in coronary heart disease, hypertension, diabetes, rheumatoid arthritis, psoriasis, ulcerative colitis, and other autoimmune diseases.

This paper will focus on omega-3 fatty acids and eicosanoid metabolism; molecular aspects and gene expression; essentiality; the hypolipidemic, antiatheromatous, antithrombotic, vascular, and antiarrhythmic effects; and the influence of omega-3 fatty acids on restenosis and in decreasing lipoprotein (a).

ELONGATION, DESATURATION, AND SOURCES

Polyunsaturated fatty acids (PUFA) consist of two families of fatty acids, omega-6 and omega-3, designated by the location of the first double bond counting from the methyl end of the fatty acid molecule (Figure 1). Linoleic acid (LA) is the parent fatty acid of the omega-6 family and alpha-linolenic acid (LNA) is the parent fatty acid of the omega-3 family. These PUFAs cannot be made by human beings. They are essential fatty acids (EFA) that must be obtained as such from the diet. Both fatty acid families are plentiful in nature. LA is found in the seeds of most plants with the exception of coconut, cocoa, and palm. LNA is found mostly in the chloroplast of green leafy vegetables instead of the seeds, with the exception of flax seed. Some nuts such as English walnuts are also rich in LNA. Purslane leaves contain more LNA than any other edible green leafy vegetable (Table 1).[1]

In the human body, both essential fatty acids are metabolized to longer chain fatty acids of 20 and 22 carbon atoms, increasing the chain length and degree of unsaturation by adding extra double bonds to the carboxyl end of the fatty acid molecule. By using deuterated material it has been shown that linoleic acid is metabolized to arachidonic acid (AA), and LNA to eicosapentanoic acid (EPA) and docosahexanoic acid (DHA) (Figure 2).[2] AA is found predominantly in the phospholipids of grain-fed animals, whereas EPA and DHA are found in the oils of fish, particularly fatty fish (Table 2),[3] and in some algae and ferns.[4] Omega-6 and omega-3 fatty acids compete for the desaturation enzymes, but both delta-5 and delta-6 desaturases prefer omega-3 to omega-6 fatty acids.[5-7] However, increased amounts of omega-6 fatty acids in the diet, as is found in the U.S. diet, interfere or slow down the metabolism of LNA to EPA and DHA. Delta-6 desaturase may decrease with age.[5] Premature infants[8] are

O
‖
C
⁄ ＼OH   Palmitic Acid (16:0)

O
‖
C
⁄ ＼OH   Oleic Acid (18:1w9)

O
‖
C
⁄ ＼OH   Linoleic Acid (18:2w6)

O
‖
C
⁄ ＼OH   Arachidonic Acid (20:4w6)

O
‖
C
⁄ ＼OH   Linolenic Acid (18:3w3)

O
‖
C
⁄ ＼OH   Eicosapentaenoic Acid (20:5w3)

O
‖
C
⁄ ＼OH   Docosahexaenoic Acid (22:6w3)

FIGURE 1. Structural formulas for omega-6 (linoleic acid, 18:2ω6) and omega-3 (alpha-linoleic acid, 18:3ω3) fatty acids. The first number (before the colon) gives the number of carbon atoms in the molecule and the second gives the number of double bonds. ω6 and ω3 indicate position of the first double bond in a given fatty acid molecule.

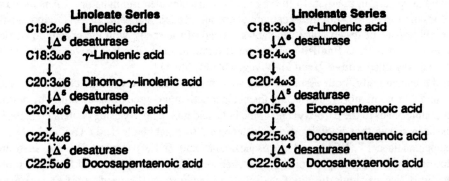

Linoleate Series	**Linolenate Series**
C18:2ω6 Linoleic acid	C18:3ω3 α-Linolenic acid
↓Δ⁶ desaturase	↓Δ⁶ desaturase
C18:3ω6 γ-Linolenic acid	C18:4ω3
↓	↓
C20:3ω6 Dihomo-γ-linolenic acid	C20:4ω3
↓Δ⁵ desaturase	↓Δ⁵ desaturase
C20:4ω6 Arachidonic acid	C20:5ω3 Eicosapentaenoic acid
↓	↓
C22:4ω6	C22:5ω3 Docosapentaenoic acid
↓Δ⁴ desaturase	↓Δ⁴ desaturase
C22:5ω6 Docosapentaenoic acid	C22:6ω3 Docosahexaenoic acid

FIGURE 2. Essential fatty acid metabolism desaturation and elongation of ω6 and ω3. Arachidonic acid is the precursor of the 2-series of prostanoids (prostaglandins and thromboxanes) and of leukotrienes of the 4-series, whereas eicosapentanoic and docosahexanoic acids are precursors of the 3-series of prostanoids and leukotrienes of the 5-series. Dihomo-gamma-linolenic acid is the precursor of the prostaglandin E₁ series.

TABLE 1
Fatty Acid Content of Plants

Fatty acid	Purslane	Spinach	Red leaf lettuce	Buttercrunch lettuce	Mustard
		(mg/g of wet weight)			
14:0	0.16	0.03	0.03	0.01	0.02
16:0	0.81	0.16	0.10	0.07	0.13
18:0	0.20	0.01	0.01	0.02	0.02
18:1ω9	0.43	0.04	0.01	0.03	0.01
18:2ω6	0.89	0.14	0.12	0.10	0.12
18:3ω3	4.05	0.89	0.31	0.26	0.48
20:5ω3	0.01	0.00	0.00	0.00	0.00
22:6ω3	0.00	0.00	0.002	0.001	0.001
Other	1.95	0.43	0.12	0.11	0.32
Total fatty acid content	8.50	1.70	0.702	0.60	1.101

Adapted from Simopoulos, A.P. and Salem, Jr., N., *N. Engl. J. Med.*, 315, 833, 1986.

limited in their ability to make EPA and DHA from LNA as are some hypertensive individuals[9] and some diabetics.[10] Evidence is also accumulating that DHA retroconverts to EPA.

Both omega-6 and omega-3 fatty acids and their longer-chain derivatives are important components of plant and animal cell membranes and are distributed selectively among lipid classes in the cells of mammals and birds. LNA is found mostly in triglycerides, in cholesteryl esters, and only in very small amounts in phospholipids. EPA is found mostly in phospholipids and in cholesteryl esters and in smaller amounts in triglycerides. DHA is found mostly in phospholipids. The cerebral cortex[11] and retina,[12] and testis and sperm[13] of all mammals, including human beings, are particularly rich in DHA. Human beings can obtain EPA and DHA either directly from the diet by eating fish or taking fish oils, or by synthesizing them from dietary LNA. Based on estimates from studies of present-day hunter-gatherer populations and paleolithic nutrition, human beings evolved on a diet that contained roughly equal amounts of omega-6 and omega-3 PUFAs (Figure 3) and low amounts of saturated fats.[14-16]

Today diets are high in omega-6 fatty acids because of the increased production of oils from oil seeds such as corn, safflower, and cottonseed, and other vegetable oils for cooking.[17] The large-scale production of vegetable oils became more efficient and economic due to the increased use of solvent extraction of oilseeds that came into increased use after World War I. The hydrogenation process contributed further to the increase of omega-6 fatty acids and the decrease of omega-3 fatty acid availability in the food supply. For example, the partial selective hydrogenation of soybean oil reduced the LNA content of the oil while leaving a high concentration of LA, because the LNA in soybean oil caused many organoleptic problems.

The hydrogenation process led to another problem; it not only decreased the LNA, but also led to increased production of trans fatty acids. It has been recently shown that trans fatty acids lead to increases in serum cholesterol concentrations, whereas LA in its liquid state lowers serum cholesterol concentration when substituted for saturated fat in the diet. In other words, trans fatty acids behave like saturated fats in that they raise serum cholesterol levels.[18] Trans fatty acids raise lipoprotein(a)[19] and lower HDL.[20] Grain-fed animals kept in stalls contain more saturated fat in their carcass and increased amounts of AA since grains are rich sources of omega-6 acids. It has been estimated that the dietary intake of AA is about 200 to 1000 mg/d,[21] whereas that of EPA and DHA is 46 and 78 mg/d, respectively.[22] Birds and animals in the wild feed on wild plants and are very lean, with a carcass content of only 3.9%.[23] They also

TABLE 2
Content of ω3 Fatty Acids and Other Fat Components in Selected Fish[a]

Fish	Total fat	Total saturated	Total mono-unsaturated	Total poly-unsaturated	18:3	20:5	22:6	Cholesterol
Anchovy, European	4.8	1.3	1.2	1.6	—	0.5	0.9	—
Bass, striped	2.3	0.5	0.7	0.8	Tr	0.2	0.6	80
Bluefish	6.5	1.4	2.9	1.6	—	0.4	0.8	59
Carp	5.6	1.1	2.3	1.4	0.3	0.2	0.1	67
Catfish, brown bullhead	2.7	0.6	1.0	0.8	0.1	0.2	0.2	75
Catfish, channel	4.3	1.0	1.6	1.0	Tr	0.1	0.2	58
Cod, Atlantic	0.7	0.1	0.1	0.3	Tr	0.1	0.2	43
Croaker, Atlantic	3.2	1.1	1.2	0.5	Tr	0.1	0.1	61
Flounder, unspecified	1.0	0.2	0.3	0.3	Tr	0.1	0.1	46
Grouper, red	0.8	0.2	0.1	0.2	—	Tr	0.2	—
Haddock	0.7	0.1	0.1	0.2	Tr	0.1	0.1	63
Halibut, Greenland	13.8	2.4	8.4	1.4	Tr	0.5	0.4	46
Halibut, Pacific	2.3	0.3	0.8	0.7	0.1	0.1	0.3	32
Herring, Pacific	13.9	3.3	6.9	2.4	0.1	1.0	0.7	77
Herring, Round	4.4	1.3	0.8	1.5	0.1	0.4	0.8	28
Mackerel, king	13.0	2.5	5.9	3.2	—	1.0	1.2	53
Mullet, striped	3.7	1.2	1.1	1.1	0.1	0.3	0.2	49
Ocean perch	1.6	0.3	0.6	0.5	Tr	0.1	0.1	42
Plaice, European	1.5	0.3	0.5	0.4	Tr	0.1	0.1	70
Pollock	1.0	0.1	0.1	0.5	—	0.1	0.4	71
Pompano, Florida	9.5	3.5	2.6	1.1	—	0.2	0.4	50
Salmon, Chinook	10.4	2.5	4.5	2.1	0.1	0.8	0.6	—
Salmon, pink	3.4	0.6	0.9	1.4	Tr	0.4	0.6	—
Snapper, red	1.2	0.2	0.2	0.4	Tr	Tr	0.2	—
Sole, European	1.2	0.3	0.4	0.2	Tr	Tr	0.1	50
Swordfish	2.1	0.6	0.8	0.2	—	0.1	0.1	39
Trout, rainbow	3.4	0.6	1.0	1.2	0.1	0.1	0.4	57
Tuna, albacore	4.9	1.2	1.2	1.8	0.2	0.3	1.0	54
Tuna, unspecified	2.5	0.9	0.6	0.5	—	0.1	0.4	—

[a] Values are given as g/100 g edible portion, raw except for cholesterol, which is given as mg. Dash (—) denotes lack of reliable data for nutrient known to be present. Tr, trace (less than 0.05 g/100 g of food).

Adapted from the U.S. Department of Agriculture Provisional Table on the Content of Omega-3 Fatty Acids and Other Fat Components in Seafoods, as presented by Simopoulos et al.[3]

contain about five times more PUFA per gram than is found in domestic livestock[24,25] and 4% of their fat is PUFA. Domestic beef obtained under modern agricultural conditions contains very small or practically undetectable amounts of LNA.[26] On the other hand, deer that forage on ferns and mosses contain more omega-3 fatty acids (LNA) in their meat.

The ratio of omega-6/omega-3 has changed in other foods as well due to modern agriculture and aquaculture and to the fact that people in the U.S. and Western Europe do not eat wild plants or fresh herbs and eat a limited number of cultivated vegetables.[1,27] Table 3 shows the different concentration of omega-6 and omega-3 fatty acids in fish obtained in the wild and in farming.[28] Fish from aquaculture contain less omega-3 and more omega-6 fatty acids than is found in the wild. Even chickens under free ranging conditions produce eggs rich in omega-

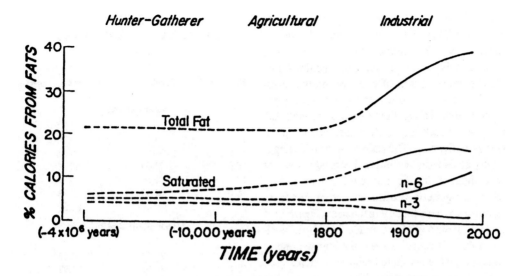

FIGURE 3. Scheme of the relative percentages of different dietary fatty acids (saturated fatty acids and omega-6 and omega-3 unsaturated fatty acids) and possible changes subsequent to industrial food processing, involving fattening of animal husbandry and hydrogenation of fatty acids.[14]

TABLE 3
Fat Content and Fatty Acid Composition of Wild and Cultured Trout, Eel, and Salmon[a]

	Trout (*Salmo gairdneri* and *S. trutta fario*)		Eel (*Anguilla anguilla*)		Salmon (*S. salar*)	
	Wild (n = 2)	Cultured (n = 9)	Wild (n = 4)	Cultured (n = 4)	Wild (n = 2)	Cultured (n = 2)
Fat (g/100 g)	5 ± 3	6 ± 1	21 ± 6	30 ± 2[b]	10 ± 0.1	16 ± 0.6[c]
Fatty acids (g/100 g fatty acid)						
18:3ω3	3 ± 2	1 ± 0.3[c]	2 ± 2	1 ± 0.3	1 ± 0.1	1 ± 0.1
20:5ω3	7 ± 0.6	4 ± 1[c]	4 ± 2	3 ± 0.6	5 ± 0.2	5 ± 0.1
22:6ω3	15 ± 2	13 ± 1[b]	4 ± 2	6 ± 0.4	10 ± 2	7 ± 0.1[b]
Other ω3[d]	5 ± 0.6	2 ± 0.7[c]	3 ± 1	2 ± 0.2[b]	3 ± 0.5	4 ± 0.1
18:2ω6	4 ± 3	9 ± 2[c]	2 ± 2	5 ± 0.3[c]	1 ± 0.1	3 ± 0.1
Other ω6[e]	1 ± 0.4	0.6 ± 0.1[c]	2 ± 0.3	0.4 ± 0.1[c]	0.2 ± 0.1	0.5 ± 0.1
Sum of ω3	30 ± 0.2	20 ± 3[c]	14 ± 3	12 ± 1	20 ± 2	17 ± 0.2
Sum of ω6	5 ± 3	9 ± 2[b]	3 ± 1	6 ± 0.3[c]	2 ± 0.1	3 ± 0.1
ω3:ω6	7 ± 5	2 ± 0.6[c]	5 ± 2	2 ± 0.3[b]	11 ± 2	6 ± 0.1[b]

[a] x ± SD; n, number of lots; each lot consisted of about six trout or eel or one or two salmon.
[b] Significantly different from wild: $p < 0.05$ compared with wild fish.
[c] Significantly different from wild: $p < 0.01$ compared with wild fish.
[d] 18:4ω3 + 20:3ω3 + 22:5ω3.
[e] 20:4ω6 + 22:4ω6.

Adapted from van Vliet, T. and Katan, M.B., *Am. J. Clin. Nutr.*, 51, 1, 1990.

3 fatty acids, whereas modern chicken feed leads to the production of eggs with much smaller amounts of omega-3 fatty acids (Table 4).[29] The egg from free-ranging chickens has an omega-6/omega-3 ratio of 1.3, whereas the USDA standard egg has an omega-6/omega-3 ratio of 19.4. An absolute and a relative increase in LA and AA, due to increases in vegetable oils

and agriculture policies with a decrease in LNA, EPA, and DHA, has occurred since the turn of the century.[30] Before the 1940s, children took 1 tsp of cod liver oil as a source of vitamins A and D. The consumption of cod liver oil was drastically decreased once these vitamins were synthesized. Today the ingestion of both LA and AA is much higher than the amount human beings evolved on. The ratio of omega-6/omega-3 fatty acids from animal and vegetable sources was about 1 during our evolutionary period when our genetic profile was programmed in response to these dietary influences (Figure 3).[14] In fact, Homo sapiens made his appearance about 40,000 years ago and the human genetic constitution has remained relatively unchanged. Agriculture began to produce changes in diet about 10,000 years ago. But it is only since the industrial revolution and particularly the last 150 years that enormous changes in total fat and types of fatty acids occurred along with other changes besides fatty acids. This very rapid change is characterized by increases in saturated fat from animal sources and trans fatty acids from the hydrogenation of vegetable oils, an increase in omega-6 fatty acids, and a decrease in omega-3 fatty acids, leading to imbalances in omega-6/omega-3 from about 1 to 20-25/1 and to Western diets that are deficient in omega-3 fatty acids compared with the diet on which humans evolved and their genetic patterns were established. The optimal intake of 18:3ω3 is estimated to be 800 to 1100 mg/d and that of EPA and DHA, 300 to 400 mg/d.[31]

OMEGA-3 FATTY ACIDS AND EICOSANOID METABOLISM

It was indicated earlier that omega-6 and omega-3 fatty acids compete for the delta-5 and delta-6 desaturases. There is even greater competition at the level of AA and EPA metabolites. Both are precursors of substances consisting of 20 carbon atoms known collectively as eicosanoids (prostaglandins, thromboxanes, and leukotrienes) (Figure 4). AA is the precursor of the 2-series of prostanoids (prostaglandins and thromboxanes) and of leukotrienes of the 4-series, whereas EPA and DHA are precursors of the 3-series of prostanoids and leukotrienes of the 5-series. There is plenty of evidence indicating that consumption of essential fatty acids has a significant effect on the production and distribution of prostanoids and leukotrienes.[32]

TABLE 4
Fatty Acid Levels in Chicken Egg Yolks[a]

Fatty acid	Greek egg[b]	Supermarket egg[b]
Saturated fats		
14:0	1.10	0.70
15:0	—	0.07
16:0	77.60	56.66
17:0	0.66	0.34
18:0	21.30	22.88
Total	100.66	80.65
Monounsaturated fats		
16:1ω7	21.70	4.67
18:1	120.50	109.97
20:1ω9	0.58	0.68
22:1ω9	—	—
24:1ω9	—	0.04
Total	142.78	115.36
ω6 Fatty acids		
18:2ω6	16.00	26.14
18:3ω6	—	0.25
20:2ω6	0.17	0.36
20:3ω6	0.46	0.47
20:4ω6	5.40	5.02
22:4ω6	0.70	0.37
22:5ω6	0.29	1.20
Total	23.02	33.81
ω3 Fatty acids		
18:3ω3	6.90	0.52
20:3ω3	0.16	0.03
20:5ω3	1.20	—
22:5ω3	2.80	0.09
22:6ω3	6.60	1.09
Total	17.66	1.73
Ratio of fatty acids to saturated fats	0.4	0.44
Ratio of ω6 to ω3	1.3	19.4

[a] The eggs were hard-boiled, and their fatty acid composition and lipid content were assessed as described elsewhere.
[b](mg/g yolk)
Adapted from Simopoulos, A.P. and Salem, Jr., N., *N. Engl. J. Med.*, 321, 1412, 1989.

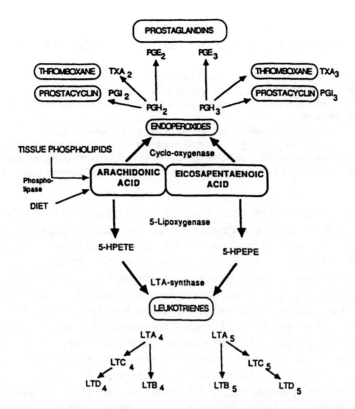

FIGURE 4. Oxidative metabolism of arachidonic acid and eicosapentanoic acid by the cyclooxygenase and 5-lipoxygenase pathways. 5-HPETE denotes 5-hydroperoxyeicosatetranoic acid and 5-HPETE denotes 5-hydroxyeicosapentanoic acid.

In 1978, two important concepts on eicosanoid formation were presented: (1) Moncada and Vane[33] showed that the vascular wall could utilize endoperoxides released by adhering platelets for the formation of prostacyclin; and (2) Dyerberg et al.[34] proposed that EPA from oily fish or fish oil could, through the formation of the active analog of prostacyclin, PGI_3, but inactive thromboxane, TXA_3, produce an antithrombotic state that protects against cardiovascular disease.

In 1979[35] Needleman discovered that prostaglandins derived from EPA have different biological properties than those derived from AA, which stimulated further research on EPA and DHA from fish and fish oils. When humans ingest fish or fish oils, the EPA and DHA from the diet partially replace the omega-6 fatty acids (especially AA) in the membrane of probably all cells and especially in the membranes of erythrocytes, platelets, neutrophils, monocytes, and liver cells. Therefore, ingestion of EPA and DHA from fish or fish oil leads to a more physiologic state characterized by the production of prostanoids and leukotrienes that have antithrombotic, antichemotactic, antivasocontrictive, and antiinflammatory properties shown in Table 5.[36,37] It is evident that omega-3 fatty acids can affect a number of chronic diseases through eicosanoids alone. As seen below, omega-3 fatty acids exert effects on many systems.

BEYOND EICOSANOIDS: MOLECULAR ASPECTS AND GENE EXPRESSION

Although most of the investigations on the role of EFA involved research into eicosanoid metabolism, other aspects of their metabolism have also been pursued. Studies indicate that the phospholipid class and fatty acid composition and cholesterol content of biomembranes

TABLE 5
Effects of Ingestion of EPA and DHA from Fish or Fish Oil

Decreased production of prostaglandin E_2 (PGE_2) metabolites
A decrease in thromboxane A_2, a potent platelet aggregator and vasoconstrictor
A decrease in leukotriene B_4 formation, an inducer of inflammation, and a powerful inducer of leukocyte
   chemotaxis and adherence
An increase in thromboxane A_3, a weak platelet aggregator and a weak vasoconstrictor
An increase in prostacyclin PGI_3, leading to an overall increase in total prostacyclin by increasing PGI_3 without a
   decrease in PGI_2, both PGI_2 and PGI_3 are active vasodilators and inhibitors of platelet aggregation
An increase in leukotriene B_5, a weak inducer of inflammation and a weak chemotactic agent[30,31]

are critical determinants of physical properties of membranes and have been shown to influence a variety of membrane-dependent functions such as integral enzyme activity, membrane transport, and receptor function. The fact that EFAs have the ability to alter membrane lipid composition and function *in vivo,* even when EFA are adequately supplied, indicates the importance of diet in growth and metabolism throughout the life cycle.

Dietary manipulation of EFA leads to complex interactions and the displacements of the omega-3 and omega-6 fatty acids in both plasma and cellular lipids. Dietary fatty acids induce cell activation such as generation of inositol phosphates affecting key processes in cell function. Nutrients, like hormones, influence and control gene expression.[38] Omega-3 fatty acids from menhaden oil have been shown to lower the enzyme fatty acid synthetase in the liver by decreasing fatty acid synthase mRNA.[39] Although the mechanisms by which omega-3 fatty acids produce hypolipidemic, antiatheromatous, antithrombotic, anti-inflammatory, and antitumor effects are not precisely known, great progress has taken place over the past 10 years due primarily to advances in methods and the interest on the role of diet in chronic diseases. First and foremost, though, omega-3 fatty acids are essential for the normal development of the premature infant and possibly the full term.

ESSENTIALITY: THE ROLE OF OMEGA-3 FATTY ACIDS IN GROWTH AND DEVELOPMENT

The fact that human milk contains both omega-6 and omega-3 fatty acids (LA, LNA, AA, EPA, and DHA), whereas cow milk does not, raised the interest of investigations in the role of the longer-chain omega-3 fatty acids, particularly DHA, in development (Table 6).[8,40]

ANIMAL STUDIES

Animal studies involving rats and rhesus monkeys showed that dietary restriction of omega-3 (LNA) during pregnancy and lactation interferes with normal visual function and may impair learning ability in offspring.[41] Connor et al.[42] carried out a series of studies in rhesus monkeys and showed that the abnormalities in the electroretinogram (ERG) induced by omega-3 fatty acid deficiency during development appeared to be irreversible. Connor concluded that omega-3 fatty acid deficiency in this model is characterized by reduced vision, abnormal ERG, and polydipsia (Table 7).[42,43] Rotstein et al.[44] concluded that in the aged rat, impairment of the delta-4 desaturase enzyme system is probably responsible for the decreased concentrations of 22:6ω3 (and 22:5ω6) in retinal lipids. DHA is required for the normal function of photoreceptors of rats and primates. DHA, therefore, may be important in the visual impairments that accompany old age. For that reason, dietary DHA rather than LNA is the omega-3 fatty acid to use in studies of visual impairment. DHA might even improve visual function in the aged.

TABLE 6
Fatty Acid Composition of Human Milk and Formulas (mol%)[a]

Common name	Fatty acid	Human milk (n = 11)	Portagen	Enfamil premature	Similac special care
Caprylic acid	8:0	0.3	60	24.5	24.1
Capric acid	10:0	1.4	24	14.1	17.7
Lauric acid	12:0	7.0	0.42	12.2	14.9
Myristic acid	14:0	8.0	Trace	4.7	5.8
Palmitic acid	16:0	19 8	0.19	7.5	6.8
Palmitoleic acid	16:1	3.2		0.1	0.2
Stearic acid	18:0	5.9	0.47	1.7	2.3
Oleic acid	18:1	34.8	4.1	12.4	10.0
Linoleic acid	18:2 (ω6)	16.0	8.1	22.4	17.4
Alpha-linolenic acid	18:3 (ω3)	0.6	Trace	0.6	0.9
Gondoic acid	20:1	1.1		0.3	0.1
	20:2 (ω6)	0.6			
Dihomo-y-linolenic acid	20:3 (ω6)	0.4			
Arachidonic acid	20:4 (ω6)	0.6			
Eicosapentanoic acid	20:5 (ω3)	0.0			
Docosenoic acid	22:1	0.1			
Docosatetranoic acid	22:4 (ω6)	0.2			
Docosapentanoic acid	22:5 (ω6)	0.2			
Docosapentanoic acid	22:5 (ω3)	0.1			
Docosahexanoic acid	22:6 (ω3)	0.2			

[a] Values from Pediatric Products Handbook, 1983 Edition, Mead Johnson Nutritional Division, Evanston, IN.

Adapted from Carlson in Simopoulos.[40]

TABLE 7
The Differing Characteristics of Omega-3 and Omega-6 Essential Fatty Acid Deficiencies

	Omega-3	Omega-6
Clinical features	Normal skin, growth, and reproduction Reduced learning Abnormal electroretinogram Impaired vision Polydipsia	Growth retardation Skin lesions Reproductive failure Fatty liver Polydipsia
Biochemical markers	Decreased 18:3ω3 and 22:6ω3 Increased 22:4ω6 and 22:5ω6 Increased 20:3ω9 (only if ω6 also low)	Decreased 18:2ω6 and 20:4ω6 Increased 20:3ω9 (only if ω3 also low)

Adapted from Connor, W.E., Neuringer, M., and Reisbick, S., *World Rev. Nutr. Diet, 66,* 118, 1991.

HUMAN STUDIES

In dietary studies involving human growth and development for the past 50 years the emphasis has been on protein-calorie malnutrition. The role of omega-6 and omega-3 fatty acids has been studied only recently. During pregnancy, the LA and LNA requirement is about

TABLE 8
Criteria for the Diagnosis of ω3 Fatty Acid Deficiency in Humans

Long-term feeding on a completely defined diet supplying alpha-linolenic acid at less than approximately 50-100 mg/d, and long-chain ω3 fatty acids less than 20-30 mg/d
Low intake of ω3 fatty acids should be confirmed by low concentration of 20:5ω3 and 22:6ω3 in plasma and/or cellular lipids
Disappearance of clinical symptoms after supplementing with pure ω3 fatty acids to an otherwise unchanged diet
Verify a concomitant increase in ω3 fatty acid concentration in plasma and/or cellular lipids

Adapted from Bjerve, K.S., *World Rev. Nutr. Diet,* 66, 133, 1991.

1% of the total energy intake of the non-pregnant woman and AA and DHA are another 0.5%. Many studies have been carried out on the fatty acid composition of human milk. Although the composition varies with the mother's diet, i.e., fish-eating women have higher concentrations of DHA in their milk than vegetarians, the range among omnivores does not differ much. Cow milk, on the other hand, is devoid of DHA. In 1973 Crawford et al.[45] analyzed 32 samples of human milk and found that it contained LA, LNA, AA, EPA, and DHA and recommended their inclusion in infant formula. EPA and DHA are higher in the erythrocytes of breast-fed infants than in those who are bottle-fed.[46] Recent studies in humans demonstrated that in preterm infants, plasma and red cell phospholipid contents of omega-3 longer-chain polyunsaturated fatty acids (LCP) are predictors of visual acuity.[47,48] Because the greatest amount of DHA accumulation occurs during the last trimester of pregnancy, the amount of DHA available to premature infants assumes critical importance.

During the first 3 to 4 months of life, preterm infants' LCP requirements appear to exceed their endogenous capacity for LCP synthesis from LNA to DHA. For this reason, during the early period of life, LCP are considered essential nutrients that should be supplied in the diet. Infants born at term and fed mother's milk had approximately twice as much DHA in their erythrocyte phospholipids as did infants fed formula containing LNA but not DHA.

Carlson[49] showed that the addition of fish oil as a source of DHA to the infant formula in doses comparable to human milk could prevent the decrease of DHA in plasma phospholipids and red cell membrane phospholipids. One month after delivery, preterm infants not fed human milk or formula supplemented with the DHA had plasma phospholipid DHA (22:6ω3) more like that of deficient monkeys at which demonstrable deficits in visual acuity occur in infant monkeys. Uauy et al.[47] provided further evidence that DHA is essential for the functional development of the eye and brain of the premature infant.

Other studies involving children, adults, and the elderly who became deficient in omega-3 fatty acids because of either complete absence or very small amounts in the diet indicate that omega-3 fatty acids are essential throughout the life cycle (Tables 8 and 9).[50-52] A number of scientific groups have recommended the establishment of an RDA for omega-3 fatty acids in order that omega-3 fatty acids be included in infant formulas, and in enteral (tube feeding) and parenteral feedings.[3,31,53] In the U.S. there is no RDA for omega-3 fatty acids, whereas the 1990 Canadian nutrition recommendations already include specific amounts for omega-3 and omega-6 fatty acids (in grams per day) for the various age groups, with additional amounts recommended for pregnancy and lactation (Table 10).[54]

In 1991 the ESPGAN Committee on Nutrition published its report on the content and composition of lipids in infant formulas.[55] The report states that "the Committee feels that the enrichment with metabolites of both linoleic and alpha-linolenic acids approximating levels typical of human milk lipids (n-6 LCP 1%, n-3 LCP 0.5% of total fatty acids) [note: LCP is defined by the ESPGAN Committee as longer chain polyunsaturated fatty acids of 20 or more carbon atoms, i.e., AA and DHA] is desirable for formulas for LBW infants. It is recommended that n-6 LCP and n-3 LCP should not exceed 2% and 1% respectively, of total formula fatty acids." The committee did not recommend the inclusion of LCP in the formula for full-term infants and asked for more data. It is unfortunate and an omission on their part. LCPs are in human milk and since human milk is the standard that formulas should follow, the inclusion for the full term is equally essential.

The plasma concentration of DHA of formula-fed, full-term infants is lower than that of breast-fed infants, indicating that the provision of LNA in formula is not sufficient or that the elongation/desaturation enzyme pathways are not adequate during early life to support accretion of DHA. A recent study suggests that term infants may be dependent on dietary DHA for optimal functional maturation of the retina and visual cortex.[56]

Earlier studies had suggested that formula-fed full-term infants may also be at higher risk than breast-fed infants for learning disabilities[57] and for lower scores on picture intelligence, word reading, sentence completion, nonverbal ability, and mathematical attainment tests administered at age 8 and 15 years.[58] A recent study by Lucas et al.[59] showed that preterm infants that had consumed mother's milk in the early weeks of life had a significantly higher intelligence quotient (IQ) at 7 1/2 to 8 years than did those who received no maternal milk. The same findings were reported when these same children were 18 months of age. The authors conclude that further work is needed to determine if the advantage in IQ is due to parenting skills, genetics, or to factors in human milk itself such as DHA, hormones, and other trophic factors that are responsible for the beneficial effect of human milk on neurodevelopment.

In summary, omega-3 fatty acids are essential for normal growth and development. During infancy the estimated requirements are based on the composition of human milk, which are 0.5% for omega-3 long chain fatty acids and 1% for omega-6 long chain fatty acids. It is recommended that total long chain omega-3 and omega-6 fatty acids do not exceed 1 and 2%, respectively. For adolescents and adults, the optimal intake of 18:3ω3 is estimated to be 800 to 1100 mg/d and that of EPA and DHA, 300 to 400 mg/d. This increase in omega-3 fatty acid intake should be accompanied by a decrease in omega-6 fatty acids and trans fatty acids. In Western societies, a decrease in saturated fat intake is recommended by dietary guidelines developed by governments.

TABLE 9

Estimates of Dietary Requirements of ω3 Fatty Acids in Humans

	Optimal requirement	Minimal requirement
Linolenic acid[a]		
mg/d	860-1220	290-390
energy %	1.0-1.2	0.2-0.3
Long-chain ω3[b]		
mg/d	350-400	100-200
energy %	0.4	0.1-0.2

Note: The estimates are based on nutritional information and fatty acid changes observed in ten patients with ω3 fatty acid deficiency after supplementing with ω3 fatty acids.

[a] In the absence of dietary long-chain ω3 fatty acids.
[b] Dietary intake of linolenic acid below 100 mg/d.

Adapted from Bjerve, K.S., *World Rev. Nutr. Diet,* 66, 133, 1991.

HYPOLIPIDEMIC EFFECTS

Because of the emphasis on LDL cholesterol as a risk factor for coronary heart disease, the effects of eating fish or fish oil supplementation on LDL cholesterol, HDL, and triglycerides have been extensively studied. Whereas the lowering of plasma triglyceride concentrations by omega-3 fatty acids has been found consistently in many studies,[53] its effect on LDL cholesterol has been found to vary from a slight decrease to no effect or up to 10% increase in LDL. Harris,[60] in a recent review, concluded that in normal volunteers, omega-3 fatty acids did not influence LDL cholesterol, but a slight rise (about 3%) occurred in HDL concentration and a 25% decrease in triglyceride concentration. When omega-3 fatty acids were substituted for saturated fat in the diet of normal volunteers, LDL cholesterol did not rise.[60] The effects of

TABLE 10
Summary of Examples of Recommended Nutrients Based on Energy
Expressed as Daily Rates

Age	Sex	Energy (kcal)	Thiamin (mg)	Riboflavin (mg)	Niacin (NE[a])	n-3 PUFA[b] (g)	n-6 PUFA (g)
Months							
0-4	Both	600	0.3	0.3	4	0.5	3
5-12	Both	900	0.4	0.5	7	0.5	3
Years							
1	Both	1100	0.5	0.6	8	0.6	4
2-3	Both	1300	0.6	0.7	9	0.7	4
4-6	Both	1800	0.7	0.9	13	1.0	6
7-9	M	2200	0.9	1.1	16	1.2	7
	F	1900	0.8	1.0	14	1.0	6
10-12	M	2500	1.0	1.3	18	1.4	8
	F	2200	0.9	1.1	16	1.1	7
13-15	M	2800	1.1	1.4	20	1.4	9
	F	2200	0.9	1.1	16	1.2	7
16-18	M	3200	1.3	1.6	23	1.8	11
	F	2100	0.8	1.1	15	1.2	7
19-24	M	3000	1.2	1.5	22	1.6	10
	F	2100	0.8	1.1	15	1.2	7
25-49	M	2700	1.1	1.4	19	1.5	9
	F	2000	0.8	1.0	14	1.1	7
50-74	M	2300	0.9	1.3	16	1.3	8
	F	1800	0.8[c]	1.0[c]	14[c]	1.1[c]	7[c]
75+	M	2000	0.8	1.0	14	1.0	7
	F[d]	1700	0.8[c]	1.0[c]	14[c]	1.1[c]	7[c]
Pregnancy (additional)							
1st trimester		100	0.1	0.1	0.1	0.05	0.3
2nd trimester		300	0.1	0.3	0.2	0.16	0.9
3rd trimester		300	0.1	0.3	0.2	0.16	0.9
Lactation (additional)		450	0.2	0.4	0.3	0.25	1.5

[a] Niacin equivalents.
[b] PUFA, polyunsaturated fatty acids.
[c] Level below which intake should not fall.
[d] Assumes moderate physical activity.

From Scientific Review Committee, Nutrition recommendations, N49-42/1990E, Minister of National Health and Welfare, Ottawa, Canada, 1990. With permission.

omega-3 fatty acids on patients with hyperlipidemias were variable and suggest that the type of patient studied determines the hypolipidemic response to omega-3 fatty acid supplementation. In patients with hypertriglyceridemia, the fall in total cholesterol is due to a decrease in VLDL synthesis. Many studies show that this effect is sustained[61-63] contrary to reports by Schectman et al.,[64] who suggested on the basis of a small group of patients that the triglyceride lowering effect of fish oil cannot be sustained. As little as 6 g of fish oil per day (2 g omega-3 fatty acids) has a triglyceride lowering effect in hypertriglyceridemic patients. The more common dose is 3 g/d of EPA and DHA.

In patients with type IIa hyperlipidemia, omega-3 fatty acids lowered triglyceride concentrations, did not change total or LDL cholesterol, and slightly increased HDL. In patients with type IIb combined hyperlipidemia, total cholesterol concentration did not change, but LDL and HDL concentrations rose by 5 to 7% and triglyceride concentration decreased by 38%. In type IV hyperlipidemia, LDL cholesterol concentration increased by 20%.[65,66] The issue is whether the increase in LDL following supplementation with omega-3 fatty acids poses an increased risk for atherosclerosis in patients with type II and IV hyperlipidemia in view of the antithrombotic, anti-inflammatory, and antivasorestrictive aspects of omega-3 fatty acids. Abbey et al.[67] noted in men given fish oil supplements significant changes in HDL characteristics. The proportion of HDL_2 rose, increasing the HDL_2:HDL_3 ratio that was reflected in an increase of $ApoA_1$:$ApoA_2$ ratio due partly to decreased activity of lipid transfer protein which normally transfers cholesteryl esters from HDL to other lipoproteins. The significance of this finding is a reduction in the redistribution of cholesterol from nonatherogenic to atherogenic proteins.

Recent reassessment of the role of triglyceride levels indicates that high triglycerides are in fact a significant risk factor for coronary heart disease.[68,69] Evidence is now forthcoming suggesting that high levels of triglycerides cause the reduced levels of HDL-C,[68] and postprandial elevations of triglycerides play a role in the process. The ability of fish oil supplements to reduce postprandial hypertriglyceridemia has been proven.[70] Thus, fish oil-induced inhibition of chylomicron production has substantial support. Feeding omega-3 fatty acids has been shown to either (1) inhibit triglyceride synthesis in animal models[71] or (2) inhibit the secretion of triglyceride-rich lipoproteins in a variety of cell culture and liver perfusion experiments[72] and in studies of VLDL kinetics in humans.[73-75] EPA has been reported to inhibit diacylglycerol acyltransferase, the final enzyme in triglyceride synthesis.[76] It is also possible that fish oil treatment alters enterocyte metabolism so as to slow the formation and secretion of chylomicrons.[77] Brown and Roberts[78] reported that low doses of fish oil (5 g/d), given to normal subjects for 6 weeks, lowered postprandial lipid levels as compared to values of matching volunteers taking olive oil. Harris and Windsor[79] showed that 2.2 g/d of omega-3 fatty acids added to the regular American diet for a month led to a decrease in serum triglycerides and VLDL, and an increase in HDL-2. Furthermore, Nestel[80] reported that when large amounts of cholesterol were fed to humans, fish oil supplementation blunted the expected rise in cholesterol.

In summary, the effects of omega-3 fatty acids on cholesterol concentrations are similar to those of other PUFAs. When omega-3 fatty acids replace saturated fatty acids in the diet, they lower serum cholesterol concentrations and have the added benefit of consistently lowering triglyceride concentration. Omega-3 fatty acids may raise or have no effect on HDL concentration. At high doses, 10 g omega-3 fatty acids per day, omega-3 fatty acids lower serum cholesterol concentrations. Human studies show that fish oils reduce the rate of hepatic secretion of VLDL-triglyceride,[61,81-83] and in normolipidemic subjects prevent and reverse rapidly the carbohydrate-induced hypertriglyceridemia.[81] Kinetic studies show that fish oils increase the fractional catabolic rate of VLDL.[61,82,83] These findings are consistent with a reduced rate of coronary artery disease in fish-eating populations (Tables 11 and 12).[84]

ANTIATHEROMATOUS ACTIONS

Despite the fact that omega-3 fatty acids in small doses have minimal effects or no effect on total and LDL cholesterol concentration, they have substantial effects on monocytes/macrophages, on endothelial and smooth-muscle cells, and on triglyceride concentrations, all of which are involved in the atherogenic process. In human monocytes, omega-3 fatty acids inhibit the production of platelet activating factor (PAF). PAF activates platelets and contributes to atherogenesis.[85] Similarly, omega-3 fatty acid supplementation reduces interleukin 1 (IL-1) and tumor necrosis factor (TNF) in humans. Both IL-1 and TNF are considered atherogenic because they stimulate the synthesis of adhesion molecules, causing monocytes to adhere to endothelial cells, and also activate platelets, neutrophils, and monocytes.

The current theory on the cellular aspects of atherogenesis combines the "response to injury hypothesis" originally formulated and updated by Ross[86] and Steinberg's[87] theory of oxidation of LDL. Atherosclerosis is a complex disease of the arteries and the arterial wall. Many cellular biochemical and physical components interact at and within the arterial wall. The first step in the formation of atherosclerosis is a nonspecific (functional) injury to endothelium followed by an accumulation of monocytes and macrophages, foam cell formation, and platelet aggregation. The platelets release growth factor, which leads to smooth muscle migration and proliferation. At this point cholesterol is deposited in the smooth muscle cells and monocyte macrophages in the vessel wall. These events further lead to the formation of ground substance and eventually to plaque formation. As seen in Tables 11 and 12, omega-3 fatty acid ingestion may be able to prevent the increase in cellular components generated by these cells and interfere at many steps in the development of the atherogenic process. Figure 5 shows the various steps in the development of atherogenesis and the points where omega-3 fatty acids block its development.[14,88]

Tables 11 and 12 summarize the many factors either inhibited or enhanced by the ingestion of fish or fish oils, that are involved in the development of coronary heart disease, hypertension, and inflammatory and autoimmune disorders such as arthritis, ulcerative colitis, psoriasis, diabetes, and possibly cancer.[84]

A number of animal studies support the antiatheromatous actions of omega-3 fatty acids. In dogs, supplementation with omega-3 fatty acids prevented intimal hyperplasia despite their diet that was high in saturated fatty acids and cholesterol.[89] Cod liver oil given as a supplement to the hyperlipidemic swine model reduced the development of atherosclerosis in the supplemented animals, despite the fact that there was no significant difference in plasma cholesterol concentrations between the supplemented animals and the controls.[90] Dietary fat substitution with omega-3 fatty acids inhibited atherogenesis in the aorta, carotid, and femoral arteries in the primate model.[91] In another primate species, similar findings were observed without significant differences in serum lipid levels.[92] In one study involving rabbits fish oil feeding led to an enhancement of cholesterol-induced atherogenesis,[93] whereas others found that atherosclerosis was inhibited by fish oils in cholesterol-fed rabbits.

In summary, omega-3 fatty acid supplementation inhibits and reduces a number of factors involved in the development of atherogenesis, while enhancing other factors that prevent atherogenesis. These recent research findings provide further support that omega-3 fatty acids are important in the prevention of atherogenesis.

TABLE 11
Functional Effects of ω3 Fatty Acids in the Cardiovascular System

Decrease postprandial lipemia
Reduce platelet aggregation
Reduce blood pressure
Decrease whole blood viscosity
Reduce vascular intimal hyperplasia
Reduce vasospastic response to vasoconstrictors
Reduce cardiac arrhythmias
Reduce albumin leakage in type I diabetes mellitus

Increase bleeding time
Increase platelet survival
Increase vascular (arterial) compliance
Increase cardia beta-receptor function
Increase postischemic coronary blood flow

Adapted from Weber, P.C. and Leaf, A., *World Rev. Nutr. Diet*, 66, 218, 1991.

ANTITHROMBOTIC EFFECTS

Omega-3 fatty acids prolong bleeding time in practically all studies reported. This prolongation in bleeding time is within the normal range and is an attempt to return to a more physiologic state. Because of the increased amounts of omega-6 fatty acids in the Western diet, the eicosanoid metabolic products from AA, specifically prostaglandins, thromboxanes, leukotrienes, hydroxy fatty acids, and lipoxins, are formed in larger quantities than those formed from omega-3 fatty acids, specifically EPA. The eicosanoids from AA are biologically active in very small quantities and if they are formed in large amounts, they contribute to the formation of thrombus and atheromas; to allergic and inflammatory disorders, particularly in

TABLE 12
Effects of ω3 Fatty Acids on Factors Involved in the Pathophysiology of Atherosclerosis and Inflammation

Factor	Function	Effect of w3 fatty acid
Arachidonic acid	Eicosanoid precursor; aggregates platelets; stimulates white blood cells	↓
Thromboxane	Platelet aggregation; vasoconstriction; increase of intracellular Ca++	↓
Prostacyclin (PGI$_{2/3}$)	Prevent platelet aggregation; vasodilation; increase cAMP	↑
Leukotriene (LTB$_4$)	Neutrophil chemoattractant increase of intracellular Ca++	↓
Tissue plasminogen activator	Increase endogenous fibrinolysis	↑
Fibrinogen	Blood clotting factor	↓
Red cell deformability	Decreases tendency to thrombosis and improves oxygen delivery to tissues	↑
Platelet activating factor (PAF)	Activates platelets and white blood cells	↓
Platelet-derived growth factor (PDGF)	Chemoattractant and mitogen for smooth muscles and macrophages	↓
Oxygen-free radicals	Cellular damage; enhance LDL uptake via scavenger pathway; stimulate arachidonic acid metabolism	↓
Lipid hydroperoxides	Stimulate eicosanoid formation	↓
Interleukin 1 and tumor necrosis factor	Stimulate neutrophil O$_2$ free radical formation; stimulate lymphocyte proliferation; stimulate PAF; express intercellular adhesion molecule-1 on endothelial cells; inhibit plasminogen activator, thus, procoagulants	↓
Endothelial-derived relaxation factor (EDRF)	Reduces arterial vasoconstrictor response	↑
VLDL	Related to LDL and HDL level	↓
HDL	Decreases the risk for coronary heart disease	↑
Lp(a)	Lipoprotein (a) is a genetically determined protein that has atherogenic and thrombogenic properties	↓
Triglycerides and chylomicrons	Contribute to postprandial lipemia	↓

Adapted from Weber, P.C. and Leaf, A., *World Rev. Nutr. Diet,* 66, 218, 1991.

susceptible people; and to proliferation of cells. Thus, a diet rich in omega-6 fatty acids shifts the physiological state to one that is prothrombotic and proaggregatory, with increases in blood viscosity, vasospasm, and vasoconstriction and decreases in bleeding time. Bleeding time is decreased in groups of patients with hypercholesterolemia,[94] hyperlipoproteinemia,[95] myocardial infarction, other forms of atherosclerotic disease, and diabetes (obesity and hypertriglyceridemia). Bleeding time is longer in women than in men and longer in young than in old people. There are ethnic differences in bleeding time that appear to be related to diet.

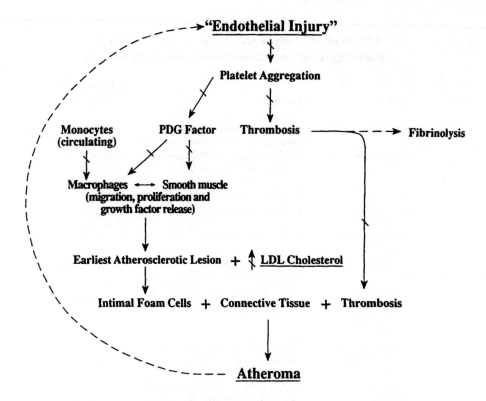

FIGURE 5. Sites of potential interventions for preventing development of atherosclerosis. Note that several of these possibilities would abort the disease before the concentration of plasma LDL cholesterol could contribute to the atherosclerotic process.[14]

In clinical trials (in both experimental and control groups) and in double-blind clinical trials (including restenosis trials and coronary artery graft surgery) there has never been increased evidence of clinical bleeding or increased perioperational blood loss despite the prolongation of bleeding time.

There is substantial agreement that ingestion of fish or fish oils has the following effects: platelet aggregation to epinephrine and collagen is inhibited, thromboxane A_2 production is decreased, whole blood viscosity is reduced, and erythrocyte membrane fluidity is increased (Tables 11 and 12).[3,31,84,96-98] Fish oil ingestion increased the concentration of plasminogen activator and decreased the concentration of plasminogen inhibitor.[99] In patients with types IIb and IV hyperlipoproteinemia and in another double-blind clinical trial involving 64 men aged 35 to 40 years, ingestion of omega-3 fatty acids decreased the fibrinogen concentration.[100] Two other studies did not show a decrease in fibrinogen, but in one a small dose of cod liver oil was used[101] and in the other the study consisted of normal volunteers and was of short duration. A recent study noted that fish and fish oil increase fibrinolytic activity, indicating that 200 g/d of lean fish or 2 g omega-3 EPA and DHA improves certain hematologic parameters implicated in the etiology of cardiovascular disease.[102]

In summary, the antithrombotic effects of EPA and DHA supplementation suggest that their ingestion leads to a return to a more physiologic state and away from a prothrombotic and atherogenic state in both animal models and human beings.

VASCULAR EFFECTS

Ingestion of omega-3 fatty acids not only increases the production of PGI_3, but also of PGI_2 in tissue fragments from the atrium, aorta, and saphenous vein obtained at surgery in patients who received fish oil 2 weeks prior to surgery.[103] Omega-3 fatty acids inhibit the production of platelet-derived growth factor (PDGF) in bovine endothelial cell.[104] PDGF is a chemoattractant for smooth muscle cells and a powerful mitogen. Thus, the reduction in its production by endothelial cells, monocytes/macrophages, and platelets could inhibit both the migration and proliferation of smooth muscle cells, monocyte/macrophages, and fibroblasts in the arterial wall. Omega-3 fatty acids increase endothelium-derived relaxing factor (EDRF).[105] EDRF, presumably nitric oxide, facilitates relaxation in large arteries and vessels. In the presence of EPA, endothelial cells in culture increase the release of relaxing factors, indicating a direct effect of omega-3 fatty acids on the cells.

ANTIARRHYTHMIC EFFECTS

Animal studies with isolated papillary muscles from either rats or marmoset monkeys indicate less susceptibility to catecholamine-induced arrhythmia in the muscles from animals fed fish oil supplements than from those on omega-6 or low fat diets.[106] In fact, adult marmosets fed fish for several months improved their cardiac function and, when subjected to ischemic stress, cardiac arrhythmia was much less. Although the mechanism is not known, it might be operating via eicosanoids, since indomethacin abolished the effect in the isolated papillary muscle experiment. Indomethacin inhibits cyclooxygenase.

Hallaq and Leaf[107] demonstrated that the enrichment of cardiac myocytes with EPA prevented the responses of neonatal cardiac myocytes to the toxic effects of ouabain, which is a potent arrhythmogenic agent. The experimental studies of Gudbjarnason[108] illustrate that the levels of omega-6 and omega-3 fatty acids in cardiac membranes are rapidly modified by stress or dietary fat. The balance between AA and DHA modulates the number and affinity of $beta_1$-adrenergic receptors in aging sarcolemma and also the incidence of ventricular fibrillation and sudden death. Sudden cardiac death is frequently a consequence of severe ventricular fibrillation or terminal cardiac arrhythmia.

In support of the animal studies is the study by Burr et al.[109,110] on the effects of dietary intervention in the secondary prevention of myocardial infarction that showed a modest intake of fatty fish two to three times per week (or 3 g fish oils per day) reduced all-cause mortality by 29% over a 2-year period, possibly by preventing sudden death from arrhythmia. The study by Gudbjarnason et al.[108] indicates that the death rate from cardiovascular disease in Iceland is lower than Sweden, Denmark, and Norway. At the age of 75 years or older, cardiovascular disease mortality decreases with increasing fish consumption. This relationship was not seen in younger age groups. However, the cardiovascular disease mortality at older age groups correlated positively with the level of AA in plasma phospholipids of the normal, healthy population at the age of 40 to 60 years, i.e., before development of cardiovascular disease. In the Nordic populations, cardiovascular disease mortality seems to be a function of both age and AA levels of plasma phospholipids, suggesting that fish consumption and plasma levels of AA have opposite effects on the development of atherosclerosis and cardiovascular disease in those populations.

EFFECTS ON RESTENOSIS

Percutaneous transluminal coronary angioplasty (PTCA) is an important treatment for selected patients with coronary heart disease. Although the success rate of angioplasty is high, restenosis occurs in 25 to 40% in the dilated lesions about 6 months after the procedure. The cause of restenosis is unknown; however, platelet aggregation, proliferation of smooth muscle cells, and coronary angiospasm are considered to be important contributors to restenosis. Most studies showed a benefit when omega-3 fatty acids supplemented the standard regimen before and after surgery.[111-114] Recently, Bairati et al.[114] reported on a double-blind randomized, controlled trial of fish oil supplements in the prevention of recurrence of stenosis after coronary angioplasty. Treatment was started 3 weeks before PTCA and continued for 6 months in 205 patients. The intervention group received 2.7 g EPA and 1.8 g DHA in 15 g of MaxEPA and the controls received 15 g olive oil. This trial documented the protective effect of fish oil on the recurrence of PTCA. Others report no benefit.[115,116] It appears that the length of time of omega-3 fatty acid supplementation prior to surgery may account for the differences as well as the amount of omega-6 fatty acids in the diet of patients.

EFFECTS ON LIPOPROTEIN (A)

Lipoprotein (a) [Lp(a)] is a genetically determined protein with a molecular structure strikingly similar to that of plasminogen. Lp(a) has atherogenic and thrombogenic properties. A number of studies indicate[117] that fish oils decrease the Lp(a) concentration only in those that have an elevated concentration and not in the normal.[118] Schmidt et al.[118] showed that omega-3 fatty acids lowered serum Lp(a) concentrations when Lp(a) concentrations were >200 mg/l, but had no effect at <200 mg/l. In another study only some of the patients lowered their Lp(a) following fish oil administration. The investigators considered the patients as responders and non-responders. The lowering of Lp(a) by fish oils is important because in patients with familial hypercholesterolemia, an elevated level of Lp(a) is a strong risk factor for CHD and the increase in risk is independent of age, sex, smoking status, and serum levels of total cholesterol.[119,120] A recent double-blind study demonstrated that 5.25 g fish oil lowers Lp(a) in primary hypertriglyceridemia.[121] Trans fatty acids, on the other hand, have been shown to increase Lp(a) levels.[19]

CONCLUSIONS

Western diets are deficient in the amount of EPA and DHA, whereas the amount of LA, AA, saturated fatty acids, and trans fatty acids are much higher than the diet on which human beings evolved. This has brought an imbalance in the ratio of omega-6/omega-3 fatty acids, which has occurred over the past 125 years, a very short period of time in human evolution.

Dietary intake of omega-3 fatty acids from fish or fish oils leads to significant incorporation of EPA (20:5ω3) and DHA (22:6ω3) into cellular phospholipids. These omega-3 fatty acids exert profound influences on eicosanoid metabolism, gene expression, and intracellular and cell-to-cell communication of practically all cells, thereby influencing numerous biochemical and pathophysiologic processes. DHA is essential for visual and cerebral function of the premature and possibly full-term infant.

Studies in animal experiments and clinical investigations indicate that omega-3 fatty acids have antiatheromatous, antithrombotic, anti-inflammatory, and beneficial hypolipidemic effects, all of which promote a more physiological state. In addition, EPA and DHA prevent ventricular fibrillation in animal models and decrease the rate of sudden death in human beings. More recently, omega-3 fatty acids have been shown to lower Lp(a) (a most atherogenic and thrombogenic genetically determined protein), to influence gene expression, and to have promising effects on the prevention of restenosis. The anti-inflammatory aspects of omega-3 fatty acids have important implications in the treatment of inflammatory and autoimmune diseases and conditions, such as rheumatoid arthritis, psoriasis, ulcerative colitis, Raynaud's disease, lupus erythematosus, and possibly cancer. The importance of omega-3 fatty acids in the diet is evident, as well as the need to return to a more physiologic omega-6/omega-3 dietary ratio of about 1 to 4:1, rather than the ratio of 20 to 30:1 provided by Western diets.

REFERENCES

1. **Simopoulos, A.P. and Salem, N., Jr.,** Purslane: a terrestrial source of omega-3 fatty acids, *N. Engl. J. Med.,* 315, 833, 1986.
2. **Emken, E.A., Adlof, R.O., Rakoff, H., et al.,** Metabolism of deuterium-labeled linolenic, linoleic, oleic, stearic and palmitic acid in human subjects, in *Synthesis and Applications of Isotopically Labelled Compounds 1988,* Baillie, T.A. and Jones, J.R., Eds., Elsevier, Amsterdam, 1989, 713.
3. **Simopoulos, A.P., Kifer, R.R., and Martin, R. E.,** Eds., *Health Effects of Polyunsaturated Fatty Acids in Seafoods,* Proc. Conf. Health Effects of Polyunsaturated Fatty Acids in Seafoods, June 24-26, 1985, Washington, D.C., Academic Press, Orlando, FL, 1986.
4. **Simopoulos, A.P.,** Terrestrial sources of omega-3 fatty acids: purslane, in *Horticulture and Human Health: Contributions of Fruits and Vegetables,* Quebedeaux, B. and Bliss, R., Eds., Prentice-Hall, Englewood Cliffs, NJ, 1987, 93.
5. **de Gomez Dumm, I.N.T. and Brenner, R.R.,** Oxidative desaturation of alpha-linolenic, linoleic, and stearic acids by human liver microsomes, *Lipids,* 10, 315, 1975.
6. **Hague, T.A. and Christoffersen, B.O.,** Effect of dietary fats on arachidonic acid and eicosapentaenoic acid biosynthesis and conversion to C_22 fatty acids in isolated liver cells, *Biochim. Biophys. Acta,* 796, 205, 1984.
7. **Hague, T.A. and Christoffersen, B.O.,** Evidence for peroxisomal retroconversion of adrenic acid (22:4n6) and docosahexaenoic acid (22:6n3) in isolated liver cells, *Biochim. Biophys. Acta,* 875, 165, 1986.
8. **Carlson, S.E., Rhodes, P.G., and Ferguson, M.G.,** Docosahexaenoic acid status of preterm infants at birth and following feeding with human milk or formula, *Am. J. Clin. Nutr.,* 44, 798, 1986.
9. **Singer, P., Jaeger, W., Voigt, S., et al.,** Defective desaturation and elongation of n-6 and n-3 fatty acids in hypertensive patients, *Prostaglandins Leukotrienes Med.,* 15, 159, 1984.
10. **Honigmann, G., Schimke, E., Beitz, J., et al.,** Influence of a diet rich in linolenic acid on lipids, thrombocyte aggregation and prostaglandins in type I (insulin-dependent) diabetes, *Diebetologia,* 23, 175, 1982.
11. **O'Brien, J.S. and Sampson, E.L.,** Fatty acid and aldehyde composition of the major brain lipids in normal gray matter, white matter and myelin, *J. Lipid Res.,* 6, 545, 1965.
12. **Anderson, R.E.,** Lipids of ocular tissues. IV. A comparison of the phospholipids from the retina of six mammalian species, *Exp. Eye Res.,* 10, 339, 1970.
13. **Poulos, A., Darin-Bennett, A., and White, I.G.,** The phospholipid bound fatty acids and aldehydes of mammalian spermatozoa, *Comp. Biochem. Physiol.,* 46B, 541, 1975.
14. **Leaf, A. and Weber, P.C.,** A new era for science in nutrition, *Am. J. Clin. Nutr.,* 45 (Suppl.), 1048, 1987.
15. **Eaton, S.B. and Konner, M.,** Paleolithic nutrition. A consideration of its nature and current implications, *N. Engl. J. Med.,* 312, 283, 1985.
16. **Simopoulos, A.P.,** Genetics and nutrition: or what your genes can tell you about nutrition, *World Rev. Nutr. Diet,* 63, 25, 1990.
17. **Kirshenbauer, H.G.,** *Fats and Oils,* 2nd ed., Reinhold Publishing, New York, 1960.
18. **Mensink, R.P. and Katan, M.B.,** Effect of dietary trans fatty acids on high-density and low-density lipoprotein cholesterol levels in healthy subjects, *N. Engl. J. Med.,* 323, 439, 1990.
19. **Nestel, P., Noakes, M., et al.,** Plasma lipoprotein lipid and Lp(a) changes with substitution of elaidic acid for oleic acid in the diet, *J. Lipid Res.,* 33, 1029, 1992.
20. **Zock, P.L. and Katan, M.B.,** Hydrogenation alternatives: effects of trans fatty acids and stearic acid versus linoleic acid on serum lipids and lipoproteins in humans, *J. Lipid Res.,* 33, 399, 1992.
21. **Phinney, S.D., Odin, R.S., Johnson, S.B., and Holman, R.T.,** Reduced arachidonate in serum phospholipids and cholesteryl esters associated with vegetarian diets in humans, *Am. J. Clin. Nutr.,* 51, 385, 1990.
22. **Raper, N.R., Cronin, F.J., and Exler, J.,** Omega-3 fatty acid content of the US food supply, *J. Am. Coll. Nutr.,* 11, 304, 1992.
23. **Ledger, H.P.,** Body composition as a basis for a comparative study of some East African mammals, *Symp. Zool. Soc. London,* 21, 289, 1968.
24. **Crawford, M.A.,** Fatty acid ratios in free-living and domestic animals, *Lancet,* 1, 1239, 1968.
25. **Wo, C.K.W. and Draper, H.H.,** Vitamin E status of Alaskan Eskimos, *Am. J. Clin. Nutr.,* 28, 808, 1975.
26. **Crawford, M.A., Gale, M.M., and Woodford, M.H.,** Linoleic acid and linolenic acid elongation products in muscle tissue of Syncerus caffer and other ruminant species, *Biochem. J.,* 115, 25, 1969.
27. **Simopoulos, A.P., Norman, H.A., Gillaspy, J.E., and Duke, J.A.,** Common purslane: a source of omega-3 fatty acids and antioxidants, *J. Am. Coll. Nutr.,* 11, 374, 1992.
28. **van Vliet, T. and Katan, M.B.,** Lower ratio of n-3 to n-6 fatty acids in cultured than in wild fish, *Am. J. Clin. Nutr.,* 51, 1, 1990.

29. **Simopoulos, A.P. and Salem, N., Jr.,** n-3 fatty acids in eggs from range-fed Greek chickens, *N. Engl. J. Med.,* 321, 1412, 1989.

30. **Hunter, J.E.,** Omega-3 fatty acids from vegetable oils, in *Dietary ω3 and ω6 Fatty Acids: Biological Effects and Nutritional Essentiality, Series A: Life Sciences,* Vol. 171, Galli, C. and Simopoulos, A.P., Eds., Plenum Press, New York, 1989, 43.

31. **Galli, C. and Simopoulos, A.P.,** Eds., *Dietary ω3 and ω6 Fatty Acids: Biological Effects and Nutritional Essentiality, Series A: Life Sciences,* Vol. 171, Plenum Press, New York, 1989.

32. **Willis, A.L.,** Nutritional and pharmacologic factors on eicosanoid biology, *Nutr. Rev.,* 39, 289, 1981.

33. **Moncada, S. and Vane, J.R.,** Unstable metabolites of arachidonic acid and their role in haemostasis and thrombosis, *Br. Med. Bull.,* 34, 129, 1978.

34. **Dyerberg, J., Bang, H.O., Stoffersen, E., Moncada, S., and Vane, J.R.,** Eicosapentaenoic acid and prevention of thrombosis and atherosclerosis, *Lancet,* 2, 117, 1978.

35. **Needleman, P., Raz, A., Minkes, M.S., et al.,** Triene prostaglandins: prostacyclin and thromboxane biosynthesis and unique biological properties, *Proc. Natl. Acad. Sci. U.S.A.,* 76, 944, 1979.

36. **Weber, P.C., Fischer, S., von Schacky, C., et al.,** Dietary omega-3 polyunsaturated fatty acids and eicosanoid formation in man, in *Health Effects of Polyunsaturated Fatty Acids in Seafoods,* Simopoulos, A.P., Kifer, R.R., and Martin, R.E., Eds., Academic Press, Orlando, FL, 1986, 49.

37. **Lewis, R.A., Lee, T.H., and Austen, K.F.,** Effects of omega-3 fatty acids on the generation of products of the 5-lipoxygenase pathway, in *Health Effects of Polyunsaturated Fatty Acids in Seafoods,* Simopoulos, A.P., Kifer, R.R., and Martin, R.E., Eds., Academic Press, Orlando, FL, 1986, 227.

38. **Rucker, R. and Tinker, D.,** The role of nutrition in gene expression: a fertile field for the application of molecular biology, *J. Nutr.,* 116, 177, 1986.

39. **Clarke, S.D. and Armstrong, M.K.,** Suppression of rat liver fatty acid synthetase mRNA level by dietary fish oil, *FASEB J.,* 2, A852, 1988.

40. **Simopoulos, A.P.,** ω-3 fatty acids in growth and development and in health and disease. I. The role of ω-3 fatty acids in growth and development, *Nutr. Today,* 23(2), 10, 1988.

41. **Walker, B.L.,** Maternal diet and brain fatty acids in young rats, *Lipids,* 2, 497, 1967.

42. **Connor, W.E., Neuringer, M., and Reisbick, S.,** Essentiality of ω3 fatty acids: evidence from the primate model and implications for human nutrition, *World Rev. Nutr. Diet,* 66, 118, 1991.

43. **Simopoulos, A.P., Kifer, R.R., and Wykes, A.A.,** Omega 3 fatty acids: research and support in the field since June 1985 (worldwide), *World Rev. Nutr. Diet.,* 66, 51, 1991.

44. **Rotstein, N.P., Ilincheta de Boschero, M.G., Giusto, N.M., and Alveldano, M.I.,** Effects of aging on the composition and metabolism of docosahexanoate-containing lipids of retina, *Lipids,* 22, 253, 1987.

45. **Crawford, M.A., Sinclair, A.J., Msuya, P.M., et al.,** Structural lipids and their polyenoic constituents in human milk, in *Dietary Lipids and Postnatal Development,* Galli, C., Jacini, G., and Pecile, A., Eds., Raven Press, New York, 1973, 41.

46. **Sanders, T.A.B. and Naismith, D.J.,** Long-chain polyunsaturated fatty acids in the erythrocyte lipids of breast-fed and bottle-fed infants, *Proc. Nutr. Soc.,* 64A, 1976.

47. **Uauy, R.D., Birch, D.G., Birch, E.E., Tyson, J.E., and Hoffman, D.R.,** Effect of dietary omega-3 fatty acids on retinal function of very-low-birth-weight neonates, *Pediatr. Res.,* 28, 485, 1990.

48. **Carlson, S.E., Werkman, S.H., Peeples, J.M., Cooke, R.J., and Wilson, W.M.,** Plasma phospholipid arachidonic acid and growth and development of preterm infants, in *Recent Advances in Infant Feeding,* van Biervielt, J.P., Koletzko, B., Okken, A., Rej, J., and Salle, B., Eds., Thieme Verlag, Stuttgart, 1991.

49. **Carlson, S.E.,** Polyunsaturated fatty acids and infant nutrition, in *Dietary ω3 and ω6 Fatty Acids: Biological Effects and Nutritional Essentiality,* Galli, C. and Simopoulos, A.P., Eds., Plenum Press, New York, 1989, 147.

50. **Holman, R.T., Johnson, S.B., and Hatch, T.F.,** A case of human linolenic acid deficiency involving neurological abnormalities, *Am. J. Clin. Nutr.,* 35, 617, 1982.

51. **Bjerve, K.S., Mostad, I.L., and Thoresen, L.,** Alpha-linolenic acid deficiency in patients on long-term gastric-tube feeding: estimation of linolenic acid and long-chain unsaturated n-3 fatty acid requirement in man, *Am. J. Clin. Nutr.,* 45, 66, 1987.

52. **Bjerve, K.S.,** ω3 fatty acid deficiency in man: implications for the requirement of alpha-linolenic acid and long-chain ω3 fatty acids, *World Rev. Nutr. Diet,* 66, 133, 1991.

53. **Simopoulos, A.P., Kifer, R.R., Martin, R.E., and Barlow, S.M.,** Eds., Health effects of ω3 polyunsaturated fatty acids in seafoods, *World Rev. Nutr. Diet,* 66, 1991.

54. Scientific Review Committee, Nutrition recommendations, H49-42/1990E, Minister of National Health and Welfare, Ottawa, Canada, 1990.

55. ESPGAN Committee on Nutrition, Committee report, Comment on the content and composition of lipids in infant formulas, *Acta Paediatr. Scand.,* 80, 887, 1991.

56. **Uauy, R., Birch, E., Birch, D., and Peirano, P.,** Visual and brain function measurements in studies on n-3 fatty acid requirements of infants, *J. Pediatr.,* 120, S168, 1992.

57. **Menkes, J.H.,** Early feeding history of children with learning disorders, *Dev. Med. Child Neurol.,* 19, 169, 1977.
58. **Rodgers, B.,** Feeding in infancy and late ability and attainment: a longitudinal study, *Dev. Med. Child Neurol.,* 20, 421, 1978.
59. **Lucas, A., Morley, R., Cole, T.J., et al.,** Breast milk and subsequent intelligence quotient in children born preterm, *Lancet,* 339, 261, 1992.
60. **Harris, W.S.,** Fish oils and plasma lipid and lipoprotein metabolism in humans: a critical review, *J. Lipid Res.,* 30, 785, 1989.
61. **Sanders, T.A.B.,** Influence of ω3 fatty acids on blood lipids, *World Rev. Nutr. Diet,* 66, 358, 1991.
62. **Miller, J.P., Heath, I.D., Choraria, S.K., et al.,** Triglyceride lowering effect of MaxEPA fish lipid concentrate: a multicentre placebo controlled double blind study, *Clin. Chem.,* 178, 215, 1988.
63. **Saynor, R., Verel, D., and Gillott, T.,** The long term effect of dietary supplementation with fish lipid concentrate on serum lipids, bleeding time, platelets and angina, *Atherosclerosis,* 50, 3, 1984.
64. **Schectman, G., Kaul, S., Cherayil, G.D., et al.,** Can the hypotriglyceridemic effect of fish oil concentrate be sustained, *Ann. Intern. Med.,* 110, 346, 1989.
65. **Harris, W.S., Dujovne, C.A., Zucker, M.L., et al.,** Effects of a low saturated fat, low cholesterol fish oil supplement in hypertriglyceridemic patients, *Ann. Intern. Med.,* 109, 465, 1988.
66. **Sanders, T.A.B., Sullivan, D.R., Reeve, J., et al.,** Triglyceride-lowering effect of marine polyunsaturates in patients with hypertriglyceridemia, *Arteriosclerosis,* 5, 459, 1985.
67. **Abbey, M., Clifton, P., Kestin, M., et al.,** Lipoproteins, lecithin: cholesterol acyltransferase and lipid transfer protein activity in human subjects consuming n-6 fatty acids and n-3 fatty acids of vegetable and marine origin, *Arteriosclerosis,* in press.
68. **Goto, A.M., Jr., Patsch, J.R., and Yamamoto, A.,** Postprandial hyperlipidemia, *Am. J. Cardiol.,* 68(3), 11A, 1991.
69. **Castelli, W.P.,** Cholesterol and lipids in the risk of coronary heart disease — the Framingham Study, *Can. J. Cardiol.,* 49 (Suppl. A), 5A, 1988.
70. **Phillipson, B.E., Rothrock, D.W., Connor, W.E., et al.,** Reduction of plasma lipids, lipoproteins, and apoproteins by dietary fish oils in patients with hypertriglyceridemia, *N. Engl. J. Med.,* 312, 1210, 1985.
71. **Parks, J.S., Johnson, F.L., Wilson, M.D., and Rudel, L.L.,** Effect of fish oil diet on hepatic lipid metabolism in nonhuman primates: lowering of secretion of hepatic triglyceride but not apoB, *J. Lipid Res.,* 31, 455, 1990.
72. **Wong, S. and Nestel, P.J.,** Eicosapentaenoic acid inhibits the secretion of triacylglycerol and of apoprotein B and the binding of LDL in Hep G2 cells, *Atherosclerosis,* 64, 139, 1987.
73. **Nestel, P.J., Connor, W.E., Reardon, M.R., Connor, S., Wong, S., and Boston, R.,** Suppression by diets rich in fish oil of very low density lipoprotein production in man, *J. Clin. Invest.,* 74, 72, 1984.
74. **Sanders, T.A.B., Sullivan, D.R., Reeve, J., and Thompson, G.R.,** Triglyceride-lowering effect of marine polyunsaturates in patients with hypertriglyceridemia, *Arteriosclerosis,* 5, 459, 1985.
75. **Harris, W.S., Connor, W.E., Illingworth, D.R., Rothrock, D.W., and Foster, D.M.,** Effects of fish oil on VLDL triglyceride kinetics in humans, *J. Lipid Res.,* 31, 1549, 1990.
76. **Rustan, A.C., Nossen, J.O., Christiansen, E.N., and Drevon, C.A.,** Eicosapentaenoic acid reduces hepatic synthesis and secretion of triacylglycerol by decreasing the activity of acyl-coenzyme A:1,2-diacylglycerol acyltransferase, *J. Lipid Res.,* 29, 1417, 1988.
77. **Murthy, S., Albright, E., Mathur, S.N., and Field, F.J.,** Effect of eicosapentaenoic acid in triacylglycerol transport in CaCo-2 cells, *Biochim. Biophys. Acta,* 1045, 147, 1990.
78. **Brown, A.J. and Roberts, D.C.K.,** Moderate fish oil intake improves lipemic response to a standard fat meal, *Arterio. Thromb.,* 11, 457, 1991.
79. **Harris, W.S. and Windsor, S.L.,** N-3 fatty acid supplements reduce chylomicron levels in healthy volunteers, *J. Appl. Nutr.,* 43, 5, 1991.
80. **Nestel, P.J.,** Fish oil attenuates the cholesterol-induced rise in lipoprotein cholesterol, *Am. J. Clin. Nutr.,* 43, 752, 1986.
81. **Harris, W.S., Connor, W.E., Inkeles, S.B., et al.,** Omega-3 fatty acids prevent carbohydrate-induced hypertriglyceridemia, *Metabolism,* 33, 1016, 1984.
82. **Nestel, P.J., Connor, W.E., Reardon, M.R., et al.,** Suppression by diets rich in fish oil of very low density lipoprotein production in man, *J. Clin. Invest.,* 74, 72, 1984.
83. **Connor, W.E.,** Hypolipidemic effects of dietary omega-3 fatty acids in normal and hyperlipidemic humans: effectiveness and mechanisms, in *Health Effects of Polyunsaturated Fatty Acids in Seafoods,* Simopoulos, A.P., Kifer, R.R., and Martin, R.E., Eds., Academic Press, Orlando, FL, 1986, 173.
84. **Weber, P.C. and Leaf, A.,** Cardiovascular effects of ω3 fatty acids. Atherosclerosis risk factor modification by ω3 fatty acids, *World Rev. Nutr. Diet,* 66, 218, 1991.
85. **Sperling, R.I., Robin, J.L., Kylander, K.A., et al.,** The effects of n-3 polyunsaturated fatty acids on the generation of platelet-activating factor-acether by human monocytes, *J. Immunol.,* 139, 4186, 1987.
86. **Ross, R.,** The pathogenesis of atherosclerosis — an update, *N. Engl. J. Med.,* 314, 488, 1986.

87. Steinberg, D., Parthasarathy, S., Carew, T.E., et al., Beyond cholesterol: modifications of low-density lipoprotein that increase its atherogenicity, *N. Engl. J. Med.*, 320, 915, 1989.

88. Simopoulos, A.P., Omega-3 fatty acids in health and disease and in growth and development, *Am. J. Clin. Nutr.*, 54, 438, 1991.

89. Landymore, R.W., MacAulay, M., Sheridan, B., et al., Comparison of cod liver oil and aspirin-dipyridamole for the prevention of intimal hyperplasia in autologous vein grafts, *Ann. Thoracic Surg.*, 41, 54, 1986.

90. Weiner, B.H., Ockene, I.S., Levine, P.H., et al., Inhibition of atherosclerosis by cod liver oil in a hyperlipidemic swine model, *N. Engl. J. Med.*, 315, 841, 1986.

91. Davis, H.R., Bridenstine, R.T., Vesselinovitch, D., et al., Fish oil inhibits development of atherosclerosis in rhesus monkeys, *Atherosclerosis*, 7, 441, 1987.

92. Hollander, W., Hong, S., Kirkpatrick, B.J., et al., Differential effects of fish oil supplements on atherosclerosis, *Circulation*, 76 (Suppl. 4), 313, 1987.

93. Thiery, J. and Seidel, D., Fish oil feeding results in an enhancement of cholesterol induced atherosclerosis in rabbits, *Atherosclerosis*, 63, 53, 1987.

94. Brox, J.H., Killie, J.E., Osterud, B., Holme, S., and Nordoy, A., Effects of cod liver oil on platelets and coagulation in familial hypercholesterolemia (type IIa), *Acta Med. Scand.*, 213, 137, 1983.

95. Joist, J.H., Baker, R.K., and Schonfeld, G., Increased in vivo and in vitro platelet function in type II- and type IV-hyperlipoproteinemia, *Thromb. Res.*, 15, 95, 1979.

96. Bottiger, L.E., Dyerberg, J., and Nordoy, A., n-3 fish oils in clinical medicine, *J. Intern. Med.*, 225 (Suppl. 1), 1, 1989.

97. Lewis, R.A., Lee, T.H., and Austen, K.F., Effects of omega-3 fatty acids on the generation of products of the 5-lipoxygenase pathway, in *Health Effects of Polyunsaturated Fatty Acids in Seafoods*, Simopoulos, A.P., Kifer, R.R., and Martin, R.E., Eds., Academic Press, Orlando, FL, 1986, 227.

98. Cartwright, I.J., Pockley, A.G., Galloway, J.H., et al., The effects of dietary ω-3 polyunsaturated fatty acids on erythrocyte membrane phospholipids, erythrocyte deformability and blood viscosity in healthy volunteers, *Atherosclerosis*, 55, 267, 1985.

99. Barcelli, U.O., Glass-Greenwalt, P., and Pollak, V.E., Enhancing effect of dietary supplementation with omega-3 fatty acids on plasma fibrinolysis in normal subjects, *Thromb. Res.*, 39, 307, 1985.

100. Radack, K., Deck, C., and Huster, G., Dietary supplementation with low-dose fish oils lowers fibrinogen levels: a randomized, double-blind controlled study, *Ann. Intern. Med.*, 111, 757, 1989.

101. Sanders, T.A.B., Vickers, M., and Haines, A.P., Effect on blood lipids and haemostasis of a supplement of cod-liver oil, rich in eicosapentaenoic and docosahexaenoic acids, in healthy young men, *Clin. Sci.*, 61, 317, 1981.

102. Brown, A.J. and Roberts, D.C.K., Fish and fish oil intake: effect on haematological variables related to cardiovascular disease, *Thromb. Res.*, 64, 169, 1991.

103. DeCaterina, R., Giannessi, D., Mazzone, A., et al., Vascular prostacyclin is increased in patients ingesting n-3 polyunsaturated fatty acids prior to coronary artery bypass surgery, *Circulation*, 82, 428, 1990.

104. Fox, P.L. and Dicorleto, P.E., Fish oils inhibit endothelial cell production of a platelet-derived growth factor-like protein, *Science*, 241, 453, 1988.

105. Shimokawa, H. and Vanhoutte, P.M., Dietary cod-liver oil improves endothelium dependent responses in hypercholesterolemic and atherosclerotic porcine coronary arteries, *Circulation*, 78, 1421, 1988.

106. Charnock, J.S., The antiarrhythmic effects of fish oils, *World Rev. Nutr. Diet*, 66, 278, 1991.

107. Hallaq, H. and Leaf, A., Effects of ω3 and ω6 polyunsaturated fatty acids on the action of cardiac glycosides on the cultured rat myocardial cells, *World Rev. Nutr. Diet*, 66, 250, 1991.

108. Gudbjarnason, S., Benediktsdottir, V.E., and Gudmundsdottir, E., Balance between ω3 and ω6 fatty acids in heart muscle in relation to diet, stress and ageing, *World Rev. Nutr. Diet*, 66, 292, 1991.

109. Burr, M.L., Fehily, A.M., Gilbert, J.F., et al., Effect of changes in fat, fish and fibre intakes on death and myocardial reinfarction: diet and reinfarction trial (DART), *Lancet*, 2, 757, 1989.

110. Burr, M.L., Fish and ischemic heart disease, *World Rev. Nutr. Diet*, 72, 1993, p49.

111. Dehmer, G.J., Pompa, J.J., Van den Berg, E.K., et al., Reduction in the rate of early restenosis after coronary angioplasty by a diet supplemented with n-3 fatty acids, *N. Engl. J. Med.*, 319, 733, 1988.

112. Milner, M.R., Gallino, R.A., Leffingwell, A., et al., High-dose omega-3 fatty acid supplementation after coronary angioplasty, *Circulation*, 78 (Suppl. 2), 634, 1988.

113. Slack, J.D., Pinkerton, C.A., Van Tassel, J., et al., Can oral fish oil supplement minimize restenosis after percutaneous transluminal coronary angioplasty?, *J. Am. Coll. Cardiol.*, 9 (Suppl.), 64a, 1987.

114. Bairati, I., Roy, L., and Meyer, F., Double-blind, randomized, controlled trial of fish oil supplements in prevention of recurrence of stenosis after coronary angioplasty, *Circulation*, 85, 950, 1992.

115. Reis, G.J., Boucher, T.M., and McCabe, C.H., Results at a randomized, double-blind placebo-controlled trial of fish oil for prevention of restenosis after PTCA, *Circulation*, 78 (Suppl. 2), 291, 1988.

116. **Grigg, L.E., Kay, I.W.H., Valentine, P.A., et al.,** Determinants of restenosis and lack of effect of dietary supplementation with eicosapentaenoic acid on the incidence of coronary artery restenosis after angioplasty, *J. Am. Coll. Cardiol.,* 13, 665, 1989.

117. **Simopoulos, A.P.,** Executive summary, in *Dietary w3 and w6 Fatty Acids: Biological Effects and Nutritional Essentiality,* Galli, C. and Simopoulos, A.P., Eds., Plenum Press, New York, 1989.

118. **Schmidt, E.B., Klausen, I.C., Kristensen, S.D., et al.,** The effect of n-3 fatty acids on lipoprotein(a) (abstract), in Simopoulos, A.P., Kifer, R.R., Martin, R.E., and Barlow, S.M., Eds.,

119. **Seed, M., Hoppichler, F., Reaveley, D., et al.,** Relation of serum lipoprotein(a) concentration and apolipoprotein(a) phenotype to coronary heart disease in patients with familial hypercholesterolemia, *N. Engl. J. Med.,* 322, 1494, 1990.

120. **Wiklund, O., Angelin, B., Olofsson, S.-O., et al.,** Apolipoprotein(a) and ischaemic heart disease in familial hypercholesterolaemia, *Lancet,* 335, 1360, 1990.

121. **Beil, F.U., Terres, W., Orgass, M., and Greten, H.,** Dietary fish oil lowers lipoprotein(a) in primary hypertriglyceridemia, *Atherosclerosis,* 90, 95, 1991.

Chapter 2.3

OMEGA-3 FATTY ACIDS

PART II: EPIDEMIOLOGICAL ASPECTS OF OMEGA-3 FATTY ACIDS IN DISEASE STATES

Artemis P. Simopoulos

INTRODUCTION

As expected, the majority of the studies on the epidemiological aspects of omega-3 fatty acids have been carried out on coronary heart disease. This chapter, therefore, begins with coronary heart disease, followed by hypertension, diabetes, cancer, and inflammatory and autoimmune disorders (arthritis, psoriasis, and ulcerative colitis). In addition, a number of intervention studies have been carried out in patients with lupus erythematosus, atopy, asthma, Raynaud's disease, and multiple sclerosis, which are only mentioned here since the information about the effects of omega-3 fatty acids on these diseases are preliminary. Large-scale double-blind controlled trials are needed to be carried out in these diseases in order to establish the benefits of fish oil supplementation.

CORONARY HEART DISEASE

In the 1950s a number of studies involving animals and human beings indicated that ingestion of fish or fish oils had a hypolipidemic effect.[1-7] However, because omega-6 fatty acids had been shown to be effective in lowering serum cholesterol,[8,9] omega-3 fatty acids were not given the attention that was due to them. In fact, the health-related effects of omega-3 fatty acids did not become apparent to the scientific community until the epidemiologic studies by Bang and Dyerberg in the 1970s.[10-15] Their work on cardiovascular diseases and dietary fat intake among Greenland Eskimos clarified the important role of EPA and DHA. Table 1[16] shows that the major difference in the dietary fat between Eskimos and Danes was in the higher intake of omega-3 fatty acids of marine origin and not in total fat.

Bang and Dyerberg studied 130 Eskimos in the Umanak district in the northern part of West Greenland. The Eskimos had lower levels of serum cholesterol, ApoB lipoproteins, and triglycerides in comparison to the Eskimos living in Denmark whose lipid levels were comparable to those of the Danes, indicating that genetics were not the major factors that could account for this difference. Coronary heart disease and diabetes were rare among the Greenland Eskimos, whereas stroke and cirrhosis of the liver were more frequent. The Eskimos exhibited a prolonged bleeding time, easy bruisability, a decreased number of platelets, and platelet aggregation consistent with a decreased rate of coronary thrombosis. Bang and Dyerberg suggested that the high dietary intake of EPA and DHA, 6 and 7 g, respectively, and the low intake of linoleic acid of about 5 g/d, compared to 24 g/d in Western diets, were responsible for the lower cholesterol and triglyceride levels, the prolongation of bleeding time, and other aspects of health and disease states of the Greenland Eskimos, particularly their low death rate from cardiovascular disease.

As mentioned previously, the hypocholesterolemic effects of fish and fatty acids of marine oils were observed by Nelson,[2,3] Pfeifer et al.,[4] Stansby,[7] and Bierenbaum et al.[17] early on. Looking at the 1972 publication by Nelson, one might wonder why it took so long to appreciate these monumental research findings. Perhaps, because Nelson's work was published in *Geriatrics,* it escaped the notice that it merited by experts in cardiology and lipid research. Dyerberg, in 1986,[18] referring to Nelson's work, said,

In the 1950s, a Seattle cardiologist interested in nutrition decided to treat heart patients with a hypocholesterolemic diet that substituted fish for meat. Encouraged by reports of the cholesterol-lowering effect of fish oils[5,7] he advised his patients to consume fish as a main course three times a week.[2] These studies continued for 16 years and showed a fourfold greater incidence of fatal heart attacks in controls compared to the diet group. Although his study was not well received when first published, it was the first positive step supporting the view that fish in the diet held promise as a preventive dietary measure against coronary heart disease.

TABLE 1

Dietary Fats in Eskimo and Danish Diets[a]

	Eskimos	Danes
Percent of total calories from fat	39%	42%
Percent of total fatty acids		
Saturated	23%	53%
Monounsaturated	58%	34%
Polyunsaturated	19%	13%
Polyunsaturated/saturated (P/S) ratio	0.84	0.24
Grams per day		
Omega-3 fatty acids	14 g	3 g
Omega-6 fatty acids	5 g	10 g
Cholesterol	0.70 g	0.42 g

[a] Daily energy intake approximately 3000 kcal.

Adapted from Dyerberg, J., *n-3 News*, 1, 1, 1986.

Stimulated by the findings of Bang and Dyerberg, Hirai et al.[19] reported similar findings in Japanese fishing villagers. Hirai et al. compared dietary intake and plasma levels of EPA and AA in a fishing village whose inhabitants consumed 250 g of fish daily, with those living in a farming village, who consumed 90 g of fish per day. The plasma EPA/AA ratio was higher in the fishing villagers. There was also a decrease in the aggregation of platelets in the fishing villagers, which Hirai et al. considered as an explanation for the relatively low incidence of cardiovascular disease in Japan.

In a subsequent more extensive study, Hirai et al.[20] demonstrated that fishing villagers had decreased platelet aggregation and decreased blood viscosity. The distribution of fatty acids (EPA and DHA) in their plasma and platelet membrane phospholipids was higher than in the farming villagers and Hirai et al. concluded,

> The results of our epidemiological investigation in Japan strongly suggest that haemostatic function in man can be manipulated with dietary fish lipids (mainly with EPA) and that the ingestion of EPA rich fish diet could have a beneficial effect on thrombotic cardiovascular disorders by reducing platelet aggregability and whole blood viscosity.

In another study, Kagawa et al.[21] demonstrated that island inhabitants in Kohana had higher serum EPA concentrations than the inhabitants of mainland Japan, and significantly lower mortality rates from cardiovascular disease, cerebrovascular disease, and cerebral infarction.

The studies by Bang and Dyerberg reported between 1972 and 1980, Hirai et al. in 1980 and 1984, Kagawa in 1982, and a series of papers published in the *New England Journal of Medicine* in 1985 accompanied by an editorial[22-25] set the stage for the worldwide interest in the role of omega-3 fatty acids in health and disease. In 1985 Kromhout et al.[22] reported that as little as 30 g of lean fish per day lowered by 50% the mortality from coronary heart disease in a group of men living in Zutphen in The Netherlands. Again in 1985 Phillipson et al. presented data indicating that a high amount of fish oil (32 g/d) supplements in both normal volunteers and patients with hypertriglyceridemia lowered both serum cholesterol and triglyceride levels.[23] The paper by Lee et al.[24] emphasized the important role of omega-3 fatty acids as anti-inflammatory agents, and the editorial of Glomset[25] pointed to the important role of omega-3 fatty acids in coronary heart disease.

In 1985 Shekelle et al.[26] showed that "consumption of fish at entry was inversely associated in a graded manner with the 25-year risk of death from coronary heart disease and from all

TABLE 2
Ethnic Differences in Fatty Acid Concentrations in Thrombocyte Phospholipids and Frequency of Cardiovascular Disorders

	Europe, U.S.	Japan	Greenland Eskimos
Arachidonic acid, C20:4ω6 (%)	26	21	8.3
Eicosapentenoic acid, C20:5ω3 (%)	0.5	1.6	8.0
ω6:ω3	50	12	1
Cardiovascular mortality (%)[a]	45	12	7

[a]Percent of all deaths.

Adapted from Weber in Simopoulos.[33]

causes combined; it was not associated with death from other cardiovascular-renal diseases, from malignant neoplasms, or from other causes combined." Not all studies reported a decrease in coronary heart disease mortality rate. Curb and Reed[27] reported that their data from the Honolulu Heart Program revealed no significant difference in the incidence of total and fatal coronary heart disease between subjects on a high fish diet vs. those on a low fish diet. However, the total incidence of coronary heart disease was higher among the men who never ate fish than among those who did. In three other studies,[28-30] fish oils did not lower mortality from coronary heart disease, probably due to a small number of subjects,[28] a high intake of saturated fatty acids,[29] or to changes due to cooking which led to increases in omega-6 fatty acids and loss of omega-3 fatty acids.[30]

In addition, the studies of Burr et al.[31] and Dolecek and Grandits[32] provide further evidence on the beneficial role of omega-3 fatty acids in the prevention of coronary heart disease mortality. Burr carried out a controlled prospective dietary intervention trial in 2033 men who had recovered from myocardial infarction. Those who were advised to eat fish or take fish oil had a 29% reduction in 2-year all-cause mortality ($p < 0.05$) compared with those who were not given this advice. Of interest is that those given advice to reduce fat and to increase the P/S ratio, and those given advice to increase fiber intake did not have any decrease in mortality. This is the first prospective dietary intervention trial for secondary prevention of coronary heart disease that demonstrated a decrease in total mortality without a decrease in coronary events, suggesting that a decrease in the rate from sudden death is most likely due to the prevention of cardiac arrhythmias by fish or fish oil.

Dolecek and Grandits[32] investigated the 24-h dietary-recall data in the usual-care group of the Multiple Risk Factor Intervention Trial (MRFIT). They distinguished between omega-3 and omega-6 fatty acid intake and their relationship to four mortality categories: coronary heart disease, total cardiovascular disease, all-cause mortality, and cancer. Significant inverse associations were found between coronary heart disease, cardiovascular disease, and all-cause mortality groups and intake of EPA and DHA. The benefit appeared to be in the highest intake quintile with a mean ingestion of about 664 mg/d of EPA and DHA. When compared with zero intake, mortality from coronary heart disease, cardiovascular disease, and all-cause mortality was 40, 41, and 24% lower, respectively.

Populations with high consumption of fish, such as the Eskimos and Japanese, have lower rates of myocardial infarction (Table 2) and practically all epidemiological studies show that ingestion of omega-3 fatty acids is associated with a decrease in coronary heart disease mortality.[33] The mechanisms involved in the hypolipidemic, hypotensive, antiatherogenic, antithrombotic, and anti-inflammatory aspects of omega-3 fatty acids have been described in the previous chapter and are summarized in Tables 3 and 4.[34] The effects of omega-3 fatty acids in decreasing the rate of restenosis noted in the majority of the studies are very encouraging and suggest that omega-3 fatty acids may prevent the development of atherosclerosis. Inconsistencies in some of the intervention trials are most likely due to differences in amount and source of omega-3 fatty acid, length of observation, disease state, and genetic differences in patients with various forms of hyperlipidemia.

Coronary heart disease is a multifactorial disorder due to many genetic and environmental factors and their interactions. Elevated serum cholesterol has been shown to be a risk factor for coronary heart disease. At the same time it is known that 50% of the serum cholesterol level is genetically determined, and that all the dyslipidemias described thus far are genetically determined. Patients who survive a myocardial infarction have one or more of four lipoprotein abnormalities: (1) increased LDL cholesterol concentrations; (2) decreased HDL cholesterol concentrations usually accompanied by increased triglyceride or VLDL concentrations; (3) increased concentration of chylomicron remnants and intermediate density lipoprotein (IDL); and (4) increased Lp(a). Omega-3 fatty acids lower triglycerides and chylomicrons, usually raise HDL, lower Lp(a), and have antithrombotic and antiatheromatous effects. Although omega-3 fatty acids alone will not lead to the universal eradication of coronary heart disease, as we begin to unravel the genetics of coronary heart disease

TABLE 3
Functional Effects of Omega-3 Fatty Acids in the Cardiovascular System

Decrease postprandial lipemia
Reduce platelet aggregation
Reduce blood pressure
Decrease whole blood viscosity
Reduce vascular intimal hyperplasia
Reduce vasospastic response to vasoconstrictors
Reduce cardiac arrhythmias
Reduce albumin leakage in type I diabetes mellitus

Increase bleeding time
Increase platelet survival
Increase vascular (arterial) compliance
Increase cardia beta-receptor function
Increase postischemic coronary blood flow

Adapted from Weber, P.C., Fischer, S., von Schacky, C., et al., "Dietary Omega-3 Polyunsaturated Fatty Acids and Eicosanoid Formation in Man," in *Health Effects of Polyunsaturated Fatty Acids in Seafoods*, Simopoulos, A.P., Kifer, R.R., and Martin, R.E., Eds., Academic Press, Orlando, FL, 1986, 49.

and the mechanisms of atherogenesis, we should be able to identify individuals with genetic susceptibility who should modify their diet early in life. The provision of a low saturated fat diet with increased amounts of omega-3 fatty acids should be beneficial in the prevention of coronary heart disease.

HYPERTENSION

The first studies on the effect of omega-3 fatty acids on blood pressure were reported in 1983 by two groups of investigators, Singer et al.[35] and Lorenz et al.[36] They showed that adding mackerel to the diet of patients with mild hypertension lowered blood pressure. Others confirmed these findings in hypertensive patients and in normal,[37-42] although others did not.[43,44] Knapp and FitzGerald[40] evaluated the effects of omega-3 and omega-6 fatty acids in patients with essential hypertension, using different doses. High doses of fish oil, 50 ml of MaxEPA, reduced blood pressure in men with essential hypertension. In this group the formation of vasodilatory prostaglandins (PGI$_2$ and PGI$_3$) increased initially, but this effect was not sustained as blood pressure fell. The concentration of thromboxane A$_2$ metabolites fell, and thromboxane A$_3$ metabolites were detected in the groups receiving fish oil. As expected, omega-6 fatty acids, 50 ml given in the form of safflower oil, led to an increase of PGE$_2$ and tended to decrease with fish oil, although no PGE$_3$ metabolite was detected (Figure 1).

In another study Bonaa et al.[45] carried out a population-based intervention trial in which decreases of 6 mmHg in systolic and 3 mmHg blood pressure occurred with fish oil supplementation. Of particular interest in this study is the observation that dietary supplementation with fish oil did not change mean blood pressure in the subjects who ate fish three or more times per week as part of their usual diet or in those who had a baseline concentration of plasma phospholipid omega-3 fatty acids >175.1 mg/l, suggesting that a relationship exists between plasma phospholipid omega-3 fatty acid concentration and blood pressure. Those subjects who habitually consumed more fish had a lower blood pressure at baseline. This

TABLE 4
Effects of ω3 Fatty Acids on Factors Involved in the Pathophysiology of Atherosclerosis and Inflammation

Factor	Function	Effect of ω3 fatty acid
Arachidonic acid	Eicosanoid precursor; aggregates; platelets; stimulates white blood cells	↓
Thromboxane	Platelet aggregation; vasoconstriction; increase of intracellular Ca⁺⁺	↓
Prostacyclin (PGI$_{2/3}$)	Prevent platelet aggregation; vasodilation; increase cAMP	↑
Leukotriene (LTB$_4$)	Neutrophil chemoattractant increase of intracellular Ca⁺⁺	↓
Tissue plasminogen activator	Increase endogenous fibrinolysis	↑
Fibrinogen	Blood clotting factor	↓
Red cell deformability	Decreases tendency to thrombosis and improves oxygen delivery to tissues	↑
Platelet activating factor (PAF)	Activates platelets and white blood cells	↓
Platelet-derived growth factor (PDGF)	Chemoattractant and mitogen for smooth muscles and macrophages	↓
Oxygen free radicals	Cellular damage; enhance LDL uptake via scavenger pathway; stimulate arachidonic acid metabolism	↓
Lipid hydroperoxides	Stimulate eicosanoid formation	↓
Interleukin 1 and tumor necrosis factor	Stimulate neutrophil O$_2$ free radical formation; stimulate lymphocyte proliferation; stimulate PAF; express intercellular adhesion molecule-1 on endothelial cells; inhibit plasminogen activator, thus, procoagulants	↓
Endothelial-derived relaxation factor (EDRF)	Reduces arterial vasoconstrictor response	↑
VLDL	Related to LDL and HDL level	↓
HDL	Decreases the risk for coronary heart disease	↑
Lp(a)	Lipoprotein(a) is a genetically determined protein that has atherogenic and thrombogenic properties	↓
Triglycerides and chylomicrons	Contribute to postprandial lipemia	↓

Adapted from Weber, P.C. and Leaf, A., *World Rev. Nutr. Diet,* 66, 218, 1991.

finding suggests that supplementation with fish oils would be important from the primary prevention standpoint. Normal subjects given large doses of fish oil do not show any change in renal function,[46] which is encouraging in terms of safety issues. Singer[38] used three cans of mackerel per week (equivalent to 1.2 g omega-3 fatty acids per day or $1.2 \times 3 = 3.6$ g of fish oil per day) for 8 months to lower blood pressure with good results and concluded that this

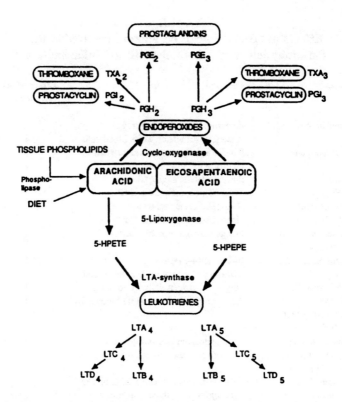

FIGURE 1. Oxidative metabolism of arachidonic acid and eicosapentanoic acid by the cyclooxygenase and 5-lipoxygenase pathways. 5-HPETE denotes 5-hydroperoxyeicosatetranoic acid and 5-HPEPE denotes 5-hydroxyeicosapentanoic acid.

amount of mackerel or equivalent amount of fish oil is considered acceptable for daily intake of the general population. The mechanisms by which omega-3 fatty acids exert their hypotensive effects have been investigated (Tables 3 and 4). Two research groups have shown that dietary EPA is converted to PGI_3 in humans without suppressing the formation of PGI_2 from AA.[47,48] Other mechanisms include effects of omega-3 fatty acids on renal function, a lowering of whole blood viscosity, and a reduction in vascular responsiveness to systemic vasoconstrictors.[49] The combination of propanolol and fish oil supplements potentiate their blood pressure-lowering effects. In addition, the increase in plasma triglycerides that is often seen during antihypertensive therapy did not occur, indicating once again the importance of omega-3 fatty acids as adjuncts to drug therapy.

DIABETES

There are two different types of diabetes: non-insulin-dependent diabetes mellitus (NIDDM) and insulin-dependent diabetes (IDDM). NIDDM is more common than IDDM. Both types are characterized by abnormalities in lipid metabolism in addition to abnormalities in carbohydrate metabolism.

Diabetes is rare in Eskimos. This observation and the fact that diabetes is a chronic disorder with complications that include hypertriglyceridemia, hypercholesterolemia, atherosclerosis, coronary heart disease, and hypertension led to the use of omega-3 fatty acids in patients with either type of diabetes.[50,51]

NON-INSULIN-DEPENDENT DIABETES MELLITUS

NIDDM is characterized by a relative lack of insulin, decreased insulin sensitivity, and impaired glucose homeostasis. The addition of fish has led to impairment in glucose homeo-

TABLE 5	TABLE 6
Effects of Omega-3 Fatty Acids on Glucose Homeostasis in Noninsulin-Dependent Diabetes Mellitus	**Effects of Omega-3 Fatty Acids on Lipoprotein Composition in Noninsulin-Dependent Diabetes Mellitus**

Fasting blood glucose	± or ↑
Meal-stimulated glucose response	↑
Glycosylated hemoglobin	± or ↑
Basal hepatic glucose output	↑
Peripheral glucose disposal	±
Insulin/glucose ratio	↓
Stimulated insulin response	± or ↓
Fasting serum insulin concentrations	± or ↓

Note: ↑ = increased; ± = unchanged; ↓ = decreased.

Adapted from Vessby, B., *World Rev. Nutr. Diet,* 66, 407, 1991.

Serum triglycerides	↓ (or ±)
Serum cholesterol	↓ or ±
VLDL triglycerides	↓
LDL cholesterol/LDL apo B	↓
Serum apo A-II	(↓)
Serum apo A-I	±
HDL cholesterol	± or ↑
LDL apo B	↑
LDL cholesterol	↑ or ±
Serum apo B	± or ↑

Note: ↑ = increased; ± = unchanged; ↓ = decreased.

Adapted from Vessby, B., *World Rev. Nutr. Diet,* 66, 407, 1991.

stasis (Table 5) and produced an elevation in some patients in LDL cholesterol, while triglycerides decreased in all cases (Table 6).[52] The magnitude of these changes was small.[53,54] The changes in glucose metabolism were associated with increased hepatic glucose output and impaired insulin secretion but unaltered glucose disposal.

INSULIN-DEPENDENT DIABETES MELLITUS

In a double-blind crossover design clinical trial Jensen et al.[51] studied the effects of cod liver oil on endothelial permeability, blood pressure, and plasma lipids in patients with IDDM. The patients' diabetic diets were supplemented for 8 weeks with cod liver oil and for 8 weeks with olive oil. The trial included 18 patients with IDDM and albuminuria >30 mg/d. Dietary supplementation with cod liver oil significantly reduced the blood pressure from 146/90 to 139/85 mmHg and there was a significant fall in transcapillary escape rate of albumin compared with baseline. During supplementation with cod liver oil, the HDL cholesterol increased and the VLDL cholesterol and triglyceride decreased, whereas no changes occurred in LDL cholesterol, glomerular filtration rate, degree of albuminuria, or glycosylated hemoglobin. In this study, omega-3 fatty acids did not have any effect on glucose metabolism. Wahlqvist et al.[55] recently reported that diabetics who eat fish have higher arterial compliance when compared to those not eating fish. Tilvis et al.[56] and Miller et al.[57] found no changes in LDL, whereas Haines et al., in 1986,[58] reported increases in LDL following 15 g of fish oil per day (MaxEPA).

In all of the above studies, different types of omega-3 fatty acids were used (cod liver oil, fish oil, MaxEPA) at different dosages, in two different types of diabetic patients. Further studies are needed under defined conditions in terms of the type of diabetes (NIDDM and IDDM), metabolic control, diet, and fatty acid composition of phospholipids. The studies thus far indicate that omega-3 fatty acids are beneficial in IDDM patients with albuminuria. These patients are at high risk for cardiovascular death since they have a number of established cardiovascular risk factors (hypertriglyceridemia, hypertension, hypercholesterolemia, atherosclerosis). Long-term studies on the effect of omega-3 fatty acids are needed.

CANCER

In 1991, Cave reviewed the effects of omega-3 fatty acids on tumorigenesis in experimental animals.[59] Experimental studies included animal tumor models in which the tumor was induced, and animal models with transplantable tumors (breast, colon, pancreas, and prostate). In both animal models the omega-3 fatty acids in the form of linseed oil (LNA) or fish oil or cod liver oil (EPA and DHA) consistently delayed tumor appearance, and decreased both the rate of growth and the size and number of tumors. Furthermore, caloric restriction potentiated the effects of omega-3 fatty acids.[59,60] Omega-3 fatty acids in the form of corn oil increased tumor formation, size, and number.[59,61] Fatty acid analysis of the transplanted tumors reflected the ingested fatty acid of the host. High dietary intake of omega-3 fatty acids prevented or delayed the development of neoplasms in the host, and PGE_2 production was decreased. Fernandes[60] used human breast cancer cells in nude mice. The mice fed omega-3 fatty acids had fewer pulmonary metastasis, decreased serum estrogen, and prolactin concentration. Both the amount of PGE_2 in the tumor and c-myc oncogene mRNA concentrations in the tumor tissue cells were reduced,[60] whereas the opposite occurred in the corn-fed mice.

Whereas in the experimental animal the beneficial effects of omega-3 fatty acids have been consistently shown, epidemiological data on the effects of omega-3 fatty acids on cancer are scarce. Kaizer et al.[62] compared breast cancer incidence and mortality rates with fish consumption and found an inverse association between percent of calories consumed from fish and the incidence of breast cancer. Dolecek[32] found that the 18:3ω3/18:2ω6 ratio, as well as the ratio of total omega-3/total omega-6 in the MRFIT study, is negatively correlated with the risk for cancer mortality. In this study, there was no correlation between the levels of dietary 18:2ω6 and cancer, which is the case with the animal experimental models. It is quite possible that the intake of 18:2ω6 in this population was above the amount that is considered critical in order to show any effect, whereas the low levels of 18:3ω3 were in the range where changes may effectively influence processes involved in the development of cancer.

In 1980, Willet et al. followed 88,745 women aged 34 to 59 years without history of cancer, inflammatory bowel disease, or familial polyposis who completed a questionnaire regarding their usual diet. By 1986, during 512,785 person-years of follow-up (98% complete), 134 incident cases of colon cancer were documented. After adjustment for total energy intake, animal fat was positively associated with risk of colon cancer, and the relative risk for the highest vs. lowest quintile was 1.84 (95% confidence interval 1.07 to 3.17). No association was found for vegetable fat or cholesterol intake. Use of beef, pork, or lamb as a main dish, processed meats, and liver were each significantly associated with increased risk, while the use of fish and chicken without skin were both related to decreased risk. The ratio of red meat divided by chicken or fish intake was strongly associated with increased incidence of colon cancer, and the relative risk for women in the highest quintile compared with those in the lowest was 2.76 (95% confidence interval 1.58 to 4.82). Fiber intake was significantly associated with a lower risk but cereal fiber was not. These prospective data provide evidence for the hypothesis that animal fat intake increases the risk of colon cancer and support recommendations to substitute fish and chicken, particularly with skin removed, for beef, pork, and lamb.[63]

It is evident that in studies involving fat and cancer, the specific types of fatty acids and the quantity need to be taken into consideration. Thus far, the limited epidemiological data relating omega-3 fatty acid intake to cancer are consistent with data from animal experiments. Long-term clinical trials are needed to define the benefits of omega-3 fatty acids using the different fatty acids, 18:3ω3, EPA, and DHA, both in cancer prevention and in decreasing the rate of cancer metastasis.

INFLAMMATORY AND AUTOIMMUNE DISORDERS

The protective effect of marine lipids on autoimmune renal disease is one of the most dramatic effects of omega-3 fatty acids on any pathology.[64] In 1983 Prickett et al.[64] demonstrated that the fatal spontaneous autoimmune disease in a genetic strain of NZB mice can be

TABLE 7
Effect of Omega-3 Fatty Acids (EPA and DHA) on Rheumatoid Arthritis — Summary of Clinical Trials

Design	Biochemical changes	Clinical outcome	Ref.
Fish oil (20 g/d) given to 12 patients for 6 weeks	As expected, decreased AA:EPA and decreased LTB_4 in neutrophils	Reduction in severity of some, but not all, measurements of disease activity	71
3.2 g/d EPA + DHA given to 23 patients for 12 weeks	As expected, decreased LTB_4 in stimulated granulocytes	Improvement achieved	72
Two different doses of fish oil (27 mg/kg EPA + 18 mg/kg DHA or 54 mg/kg EPA + 36 mg/kg DHA) given to 20 patients for 24 weeks. Olive oil (6.8 g/d) given to 12 patients for 24 weeks	As expected, decrease in interleukin-1 and LTB_4 production in macrophages and neutrophils	Subjective and objective improvement in joint inflammation, particularly in higher dose group	73
Double-blind, placebo-controlled crossover trial; EPA (2.04 g/d) + DHA (1.32 g/d) given to 16 patients for 12 weeks	As expected, increase in membrane EPA and DHA. Decrease in LTB_4; increase in LTB_5	Modest clinical improvement; reduction in morning stiffness and joint swelling	74

Adapted from the British Nutrition Foundation Task Force, *Unsaturated Fatty Acids. Nutritional and Physiological Significance,* Chapman & Hall, London, 1992.

largely prevented by changing the fat in the diet from beef tallow to fish oil. These findings led to studies in humans with arthritis and other diseases of autoimmune nature such as psoriasis, ulcerative colitis, lupus erythematosus, and asthma. Many experimental studies have shown that supplementing the diet with omega-3 fatty acids (3.2 g EPA and 2.2 g DHA) in normal subjects increased the EPA content in neutrophils and monocytes more than sevenfold without changing the quantities of AA and DHA. The anti-inflammatory effects of fish oil are mediated by: (1) inhibiting the 5-lipoxygenase pathway in neutrophils and monocytes which leads to the inhibition of LTB_4 while increasing the production of LTB_5 (Figure 1);[24,65,66] and (2) by decreasing IL-1 production in patients with elevated IL-1 levels.[67-69] In normal volunteers, marine lipids suppress production of PAF from monocytes.[70] The chemotactic response of neutrophils to transmembrane agonists is also suppressed in normals by dietary fish oils, but not in patients with rheumatoid arthritis.

ARTHRITIS

In patients with rheumatoid arthritis, omega-3 fatty acids decrease LTB_4 levels and IL-1 while increasing IL-2.[68] Kremer et al., in a prospective randomized double-blind placebo-controlled study, investigated three groups of patients for 24 weeks with two different doses of fish oil and one dose of olive oil. Fish oils led to a decrease in LTB_4 and IL-1 and an increase in LTB_5 and IL-2. LTB_4 exerts a positive modulating effect on the genetic control of IL-1 probably at the translational level within the cytoplasm. The decrease in LTB_4 was accompanied by a decrease in pain and swelling of the joints, along with a decrease in the number of joints involved.[71-74] Table 7 includes a summary of clinical trials in patients with rheumatoid arthritis treated with fish oil (EPA and DHA).[75]

PSORIASIS

Eskimos exhibit a low incidence of psoriasis.[76] The absence of psoriasis in Eskimos and their high omega-3 fatty acid intake prompted the use of omega-3 fatty acids in patients with psoriasis.[77-81] In patients with psoriatic lesions, LTB_4 and 12-hydroxyeicosatetranoic acid (12-

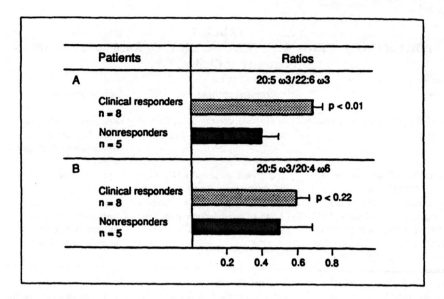

FIGURE 2. Polyunsaturated ratios in psoriatic lesions in clinical responders and nonresponders. (A) EPA (20:3ω5)/ DHA (22:6ω3) ratio; (B) EPA (20:5ω3)/AA (20:4ω6) ratio.[81]

HETE) are markedly elevated in the lesion.[77] Since omega-3 fatty acids have been shown to decrease LTB_4, it was therefore obvious to test the effect of MaxEPA to the standard treatment in patients with psoriasis. As expected, the addition of MaxEPA produced further improvement in skin lesions, it reduced itching and erythema, and decreased both LTB_4 and 12-HETE. However, not all patients responded to the administration of MaxEPA. Ziboh[77] referred to those who benefited from MaxEPA administration as responders and those who did not as nonresponders (Figure 2). About 60% of patients demonstrated mild to moderate improvement of their lesions. The improved clinical response correlated with a high EPA/AA ratio attained in the epidermal tissue. Although EPA and DHA were rapidly incorporated into the serum lipids, neutrophil lipids, and epidermal lipids, it was the amount of EPA and DHA in the epidermal lipids that correlated significantly with the activity of the disease. Further studies showed additional benefits of fish oil. In patients receiving etretinate, fish oil reduced the hyperlipidemia that results from this drug.[82] Fish oil also prolonged the beneficial effects of phototherapy when used in combination with ultraviolet B (UVB).[82]

ULCERATIVE COLITIS

Leukotriene B_4 and prostaglandin E_2, both products of AA metabolism, are increased in patients with ulcerative colitis. In ulcerative colitis LTB_4 is an important mediator of inflammation and has the ability to recruit additional neutrophils from the bloodstream into the mucosa, further exacerbating the disease process with the further increase of LTB_4. Since omega-3 fatty acids decrease the production of LTB_4, Stenson et al.[83] investigated the effects of dietary supplementation with fish oil in ulcerative colitis. Four months of diet supplementation with fish oils resulted in reductions in rectal dialysate LTB_4 levels, improvements in histologic findings, weight gain, and a reduction in the dose of prednisone. None of these changes occurred in the placebo group in which the dose of prednisone had to be raised during the period of the study. Encouraging results have been reported by others.[84,85] Dietary fish oil supplementation might have an adjunctive role in the treatment of patients with ulcerative colitis (Table 8).[86-90]

TABLE 8
Conditions in Which Omega-3 Fatty Acids
Have Been Shown to Have Synergistic Effects
with Drugs

Human studies	Ref.	Animal studies	Ref.
Hypertension	86	Autoimmune disorders	89
Arthritis	68, 87		
Psoriasis	82		
Ulcerative colitis	83		
Restenosis	88		

Adapted from Simopoulos, A.P., Kifer, R.R., and Wykes, A.A., *World Rev. Nutr. Diet*, 66, 51, 1991.

FACTORS TO BE CONSIDERED IN PROTOCOL DEVELOPMENT IN CLINICAL INVESTIGATIONS AND THE BIOMEDICAL TEST MATERIALS PROGRAM

A number of randomized double-blind controlled studies have been carried out in disease states such as following angioplasty, rheumatoid arthritis, psoriasis, atopic dermatitis, Raynaud's disease, ulcerative colitis, and bronchial asthma. Inconsistencies in the results of some of the above studies appear to be due to variations in the design and duration of the studies; selection of proper controls; failure to determine omega-3 and omega-6 fatty acid status along with the status of other dietary factors, e.g., vitamin fatty acid status along with the status of other dietary factors, e.g., vitamin E; and also failure to consider the genetic variations of multifactorial diseases and consequently the need to study subgroups for the various hypercholesterolemias, as well as hypertension, diabetes (whether NIDDM or IDDM), and arthritis; and failure to control for the amount of exercise and for body mass index.

In undertaking studies it is therefore essential to know the initial phospholipid content of cells and to know about related factors that pertain to the disease under study. Also, genetic variations must always be considered in designing experimental protocols. Patients and controls must always be matched for age and sex, since there is evidence that women respond to omega-3 fatty acids differently than men. To illustrate, the same dose of omega-3 fatty acids lowers thromboxane 2 (TXA$_2$) more in women than in men.[91] Dietary control of saturated and polyunsaturated fatty acids during the control and intervention period, as well as differences in lifelong dietary intake of fatty acids must be carefully considered and differences in the ratio of omega-6/omega-3 noted.

In December 1986 the U.S. Department of Commerce developed a special program, the Biomedical Test Materials (BTM) program, which provides standardized test materials of known composition of EPA and DHA nationally and internationally.[92] The objectives of the BTM program are to develop and provide test materials necessary to attain a thorough understanding of the mechanisms and interactions of omega-3 fatty acids, and to stimulate the conduct of well-designed clinical studies in order to assist in the interpretation of the action of omega-3 fatty acids. These test materials are being utilized by U.S. and foreign investigators for research principally on arthritis, cardiovascular diseases, diabetes, blood coagulation, hyperlipidemias, autoimmune disorders, kidney disorders, lipid metabolism, malaria, cancer, and skin disorders. Table 9 shows the types of test materials available.

TABLE 9
Test Materials Currently Available

ω-3 ethyl ester concentrate, prepared from menhaden oil, bulk packed or soft-gel encapsulated (80% ω-3 fatty acids including EPA and DHA)

Ethyl esters of corn, olive, or safflower oil, bulk packed or soft-gel encapsulated

Deodorized menhaden oil, bulk packed or soft-gel encapsulated

Commercial preparations of corn, olive, or safflower oil, soft-gel encapsulated only

EPA ethyl ester (>95% ethyl esters), prepared from menhaden oil, packaged in 1- to 5-g aliquots

DHA ethyl ester (>95% ethyl esters), prepared from menhaden oil, packaged in 1- to 5-g aliquots

Reproduced from Simopoulos, A.P., Kifer, R.R., and Wykes, A.A., *World Rev. Nutr. Diet,* 66, 51, 1991.

CONCLUSION

The establishment of the BTM program has made it possible to carry out intervention trials, clinical trials, and animal experimentation with the standardized test materials.

At present, there are collaborative studies on restenosis, hypertension, arthritis, and other autoimmune disorders that will define the prevention and therapeutic role(s) of omega-3 fatty acids (EPA, DHA, and LNA) in chronic diseases. Today we know more about the effects of omega-3 fatty acids in health and disease and in development than any other family of fatty acids.

It is obvious that the ratio of omega-6/omega-3 fatty acids in cell membrane phospholipids and plasma phospholipids plays a pivotal role in determining membrane fluidity, gene expression, cytokine formation, lipid levels, and immune responses, all of which trigger various aspects and may prevent or contribute to coronary heart disease, hypertension, diabetes, cancer, arthritis, psoriasis, ulcerative colitis, Raynaud's disease, atopy, asthma, lupus erythematosus, multiple sclerosis, and other autoimmune disorders. Omega-3 fatty acids play a role as adjuvants to drug therapy. This property ranges from the control of side effects of drugs to lower doses of drugs.

Finally, research on the nutritive aspects of omega-3 fatty acids has contributed enormously to the scientific basis of nutrition and, since they influence gene expression and the development of molecules involved in intracellular as well as cell-to-cell communication, has helped define and further enhance the molecular basis of nutrition.

REFERENCES

1. **Simopoulos, A.P.,** Historical perspective, conference conclusions and recommendations, and actions by federal agencies, in *Health Effects of Polyunsaturated Fatty Acids in Seafoods,* Simopoulos, A.P., Kifer, R.R., and Martin, R.E., Eds., Academic Press, Orlando, FL, 1986, 3.
2. **Nelson, A.M.,** Diet therapy in coronary disease. Effect on mortality of high-protein, high-seafood, fat-controlled diet, *Geriatrics,* 27, 103, 1972.
3. **Nelson, A.M. and Douglas, C.,** The effect of a fat-controlled diet on patients with coronary disease, *Med. Welt,* 36, 2063, 1965.
4. **Pfeifer, J.J., Janssen, F., and Muesing, R.,** The lipid depressant activities of whole fish and their component oils, *J. Am. Oil Chem. Soc.,* 39, 292, 1962.
5. **Bronte-Stewart, B., Antonis, A., Eales, L., and Brock, J.F.,** Effects of feeding different fats on serum-cholesterol level, *Lancet,* 1, 521, 1956.
6. **Kinsell, L.W. and Sinclair, H.M.,** Fats and disease, *Lancet,* 1, 883, 1957.
7. **Stansby, M.E.,** Nutritional properties of fish oils, *World Rev. Nutr. Diet,* 11, 46, 1969.
8. **Keys, A., Anderson, J.T., and Grande, F.,** "Essential" fatty acids, degree of unsaturation and effect of corn (maize) oil on the serum cholesterol level in man, *Lancet,* 1, 66, 1957.

9. **Ahrens, E.H., Insull, W., Hirsh, J., et al.,** The effects on human serum lipids of a dietary fat, highly unsaturated, but poor in essential fatty acids, *Lancet*, 1, 115, 1959.

10. **Bang, H.O. and Dyerberg, J.,** Plasma lipids and lipoproteins in Greenlandic West-coast Eskimos, *Acta Med. Scand.*, 192, 85, 1972.

11. **Dyerberg, J., Bang, H.O., and Hjorne, N.,** Fatty acid composition of the plasma lipids in Greenland Eskimos, *Am. J. Clin. Nutr.*, 28, 958, 1975.

12. **Bang, H.O., Dyerberg, J., and Hjorne, N.,** The composition of food consumed by Greenland Eskimos, *Acta Med. Scand.*, 200, 69, 1976.

13. **Dyerberg, J., Bang, H.O., and Stoffersen, E.,** Eicosapentaenoic acid and prevention of thrombosis and atherosclerosis, *Lancet*, 2, 117, 1978.

14. **Dyerberg, J. and Bang, H.O.,** Haemostatic function and platelet polyunsaturated fatty acids in Eskimos, *Lancet*, 2, 433, 1979.

15. **Bang, H.O. and Dyerberg, J.,** Lipid metabolism and ischemic heart disease in Greenland Eskimos, in *Advanced Nutrition Research*, Vol. 3, Draper, H.H., Ed., Plenum Press, New York, 1980, 1.

16. **Dyerberg, J.,** The Eskimo experience, *n-3 News*, 1, 1, 1986.

17. **Bierenbaum, M.L., Fleischman, A.I., and Green, D.P.,** The 5-year experience of modified fat diets on younger men with coronary heart disease, *Circulation*, 42, 943, 1970.

18. **Dyerberg, J.,** Linolenate-derived polyunsaturated fatty acids and prevention of atherosclerosis, *Nutr. Rev.*, 44, 125, 1986.

19. **Hirai, A., Hamazaki, T., Terano, T., et al.,** Eicosapentaenoic acid and platelet function in Japanese, *Lancet*, 2, 1132, 1980.

20. **Hirai, A., Terano, T., Saito, H., et al.,** Eicosapentaenoic acid and platelet function in Japanese, in *Nutritional Prevention of Cardiovascular Disease*, Goto, Y. and Homma, Y., Eds., Academic Press, New York, 1984, 231.

21. **Kagawa, Y., Nishizawa, M., Suzuki, M., et al.,** Eicosapolyenoic acids of serum lipids of Japanese Islanders with low incidence of cardiovascular disease, *J. Nutr. Sci. Vitaminol.*, 28, 441, 1982.

22. **Kromhout, D., Bosschieter, E.B., and Coulander, C. deL.,** The inverse relation between fish consumption and 20-year mortality from coronary heart disease, *N. Engl. J. Med.*, 312, 1205, 1985.

23. **Phillipson, B.E., Rothrock, D.W., Connor, W.E., Harris, W.S., and Illingworth, D.R.,** Reduction of plasma lipids, lipoproteins, and apoproteins by dietary fish oils in patients with hypertriglyceridemia, *N. Engl. J. Med.*, 312, 1210, 1985.

24. **Lee, T.H., Hoover, R.L., Williams, J.D., et al.,** Effect of dietary enrichment with eicosapentaenoic and docosahexaenoic acids on in vitro neutrophil and monocyte leukotriene generation and neutrophil function, *N. Engl. J. Med.*, 312, 1217, 1985.

25. **Glomset, J.A.,** Fish, fatty acids, and human health (editorial), *N. Engl. J. Med.*, 312, 1253, 1985.

26. **Shekelle, R.B., Missell, L.V., Oglesby, P., Shryock, A.M., and Stamler, J.,** Letter to the editor, *N. Engl. J. Med.*, 313, 820, 1985.

27. **Curb, J.D. and Reed, D.M.,** Letter to the editor, *N. Engl. J. Med.*, 313, 821, 1985.

28. **Vollset, S.E., Hench, I., and Bjelke, E.,** Fish consumption and mortality from coronary heart disease, *N. Engl. J. Med.*, 313, 820, 1985.

29. **Simonsen, T., Vartun, A., Lyngmo, V., and Nordoy, A.,** Coronary heart disease, serum lipids, platelets and dietary fish in two communities in northern Norway, *Acta Med. Scand.*, 222, 237, 1987.

30. **Hunter, D.J., Kazda, Chockallngam, A., and Fodor, J.G.,** Fish consumption and cardiovascular mortality in Canada: an interregional comparison, *Am. J. Prev. Med.*, 4, 5, 1988.

31. **Burr, M.L., Fehily, A.M., Gilbert, J.F., et al.,** Effect of changes in fat, fish and fibre intakes on death and myocardial reinfarction: diet and reinfarction trial (DART), *Lancet*, 2, 757, 1989.

32. **Dolecek, T.A. and Grandits, G.,** Dietary polyunsaturated fatty acids and mortality in the Multiple Risk Factor Intervention Trial (MRFIT), *World Rev. Nutr. Diet*, 66, 205, 1991.

33. **Simopoulos, A.P.,** Omega-3 fatty acids in health and disease and in growth and development, *Am. J. Clin. Nutr.*, 54, 438, 1991.

34. **Weber, P.C., Fischer, S., von Schacky, C., et al.,** Dietary omega-3 polyunsaturated fatty acids and eicosanoid formation in man, in *Health Effects of Polyunsaturated Fatty Acids in Seafoods*, Simopoulos, A.P., Kifer, R.R., and Martin, R.E., Eds., Academic Press, Orlando, FL, 1986, 49.

35. **Singer, P., Jaeger, W., Wirth, M., et al.,** Lipid and blood pressure-lowering effect of mackerel diet in man, *Atherosclerosis*, 49, 99, 1983.

36. **Lorenz, R., Spengler, U., Fisher, S., Duhm, J., and Weber, P.C.,** Platelet function, thromboxane formation and blood pressure control during supplementation of the Western diet with cod liver oil, *Circulation*, 67, 504, 1983.

37. **Mortensen, J.Z., Schmidt, E.B., Nielsen, A.H., and Dyerberg, J.,** The effect of n-6 and n-3 polyunsaturated fatty acids on hemostasis, blood lipids and blood pressure, *Thromb. Haemost.*, 50, 543, 1983.

38. **Singer, P., Berger, I., Luck, K., Taube, C., Naumann, E., and Godicke, W.,** Long-term effect of mackerel diet on blood pressure, serum lipids and thromboxane formation in patients with mild essential hypertension, *Atherosclerosis,* 62, 259, 1986.

39. **Rogers, S., James, K.S., Butland, B.K., Etherington, M.D., O'Brien, J.B., and Jones, J.G.,** Effects of a fish oil supplement on serum lipids, blood pressure, bleeding time, haemostatic and rheological variables: a double blind randomised controlled trial in healthy volunteers, *Atherosclerosis,* 63, 137, 1987.

40. **Knapp, H.R. and FitzGerald, G.A.,** The antihypertensive effects of fish oil. A controlled study of polyunsaturated fatty acid supplements in essential hypertension, *N. Engl. J. Med.,* 320, 1037, 1989.

41. **Singer, P., Wirth, M., Voigt, S., et al.,** Blood pressure- and lipid-lowering effect of mackerel and herring diets in patients with mild essential hypertension, *Atherosclerosis,* 56, 223, 1985.

42. **Norris, P.G., Jones, C.J.H., and Weston, M.J.,** Effect of dietary supplementation with fish oil on systolic blood pressure in mild essential hypertension, *Br. Med. J.,* 293, 104, 1986.

43. **von Houwelingen, R., Nordoy, A., van der Beek, E., Houtsmuller, U.M.T., de Metz, M., and Hornstra, G.,** Effect of a moderate fish intake on blood pressure, bleeding time, hematology, and clinical chemistry in healthy males, *Am. J. Clin. Nutr.,* 46, 424, 1987.

44. **Demke, D.M., Peters, G.R., Linet, O.I., Metzler, C.M., and Klott, K.A.,** Effects of a fish oil concentrate in patients with hypercholesterolemia, *Atherosclerosis,* 70, 73, 1988.

45. **Bonaa, K.H., Bjerve, K.S., Straume, B., Gram, I.T., and Thelle, D.,** Effect of eicosapentaenoic and docosahexaenoic acids on blood pressure in hypertension. A population-based intervention trial from the Tromso study, *N. Engl. J. Med.,* 322, 795, 1990.

46. **Dusing, R., Struck, A., Scherg, H., Pietsch, R., and Kramer, H.J.,** Dietary fish oil supplements: effects on renal hemodynamics and renal excretory function in healthy volunteers, *Kidney Int.,* 31, 268, 1987.

47. **Fischer, S. and Weber, P.C.,** Prostaglandin I_3 is formed in vivo in man after dietary eicosapentaenoic acid, *Nature,* 307, 165, 1984.

48. **Knapp, H.R., Reilly, I.A.G., Alessandrini, P., and FitzGerald, G.A.,** In vivo indexes of platelet and vascular function during fish-oil administration in patients with atherosclerosis, *N. Engl. J. Med.,* 314, 937, 1986.

49. **McMillan, D.E.,** Antihypertensive effects of fish oil, *N. Engl. J. Med.,* 321, 1610, 1989.

50. **Jensen, T.,** Dietary supplementation with ω3 fatty acids in insulin-dependent diabetes mellitus, *World Rev. Nutr. Diet,* 66, 417, 1991.

51. **Jensen, T., Stender, S., Goldstein, K., Holmer, G., and Deckett, T.,** Partial normalization by dietary cod-liver oil of increased microvascular albumin leakage in patients with insulin-dependent diabetes and albuminuria, *N. Engl. J. Med.,* 321, 1572, 1989.

52. **Vessby, B.,** Effects of ω3 fatty acids on glucose and lipid metabolism in non-insulin-diabetes mellitus, *World Rev. Nutr. Diet.,* 66, 407, 1991.

53. **Glauber, H., Wallace, P., Griver, K., and Brechtel, G.,** Adverse metabolic effect of omega-3 fatty acids in non-insulin-dependent diabetes mellitus, *Ann. Intern. Med.,* 108, 663, 1988.

54. **Schechtman, G., Kaul, S., and Kissebah, A.H.,** Effect of fish oil concentrate on lipoprotein composition in NIDDM, *Diabetes,* 37, 1567, 1988.

55. **Wahlqvist, M.L., Lo, C.S., and Myers, K.A.,** Fish intake and arterial wall characteristics in healthy people and diabetic patients, *Lancet,* 1, 944, 1989.

56. **Tilvis, R.S., Rasi, V., Viinikka, L., Ylikorkala, O., and Miettinen, T.A.,** Effects of purified fish oil on platelet lipids and function in diabetic women, *Clin. Chim. Acta,* 1654, 315, 1987.

57. **Miller, M.E., Anagnostou, A.A., Ley, B., Marshall, P., and Steiner, M.,** Effects of fish oil concentration on hemorheological and hemostatic aspects of diabetes mellitus: a preliminary study, *Thromb. Res.,* 47, 201, 1987.

58. **Haines, A.P., Sanders, T.A.B., Imeson, J.D., et al.,** Effects of a fish oil supplement on platelet function, haemostatic variables and albuminuria in insulin-dependent diabetics, *Thromb. Res.,* 43, 643, 1986.

59. **Cave, W.T.,** ω3 fatty acid diet effects on tumorigenesis in experimental animals, *World Rev. Nutr. Diet,* 66, 462, 1991.

60. **Fernandes, G. and Venkatraman, J.T.,** Modulation of breast cancer growth in nude mice by ω3 lipids, *World Rev. Nutr. Diet,* 66, 488, 1991.

61. **Karmali, R.A.,** Dietary ω-3 and ω-6 fatty acids in cancer, *Dietary ω3 and ω6 Fatty Acids: Biological Effects and Nutritional Essentiality,* Vol. 171, Galli, C. and Simopoulos, A.P., Eds., Plenum Press, New York, 1989, 351.

62. **Kaizer, L., Boyd, N.F., Kriukov, V., and Tritcher, D.,** Fish consumption and breast cancer risk: an ecological study, *Nutr. Cancer,* 12, 61, 1989.

63. **Willet, W.C., Stampfer, M.J., Colditz, G.A., Rosner, B.A., and Speizer, F.E.,** A prospective study of diet and colon cancer in women, *Am. J. Epidemiol.,* 130, 820, 1989.

64. **Prickett, J.D., Robinson, D.R., and Steinberg, A.D.,** Effects of dietary enrichment with eicosapentaenoic acid upon autoimmune nephritis in female NZB×NZW/F$_1$ mice, *Arthritis Rheum.,* 26, 133, 1983.

65. **Lewis, R.A., Lee, T.H., and Austen, K.F.,** Effects of omega-3 fatty acids on the generation of products of the 5-lipoxygenase pathway, in *Health Effects of Polyunsaturated Fatty Acids in Seafoods,* Simopoulos, A.P., Kifer, R.R., and Martin, R.E., Eds., Academic Press, Orlando, FL, 1986, 227.

66. **Kremer, J.M., Jubiz, W., and Michalek, A.,** Fish-oil fatty acid supplementation in active rheumatoid arthritis, *Ann. Intern. Med.,* 106, 497, 1987.

67. **Endres, S., Ghorbani, R., Kelley, V.E., et al.,** The effect of dietary supplementation with n-3 polyunsaturated fatty acids on the synthesis of interleukin-1 and tumor necrosis factor by mononuclear cells, *N. Engl. J. Med.,* 320, 265, 1989.

68. **Kremer, J.M., Lawrence, D.A., and Jubiz, W.,** Different doses of fish-oil fatty acid ingestion in active rheumatoid arthritis: a prospective study of clinical and immunological parameters, in *Dietary ω3 and ω6 Fatty Acids: Biological Effects and Nutritional Essentiality,* Vol. 171, Galli, C. and Simopoulos, A.P., Eds., Plenum Press, New York, 1989, 343.

69. **Robinson, D.R. and Kremer, J.M.,** Summary of panel G: rheumatoid arthritis and inflammatory mediators, *World Rev. Nutr. Diet,* 66, 44, 1991.

70. **Sperling, R.I., Robin, J.L., Kylander, K.A., Lee, T.H., Lewis, R.A., and Austin, K.F.,** The effects of N-3 polyunsaturated fatty acids on the generation of platelet-activating factor-acether by human monocytes, *J. Immunol.,* 139, 4186, 1987.

71. **Sperling, R.I., Weinblatt, M.E., Robin, J.L., et al.,** Effects of dietary supplementation with marine fish oil on leukocyte lipid mediator generation and function, *Arthritis Rheum.,* 30, 987, 1987.

72. **Cleland, L.G., French, J.K., Betts, W.H., Murphy, G.A., and Elliot, M.J.,** Clinical and biochemical effects of dietary fish oil supplements in rheumatoid arthritis, *J. Rheumatol.,* 15, 1471, 1988.

73. **Kremer, J.M., Lawrence, D.A., Jubiz, W., et al.,** Dietary fish oil and olive oil supplementation in patients with rheumatoid arthritis; clinical and immunological effects, *Arthritis Rheum.,* 33, 810, 1990.

74. **van der Tempel, H., Tulleken, J.E., Limbuerg, P.C., Muskiet, F.A.J., and van Rijswijk, M.H.,** Effects of fish oil supplementation in rheumatoid arthritis, *Ann. Rheum. Dis.,* 49, 76, 1990.

75. British Nutrition Foundation Task Force, *Unsaturated Fatty Acids. Nutritional and Physiological Significance,* Chapman & Hall, London, 1992.

76. **Kromann, N. and Green, A.,** Epidemiological studies in the Uppernavik District, Greenland, *Acta Med. Scand.,* 200, 401, 1980.

77. **Ziboh, V.A., Miller, C., Kragballe, K., et al.,** Effects of dietary supplementation of fish oil on neutrophil and epidermal fatty acids: modulation of the clinical course of psoriatic subjects, *Arch. Dermatol.,* 122, 1277, 1986.

78. **Bittinger, S.B., Cartwright, I., Tucker, W.F.G., et al.,** A double-blind, randomized, placebo-controlled trial of fish oil in psoriasis, *Lancet,* 1, 378, 1988.

79. **Maurice, P.D.L., Allen, B.R., Barkeley, A.S.J., et al.,** The effects of dietary supplementation with fish oil in patients with psoriasis, *Br. J. Dermatol.,* 117, 599, 1987.

80. **Bjornboe, A., Smith, A.K., Bjornboe, G.-E.A.A., et al.,** Effect of dietary supplementation with n-3 fatty acids on clinical manifestations of psoriasis, *Br. J. Dermatol.,* 118, 77, 1988.

81. **Ziboh, V.A.,** ω3 polyunsaturated fatty acid constituents of fish oil and the management of skin inflammatory and scaly disorders, *World Rev. Nutr. Diet,* 66, 425, 1991.

82. **Allen, B.R.,** Fish oil in combination with other therapies in the treatment of psoriasis, *World Rev. Nutr. Diet,* 66, 436, 1991.

83. **Stenson, W.F., Cort, D., Rodgers, J., et al.,** Dietary supplementation with fish oil in ulcerative colitis, *Ann. Intern. Med.,* 116, 609, 1992.

84. **McCall, T.B., O'Leary, D., Bloomsield, J., and O'Morain, C.A.,** Therapeutic potential of fish oil in the treatment of ulcerative colitis, *Aliment Pharmacol. Ther.,* 3, 415, 1989.

85. **Salomon, P., Kornbluth, A.A., and Janowitz, H.D.,** Treatment of ulcerative colitis with fish oil in 3 omega fatty acids: an open trial, *J. Clin. Gastroenterol.,* 12, 157, 1990.

86. **Singer, P. and Hueve, J.,** Blood pressure-lowering effect of fish oil, propranolol and the combination of both in mildly hypertensive patients (abstract), *World Rev. Nutr. Diet,* 66, 522, 1991.

87. **Kremer, J.M. and Robinson, D.R.,** Studies of dietary supplementation with ω3 fatty acids in patients with rheumatoid arthritis, *World Rev. Nutr. Diet,* 66, 367, 1991.

88. **Dehmer, G.J., Pompa, J.J., Van den Berg, E.K., et al.,** Reduction in the rate of early restenosis after coronary angioplasty by a diet supplemented with n-3 fatty acids, *N. Engl. J. Med.,* 319, 733, 1988.

89. **Fernandes, G., Venkatraman, J.T., Fernandes, A., et al.,** Effect of ω3 fatty acid diet therapy on autoimmune disease with or without calorie restriction (abstract), *World Rev. Nutr. Diet,* 66, 568, 1991.

90. **Simopoulos, A.P., Kifer, R.R., and Wykes, A.A.,** ω3 fatty acids: research advances and support in the field since June 1985 (worldwide), *World Rev. Nutr. Diet,* 66, 51, 1991.

91. **Hansen, J.-B., Olsen, J.O., Wilsgard, L., et al.,** Effects of dietary supplementation with cod liver oil on monocyte thromboplastin synthesis, coagulation and fibrinolysis, *J. Intern. Med.,* 225 (Suppl.), 133, 1989.

92. NIH, Availability of Fish Oil Test Materials. NIH Guide for Grants and Contracts, Vol. 18, No. 24, U.S. Department of Health and Human Services, Washington, D.C., 1989.

Chapter 2.4

TRANS FATTY ACIDS

Artemis P. Simopoulos

INTRODUCTION

Trans fatty acids are formed during hydrogenation, a process that solidifies liquid vegetable oils by adding hydrogen atoms to the double bonds of the unsaturated fatty acids in the trans instead of the cis position on the fatty acid molecule. The degree of hydrogenation dictates how much trans fatty acid is in the product. The concentration of trans fatty acids varies with the extent and type of processing of oil. Salad oils contain 8 to 17% trans, shortenings 14 to 60%, and margarines 16 to 70%.[1] In food composition tables, the trans fatty acids are listed as monounsaturated fatty acids and this has led to confusion about the amount of saturated fat in the U.S. diet. In the U.S. trans fatty acids contribute 3 to 7% of the fat consumed.[2] Trans fatty acids occur rarely or in small amounts in nature.

In the U.S., the principal vegetable oil used for hydrogenation is linoleic acid (LA, 18:2ω6) and the main fatty acids produced from hydrogenation are oleic, elaidic, and stearic acids. Oleic acid (18:1) has one double bond of the cis configuration, in which the two hydrogen atoms attached to the double bond lie on the same side, which makes the molecule flexible and liquid.[3] Elaidic acid is trans 18:1; its hydrogen atoms lie on opposite sides of the double bonds. It is a rigid molecule whose structure is similar to saturated fats. Its presence solidifies fat. Other trans monounsaturated fatty acids are formed during the hydrogenation process by moving the double bond up and down the molecule, and they most likely behave biologically very much like elaidic acid. The third fatty acid resulting from the hydrogenation process is stearic acid (18:0), which is a saturated fatty acid and has no double bonds (Figure 1).[3]

The invention of the continuous screw press, named Expeller® by V.D. Anderson, and the steam-vacuum deodorization process by D. Wesson made possible the industrial production of cottonseed oil and other vegetable oils for cooking.[4] Solvent extraction of oilseeds came into increased use after World War I and the large-scale production of vegetable oils became more efficient and economic. Subsequently, hydrogenation was applied to oils to solidify them and improve stability and texture. Soybean oil constitutes about 65% of the fat used in the manufacture of shortenings in the U.S.[5] The partial selective hydrogenation of soybean oil reduces the linolenic acid (LNA) content of the oil while leaving a high concentration of linoleic acid (LA). LNA content was reduced because LNA in soybean oil caused many organoleptic problems.

The hydrogenation process raises the melting point of oils and improves their use in frying of foods and other aspects of food processing, such as in the production of margarines, salad oils, and shortening for baking. Currently, soybean oil accounts for 84% of the fat in margarines.[6] Shortenings are semisolid fats that confer a tender quality to baked goods to make flaky pastries and delicious donuts. Shortening also modifies gluten and adds richness to the product. The so-called "health" margarines include 15 to 20% of a hydrogenated oil.[7] There is evidence that the consumption of trans fatty acids is increasing (Table 1).[2,8] Enig[8] has determined that the amount of trans fatty acids in baked goods and fast foods was higher in 1988 than in 1978. Specifically, the amount of trans fatty acids in cheese corn chips was 33.4% in 1978 and had risen to 53.9% in 1988. This is most likely the result of substituting hydrogenated vegetable oils for tropical oils in some products.

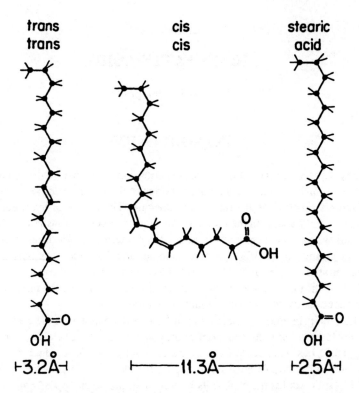

FIGURE 1. The cross-sectional area of two steroisomeric linoleic acids. The fully saturated analog stearic acid is included for comparison. The lateral cross-sectional area of the normally occurring cis-cis linoleic acid is much greater than that of the trans-trans linoleic acid produced by *in vitro* hydrogenation. The latter is not only structurally similar to stearic acid, but behaves metabolically in an anomalous way.[3]

EFFECTS OF TRANS FATTY ACIDS ON ESSENTIAL FATTY ACID METABOLISM AND GROWTH

The physical properties of trans fatty acids are like those of saturated fatty acids (SFA). Trans fatty acids have different effects on different enzyme systems in lipid metabolism of rats.[9] Trans isomers of C18:2 decrease prostaglandin (PG) synthesis[10] and increase the requirement for linoleic acid for PG functions,[11] as well as changing the fatty acid composition of heart cardiolipins[12] in rats.

Trans fatty acids (cis,trans and trans,trans linoleates 18:2ω6) are devoid of essential fatty acid (EFA) activity, and as the sole dietary source of fat they retard growth to a greater degree than an EFA-deficient diet in rats[11] (Figure 1).[3] Mahfouz et al.[13] observed inhibition of delta-9 and delta-6 desaturases by various isomers of trans 18:1ω9. Trans,trans 18:2ω6 decreased the conversion of linoleic to gamma-linolenic in rat liver microsomal preparation.[14] Trans,cis 18.2ω6 also inhibited the desaturation of 18:2ω6. Various studies involving rodents both *in vivo*[15-19] and *in vitro*[13,14,20-24] and human fibroblasts *in vitro*[25] have adequately demonstrated that trans fatty acids impair the microsomal desaturation and chain elongation of both LA and LNA to their long chain metabolites in the 20 and 22 carbon atoms, specifically arachidonic acid and docosahexanoic acid, which are of great importance during perinatal development as essential components of cell membranes and as precursors of prostaglandins and other eicosanoids. Studies in which trans isomers of linoleic or oleic acids[26-28] and of partially hydrogenated soybean oil were fed to newborn rats and mice also impaired postnatal weight gain.

TABLE 1
Partial List of Foods Containing Trans Fatty Acids in the American Diet[2,8]

Margarines
Shortening
Cooking oils
Salad dressings
Mayonnaise
Foods prepared with fats and oils high in trans fatty
  acids generally consumed by Americans
    Bread/rolls
    Cakes
    Cheese corn snacks/corn snacks
    Cookies
    Crackers
    Doughnuts
    French fried potatoes
    Fried chicken
    Fried fish
    Potato chips
    Snacks (miscellaneous)

Since the late 1960s and early 1970s concern was expressed about the metabolic effects of trans fatty acids, because animal experiments referred to earlier indicated that trans fatty acids interfere with the desaturation and elongation of linoleic ($18:2\omega6$) and linolenic acid ($18:3\omega3$) and in prostaglandin formation and metabolism. Despite these findings in animals, two reports, one prepared at the request of the Food and Drug Administration[29] and the other by the British Nutrition Foundation,[30] concluded that there is no evidence of a harmful effect in humans at the dietary level estimated by them. Furthermore, the Food and Drug Administration concluded[29] that the placenta would act as a barrier and any effects of maternal trans fatty acid consumption on the fetus were excluded. Koletzko, in a series of papers,[31-33] showed that trans fatty acids constitute the same percentage of fatty acids in maternal and cord blood, indicating that transfer does occur across the placenta. Trans fatty acids are also secreted in human milk.[31,34]

Recent studies by Holman et al. with rodents indicate that partially hydrogenated soybean oil leads to the formation of unusual isomers in rodents.[35] Holman et al. suggested that these unusual isomers may compete in the metabolism of normal PUFA and that these unusual isomers are substrates for oxidative formation of autocoids of unknown structure and function. Holman et al. concluded:

It is now clear that uncommon isomers of PUFA occur in the lipids of animals fed partially hydrogenated fat and that they inhibit the metabolism of PUFA at many steps in the normal metabolic cascade. It would, therefore, seem wise to avoid foods that contain unusual or unnatural isomeric PUFA or their isomeric monoenoic FA precursors. The latter, both cis and trans positional isomers of 18:1, occur abundantly in partially hydrogenated vegetable oils now commonly consumed by Western populations. The large-scale hydrogenation of vegetable oils reduces $\omega3$ and $\omega6$ EFA and replaces them by saturated and isomeric 18:1 acids that interfere with the $\omega3$ and $\omega6$ metabolism, inducing significant partial deficiencies of EFA. It would seem wise to preserve the essential nutrients and to avoid producing inhibitors of their metabolism by hydrogenation. Evidence is growing for the essentiality of $\omega3$ PUFA and the occurrence of deficiencies of $\omega3$ acids in humans under stress conditions. It would, therefore, be wise economy to use oils containing linolenic acid directly as foods and to avoid their hydrogenation.

A recent study by Koletzko involving premature infants of 33.6 ± 1.4 weeks gestational age showed that the amount of trans octadecenoic acid and total trans fatty acid in plasma lipids correlated inversely to the omega-3 and omega-6 long chain PUFA (20 and 22 carbon atoms) and to the product substrate ratio of long-chain PUFA biosynthesis.[36] Trans fatty acids were also inversely correlated to the birth weight but not to gestational age. The authors concluded that "these data indicate a potential impairment of essential fatty acid metabolism and early growth by trans isomers in man, and question the safety of high dietary trans isomer intakes during pregnancy and the perinatal period."[36] Whereas in studies with rodents the induction of growth failure was not seen when very high amounts of EFA were supplied,[37] premature infants are especially at risk since they have very limited reserves of EFA and may not be able to overcome the effects of trans isomers on EFA metabolism. The amount of trans fatty acids ingested by the mother during pregnancy influences the intrauterine trans fatty acid exposure of the fetus. Similarly, maternal diet influences the amount of trans fatty acids in human milk.[31,34] Some infant formulas contain trans fatty acids related to the use of butter fat and or oleo oils included in the formula.[38] Holman, in his study with pregnant and lactating women, recommended supplementation with EFA during pregnancy and lactation.[39] Thus there is serious concern about the deleterious effects of trans fatty acids on EFA metabolism and growth during intrauterine life. Certainly this is an area that needs to be carefully investigated.

EFFECTS OF TRANS FATTY ACIDS ON LIPID LEVELS, PLATELET AGGREGATION, AND OTHER ASPECTS OF ATHEROSCLEROSIS

As early as 1969, Spritz and Mishkel suggested that many of the effects of unsaturated or polyunsaturated fatty acids were due to their cross-sectional area[40] (Figure 1). The normal double bond bends the fatty acid at a 120° angle and increases the area occupied by that chain. The presence of more double bonds increases the cross-sectional area even more. These authors further suggested that the cholesterol-lowering effects of PUFA were related to their "bulk" and that the ratio of the cross-sectional area of PUFA to that of the saturated dietary fatty acids fed controls should be directly related to the percent lowering of serum cholesterol. Spritz and Mishkel carried out feeding experiments and, as expected, both LDL and HDL cholesterol were decreased following PUFA feedings. They further suggested and showed that the trans fatty acids should have the same effect as saturated fats in raising serum cholesterol concentrations. It was the first demonstration that changes in the structure of PUFA changed their biologic effects.

Studies by Vergroesen also suggested that trans fatty acids, compared to their cis isomer oleic acid, raised serum total cholesterol levels,[41,42] but other studies could not confirm it.[43] There are now many studies that have investigated the effects of trans fatty acids on serum lipid levels. In the aggregate, these studies show that trans fatty acids raise triglycerides in comparison to butterfat[44] and raise LDL cholesterol in comparison to oleic acid, while they also lower HDL, whereas oleic acid does not lower HDL.[45] Nestel et al.[46] showed also that a diet consisting of elaidic at 7% energy intake (about two times as high as the Australian diet, but within the range of the U.S. diet) increased LDL cholesterol and significantly elevated Lp(a) compared to the other three diets: (1) enriched with butter fat (lauric, myristic, palmitic), (2) oleic acid-rich, and (3) palmitic-rich. Trans fatty acids lower HDL and increase triglycerides, Lp(a), and platelet aggregation[47] (Table 2). Furthermore, these effects occurred even when trans fatty acids comprised 7% of energy, which is within the range of trans fatty acid intake in the American diet.[2]

TABLE 2
Effect of Different Fatty Acids on Lipid Levels, and Platelet Aggregation

Fatty acid[a]		LDL	HDL	LDL triglycerides	Lp(a)	Platelet aggregation
Lauric acid	12:0	↑		↑		
Myristic acid	14:0	↑		↑		
Palmitic acid[b]	16:0	↑	↑	↑	↓	
Stearic acid	18:0	↑	↓			↑
Oleic acid (olive oil)	18:1	↓	↑	↓	—	
Linoleic acid	18:2 (ω6)	↓	↓	—	—	?
Linolenic acid	18:3 (ω3)	↓	—	↓		↓
Arachidonic acid	20:4 (ω6)					↑
Eicosapentanoic acid	20:5 (ω3)	↓ or ↑[c]	—	↓	↓	↓
Docosahexanoic acid	22:6 (ω3)	↓ or ↑[c]		↓	↓	↓
Trans fatty acids (technologically developed)		↑	↓	↑	↑	↑

Note: ↑ = increase; ↓ = decrease; ? = variable or questionable; — = no effect or neutral.

[a] Medium chain fatty acids (8 to 10 carbon atoms) probably have no effect on serum cholesterol.
[b] May be neutral in normocholesterolemic individuals. (From Hayes, K.C., Chewing the fat, *Fat Nutr.*, 1, 1, 1992.)
[c] Decreases LDL at high doses.

Zock and Katan carried out a study, the objective of which was to compare the effects of linoleic acid (cis, cis-cis18.2ω6) and its hydrogenation products elaidic (trans-C18:1ω9) and stearic acid (C18:0) on serum lipoprotein levels in humans.[48] They found that 7.7% of energy (mean, 24 g/d) of trans fatty acids in the diet significantly lowered HDL and raised LDL cholesterol relative to linoleic acid. They concluded that the results of this study and previous findings suggest a linear dose-response relationship.[45] Replacement of linoleic acid by stearic acid, another product of the hydrogenation process, also caused somewhat lower HDL cholesterol and higher LDL cholesterol. Thus the hydrogenation of linoleic to either stearic or trans fatty acids produces fatty acids that may lower HDL and raise LDL, and the rise in total cholesterol by stearic acid was higher in men than in women. The lipoprotein levels of men and women respond similarly to dietary trans fatty acids; stearate, however, increased total and LDL cholesterol and triglycerides in men, but not in women, but HDL cholesterol decreased to a similar extent in men and women.

The 1990 study by Mensink and Katan[45] and the 1992 study by Zock and Katan[48] clearly show that a linear dose-response relationship exists between trans monounsaturated fatty acid intake and LDL and HDL cholesterol concentrations. Although more studies are needed, the results of the above two studies indicate that at least in their laboratories the effects of trans fatty acids on LDL and HDL cholesterol are proportional to the amounts of trans fatty acids consumed. In both studies the trans fatty acids supplied to the subjects consisted largely of C18:1ω9 (elaidic acid) and therefore their results are applicable to most monounsaturated trans C18:1ω9 fatty acid isomers as present in commercial foods. The 1992 study showed that trans fatty acid intake of 7.7% of energy (24 g/d) in the diet raised LDL cholesterol and lowered HDL cholesterol relative to linoleic. Similar results were obtained by Nestel et al.[46] Thus it is evident that 7.7% of energy as trans, the upper range of trans fatty acid intake in the U.S. diet, increases the risk of atherosclerosis just as saturated fats do.

The biological mechanism by which trans fatty acids raise serum LDL cholesterol and lower HDL cholesterol are not known. Studies in rats[49,50] indicate a reduction in lecithin/ cholesterol transferase (LCAT) with increased ingestion of trans fatty acids. The effects of LCAT deficiency lead to decreased ability of HDL cholesterol to absorb cholesterol, resulting in reduced cholesterol transport to the liver,[51] leading to decreased HDL cholesterol in serum. Another factor could be the ability of trans fatty acids to inhibit desaturation and elongation of PUFAs.[52]

Troisi et al. assessed the relationship of trans fatty acid intake to serum lipid concentration in a large, cross-sectional population-based cohort of U.S. men.[53] Trans fatty acid intake was directly related to total serum and LDL cholesterol and conversely related to HDL cholesterol. Trans fatty acid intake was positively associated with the ratios of total to HDL cholesterol and LDL to HDL cholesterol. The estimated ratios of total cholesterol to HDL cholesterol were 4.4 for persons at the 10th percentile of trans fatty acid intake of 2.1 g/d and 4.9 for persons on the 90th percentile of intake of 4.9 g/d, corresponding to a 27% increase in risk of myocardial infarction based on estimates from the Stampfer study.[54] The ratio of total serum cholesterol to HDL cholesterol is strongly associated with coronary heart disease.[55,56] The study by Troisi et al.[52] and the study by Mensink and Katan[45] suggest a strong influence of trans fatty acid intake on the ratio of total cholesterol, or LDL cholesterol to HDL cholesterol.

The results of these studies suggest that margarines should not be promoted for the prevention of atherosclerosis and coronary heart disease, and trans fatty acids should not be recommended as substitutes for saturated fats (Table 2).

CONCLUSIONS

Hydrogenation of fats has developed as a result of the need to convert liquid oils to the semi-solid form for greater utility in certain food uses and to increase the thermal stability of the fat or oil. Trans fatty acids are found in small amounts in nature. In the U.S., trans fatty acids account for at least 7 to 9% of energy intake. They are found in margarines, shortening, spread oils, and in bakery products and other processed foods made from them.

In animal experiments and clinical investigations, trans fatty acids interfere with the metabolism of essential fatty acids. Trans fatty acids cross the placenta and are secreted in human milk. Premature newborns exposed to trans fatty acids in their intrauterine environment due to high trans fatty acid intake of their mothers exhibit growth retardation.

Trans fatty acids have the physical properties of saturated fats. The amount of trans fatty acids in the diet directly relates to the amount of trans fatty acids in plasma phospholipids. Functionally, at 7% of energy intake, trans fatty acids increase LDL, decrease HDL, and increase triglycerides and Lp(a). Lp(a) is an atherogenic and thrombogenic lipoprotein that contributes to atherosclerosis. Trans fatty acids also increase platelet aggregation contributing to thrombosis.

RECOMMENDATIONS

- Because information about intake of trans fatty acids in the U.S. is controversial, there is a need for accurate determination of the trans fatty acid content of the food supply.
- In determining the fatty acid composition of the food supply, trans fatty acids should be labeled as a separate category of fats and proposed food labeling regulations should identify trans fatty acids as a separate category of fats.
- The trans fatty acid intake of various segments of the population needs to be assessed, and groups with high intake need to be identified.

- Every effort should be made to decrease the intake of trans fatty acids in the diet by avoiding margarines and shortening and substituting olive oil and other monounsaturated oils that have not been hydrogenated.
- Industry should produce soft margarines free of trans fatty acids.
- Particular attention should be given to lower the trans fatty acid intake in the diet of pregnant and lactating women and in infant formula.
- Individuals with a family history of coronary heart disease or those who have coronary heart disease should decrease their intake of trans fatty acids and saturated fats.
- Advances in technology should not be the driving force in food development without knowledge of their effects on growth and development and in health and disease; that is, the introduction of new techniques and food products should be preceded by a health impact statement.

REFERENCES

1. **Kinsella, J.E., Bruckner, G., Mai, J., and Shimp, J.,** Metabolism of trans fatty acids with emphasis on the effects of trans,trans-octadecadienoate on lipid composition, essential fatty acid, and prostaglandins: an overview, *Am. J. Clin. Nutr.,* 34, 2307, 1981.
2. **Dupont, J., White, P.J., and Feldman, E.B.,** Saturated and hydrogenated fats in food in relation to health, *J. Am. Coll. Nutr.,* 10, 577, 1991.
3. **Lees, R.S., Lees, R.S., and Karel, M., Eds.,** Impact of dietary fat on human health, in *Omega-3 Fatty Acids in Health and Disease,* Marcel Dekker, New York, 1990, 1.
4. **Kirshenbauer, H.G.,** *Fats and Oils,* 2nd ed., Reinhold Publishing, New York, 1960.
5. **USDA,** Shortening: fats and oils used in manufacture, U.S. 1973-1987, Table 188, in Agricultural Statistics, U.S. Government Printing Office, Washington, D.C., 1988.
6. **USDA,** Margarine: selected reported fats and oils used in manufacture, U.S., 1973-1987, Table 186, in Agricultural Statistics, U.S. Government Printing Office, Washington, D.C., 1988.
7. **Lefebvre, J.,** Finished product formulation, *J. Am. Oil Chem. Soc.,* 60, 295, 1983.
8. **Enig, M.G., Atal, S., Keeney, M., and Sampugna, J.,** Isomeric trans fatty acids in the U.S. diet, *J. Am. Coll. Nutr.,* 9, 471, 1990.
9. **Emken, E.F.,** Nutrition and biochemistry of trans and positional fatty acid isomers in hydrogenated oils, *Annu. Rev. Nutr.,* 4, 339, 1984.
10. **Hwang, D.H. and Kinsella, J.E.,** The effects of trans linoleic acid on the concentration of serum prostaglandins $F_{2\alpha}$ and platelet aggregation, *Prostaglandins Med.,* 1, 121, 1978.
11. **Privett, O.S., Phillips, F., Shimasaki, H., Nozawa, T., and Nickell, E.C.,** Studies of effects of trans fatty acids in the diet on lipid metabolism in essential fatty acid deficient rats, *Am. J. Clin. Nutr.,* 30, 1009, 1977.
12. **Hoy, C.-E. and Holmer, G.,** Influence of dietary linoleic acid and trans fatty acids on the fatty acid profile of cardiolipins in rats, *Lipids,* 25, 455, 1990.
13. **Mahfouz, M.M., Johnson, S., and Holman, R.T.,** The effect of isomeric trans 18:1 acids on the desaturation of palmitic, linoleic and eicosa-8,11,14-trienoic acids by rat liver microsomes, *Lipids,* 15, 100, 1980.
14. **Brenner, R.R. and Peluffo, R.O.,** Regulation of unsaturated fatty acid biosynthesis: effect of unsaturated fatty acids of 18 carbons on the microsomal desaturation of linoleic acid into gamma-linolenic acid, *Biochim. Biophys. Acta,* 176, 471, 1969.
15. **Anderson, R.L., Fullmer, C.S., and Hollenbach, E.J.,** Effects of the trans isomers of linoleic acid on the metabolism of linoleic acid in rats, *J. Nutr.,* 105, 393, 1975.
16. **Kinsella, J.E., Hwang, D.H., Yu, P., Mai, J., and Shimp, J.,** Prostaglandins and their precursors in tissues from rats fed on trans,trans-linoleate, *Biochem. J.,* 184, 701, 1979.
17. **Hwang, D.H., Chanmugam, P., and Anding, R.,** Effects of dietary 9-trans, 12-trans linoleate on arachidonic acid metabolism in rat platelets, *Lipids,* 17, 307, 1982.
18. **Lawson, L.D., Hill, E.G., and Holman, R.T.,** Suppression of arachidonic acid in lipids of rat tissues by dietary mixed isomeric cis and trans octadecenoates, *J. Nutr.,* 113, 1827, 1983.
19. **Bruckner, G., Goswami, S., and Kinsella, J.E.,** Dietary trilinolelaidate: effects on organ fatty acid composition, prostanoid biosynthesis and platelet function in rats, *J. Nutr.,* 114, 58, 1984.
20. **Lawson, L.D. and Kummerow, F.A.,** Beta-oxidation of the coenzyme A esters of elaidic, oleic, and stearic acids and their full-cycle intermediates by rat heart mitochondria, *Biochim. Biophys. Acta,* 573, 245, 1979.

21. **Mahfouz, M.M., Smith, T.L., and Kummerow, F.A.,** Effect of dietary fats on desaturase activities and the biosynthesis of fatty acids in rat-liver microsomes, *Lipids,* 19, 214, 1984.
22. **de Schrijver, R. and Privett, O.S.,** Interrelationship between dietary trans fatty acids and the 6- and 9 desaturases in the rat, *Lipids,* 17, 27, 1982.
23. **Shimp, J.L., Bruckner, G., and Kinsella, J.E.,** The effects of dietary trilinolelaidin on fatty acid and acyl desaturases in rat liver, *J. Nutr.,* 112, 722, 1982.
24. **Chern, J. and Kinsella, J.E.,** The effects of unsaturated fatty acids on the synthesis of arachidonic acid in rat kidney cells, *Biochim. Biophys. Acta,* 750, 465, 1983.
25. **Rosenthal, M.D. and Doloresco, M.A.,** The effects of trans fatty acids on fatty acyl delta 5 desaturation by human skin fibroblasts, *Lipids,* 19, 869, 1984.
26. **Aaes-Jorgensen, E.,** Essential fatty acid deficiency. III. Effects of conjugated isomers of dienoic and trienoic fatty acids in rats, *J. Nutr.,* 66, 465, 1958.
27. **Hwang, D.H. and Kinsella, J.E.,** The effect of trans,trans methyl linoleate on the concentration of prostaglandins and their precursors in rat, *Prostaglandins,* 17, 5434, 1979.
28. **Yu, P.H., Mai, J., and Kinsella, J.E.,** The effects of dietary trans,trans methyl octadienoate acid on composition and fatty acids of the heart, *Am. J. Clin. Nutr.,* 33, 598, 1980.
29. **Senti, F.R., Ed.,** Health Aspects of Dietary Trans Fatty Acids: Report Prepared for Food Safety and Applied Nutrition, Food and Drug Administration, Department of Health and Human Services, Washington, D.C. 20204, Life Sciences Research Office, Federation of the American Societies for Experimental Biology, Bethesda, MD, 1985.
30. British Nutrition Foundation Task Force, Report on Trans Fatty Acids, British Nutrition Foundation, London, 1987.
31. **Koletzko, B.,** Zufuhr, stoffwechsel und biologische wirkungen trans-isomerer fettsauren bei sauglingen, *Nahrung Food,* 35, 229, 1991.
32. **Koletzko, B., Mrotzek, M., and Bremer, H.J.,** Trans fatty acids in human milk and infant plasma and tissue, in *Human Lactation. Vol. 3. Effect of Human Milk on the Recipient Infant,* Goldman, A.S., Atkinson, S., and Hanson, L.A., Eds., Plenum Press, New York, 1987, 323.
33. **Koletzko, B. and Muller, J.,** Cis- and trans-isomeric fatty acids in plasma lipids of newborn infants and their mothers, *Biol. Neonate,* 57, 172, 1990.
34. **Koletzko, B., Mrotzek, M., and Bremer, H.J.,** Fatty acid composition of mature human milk in Germany, *Am. J. Clin. Nutr.,* 47, 954, 1988.
35. **Holman, R.T., Pusch, F., Svingen, B., and Dutton, H.J.,** Unusual isomeric polyunsaturated fatty acids in liver phospholipids of rats fed hydrogenated oil, *Proc. Natl. Acad. Sci. U.S.A.,* 88, 4830, 1991.
36. **Koletzko, B.,** Trans fatty acids may impair biosynthesis of long-chain polyunsaturates and growth in man, *Acta Paediatr.,* 81, 302, 1992.
37. **Alfin-Slater, R.B., Wells, P., and Aftergood, L.,** Dietary fat composition and tocopherol requirement. IV. Safety of polyunsaturated fats, *J. Am. Oil Chem. Soc.,* 50, 479, 1973.
38. **Koletzko, B. and Bremer, H.J.,** Fat content and fatty acid composition of infant formulae, *Acta Paediatr. Scand.,* 78, 513, 1989.
39. **Holman, R.T., Johnson, S.B., and Ogburn, P.L.,** Deficiency of essential fatty acids and membrane fluidity during pregnancy and lactation, *Proc. Natl. Acad. Sci. U.S.A.,* 88, 4835, 1991.
40. **Spritz, N. and Mishkel, M.A.,** Effects of dietary fats on plasma lipids and lipoproteins: an hypothesis for the lipid-lowering effect of unsaturated fatty acids, *J. Clin. Invest.,* 48, 78, 1969.
41. **Vergroesen, A.J.,** Dietary fat and cardiovascular disease: possible modes of action of linoleic acid, *Proc. Nutr. Soc.,* 31, 323, 1972.
42. **Vergroesen, A.J., Ed.,** *The Role of Fats in Human Nutrition,* Academic Press, New York, 1975.
43. **Mattson, F.H., Hollenbach, E.J., and Kligman, A.M.,** Effect of hydrogenated fat on the plasma cholesterol and triglyceride levels of man, *Am. J. Clin. Nutr.,* 28, 726, 1975.
44. **Anderson, J.T., Grande, R., and Keys, A.,** Hydrogenated fats in the diet and lipids in the serum of man, *J. Nutr.,* 75, 388, 1961.
45. **Mensink, R.P. and Katan, M.B.,** Effect of dietary trans fatty acids on high-density and low-density lipoprotein cholesterol levels in healthy subjects, *N. Engl. J. Med.,* 323, 439, 1990.
46. **Nestel, P., Noakes, M., Belling, B., McArther, R., Clifton, P., Janus, E., and Abbey, M.,** Plasma lipoprotein lipid and Lp(a) changes with substitution of elaidic acid for oleic acid in the diet, *J. Lipid Res.,* 33, 1029, 1992.
47. **Gautheron, P. and Renaud, S.,** Hyperlipidemia induced hypercoagulable state in rat. Role of an increased activity of platelet phosphatidylserine in response to certain dietary fatty acids, *Thromb. Res.,* 1, 353, 1972.
48. **Zock, P.L. and Katan, M.B.,** Hydrogenation alternatives: effects of trans fatty acids and stearic acid versus linoleic acid on serum lipids and lipoproteins in humans, *J. Lipid Res.,* 33, 399, 1992.
49. **Takatori, T., Phillips, F.C., Shimasaki, H., and Privett, O.S.,** Effects of dietary saturated and trans fatty acids on tissue lipid composition and serum LCAT activity in the rat, *Lipids,* 85, 272, 1976.

50. Moore, C.E., Alfin-Slater, R.B., and Aftergood, L., Effect of trans fatty acids on serum lecithin: cholesterol acyltransferase in rats, *J. Nutr.*, 110, 2284, 1980.
51. Murray, R.K., Granner, D.K., Mayes, P.A., and Rodwell, V.W., Eds., *Harper's Biochemistry*, 27th ed., Appleton and Lange, Norwalk, CT, 1990.
52. Emken, E.A., Nutrition and biochemistry of trans and positional fatty acid isomers in hydrogenated oils, *Annu. Rev. Nutr.*, 4, 339, 1984.
53. Troisi, R., Willett, W.C., and Weiss, S.T., Trans-fatty acid intake in relation to serum lipid concentrations in adult men, *Am. J. Clin. Nutr.*, 56, 1019, 1992.
54. Stampfer, M.J., Sacks, F.M., Salvini, S., Willett, W.C., and Hennekens, C.H., A prospective study of lipids, apolipoproteins and risks of myocardial infarction, *N. Engl. J. Med.*, 325, 373, 1991.
55. Shekelle, R.B., Shryock, A.M., Paul, O., et al., Diet, serum cholesterol, and death from coronary heart disease. The Western Electric Study, *N. Engl. J. Med.*, 304, 65, 1981.
56. Arntzenius, A.C., Kromhout, D., Barth, J.D., et al., Diet, lipoproteins, and the progression of coronary atherosclerosis, *N. Engl. J. Med.*, 312, 805, 1985.

Chapter 2.5

PLANT STEROLS: THEIR BIOLOGICAL EFFECTS IN HUMANS

John W. Farquhar

INTRODUCTION

Plant sterols (phytosterols) are complex alcohols comprised of nine currently identified C28 or C29 sterols, differing in most instances from cholesterol (C27) by possession of an extra methyl or ethyl group on cholesterol's 8 carbon side chain.[1]

They are present in higher plants and share the biological functions of cholesterol in higher animals as components in their unesterified state of the lipid bilayer of cell membranes.[2] Cholesterol in animals also serves as a precursor of steroid hormones and bile acids, but less is known of possible analogous roles for phytosterols. In edible oils and human diets beta sitosterol (C29), campesterol (C28), and stigmasterol (C29)[3] are the major plant sterols. Phytosterols usually comprise less than half of dietary sterol intake in the U.S., with the remainder being dietary cholesterol;[3] however, in many human populations that consume predominately foods of vegetable origin, the cholesterol/phytosterol ratio is reversed, as was shown in Mexico's Tarahumara Indians, who consume diets consisting largely of corn and beans.[4]

ABSORPTION AND METABOLISM OF PHYTOSTEROLS IN HUMANS

Human evolution has apparently resulted in selective gut absorption mechanisms that exclude almost all phytosterols. Although higher estimates of their absorption have been made,[3] the most convincing isotopic sterol balance experiments have shown that less than 5% of ingested plant sterols are absorbed, due to interference with cholesterol esterification in the gut, a step needed for efficient absorption.[5] That which is absorbed has a shorter half-life than endogenous cholesterol, and is either converted to bile acids or excreted in bile as the free sterol. On diets approximating the usual sterol intakes in the U.S. (cholesterol — 400 mg/d, phytosterol — 200 mg/d), plasma cholesterol levels are 300 to 800 times greater than the plasma phytosterol levels, attesting further to a "metabolic rejection" of phytosterols by humans. The sophisticated isotopic sterol balance study cited[5] confirms much earlier work that had shown very low absorption rates. The first studies, done without the benefit of isotopic labeling, were those reported by Schonheimer in 1930. From studies on five animal species, he concluded that little, if any, plant sterols were absorbed.[6]

EFFECTS OF PLANT STEROLS ON CHOLESTEROL METABOLISM IN HUMANS

Since earlier work had suggested an effect of plant sterols on cholesterol absorption, this led next to the question of their effect on blood cholesterol levels. In 1951 Peterson found that soy sterols prevented the expected increase in serum cholesterol of cholesterol-fed chicks.[7] Pollak then reported that both hypercholesterolemia and atherosclerosis could be prevented in cholesterol-fed rabbits which were also fed plant sterols.[8] Given the growing evidence during the 1950s for hypercholesterolemia as a risk factor for coronary heart disease, plant sterols captured the attention of clinical researchers as substances of potential use in treatment of hypercholesterolemia. Accordingly, approximately nine papers on this issue were published

0-8493-4248-1/96/$0.00+$.50
© 1996 by CRC Press Inc.

between 1953 and 1958 and all but one reported decreases in blood cholesterol levels, in most instances with use of mixtures of plant sterols that contained primarily beta-sitosterol.

The largest and most carefully controlled studies during this period by Farquhar and colleagues showed that adult patients with high cholesterol levels achieved an average 20% reduction of blood cholesterol and of beta lipoprotein cholesterol (now termed low density lipoprotein) on a total dose of 12 to 18 g of beta-sitosterol taken three times daily with meals.[9,10] In the second study, addition of a liquid vegetable oil supplement, acting by a different mechanism, produced a further reduction in blood cholesterol.[10] A full exploration of beta-sitosterol's effect on those with lower cholesterol levels, such as between the 20th to 60th percentile of cholesterol distribution, was not addressed in any of the studies, nor did an adequate test of dose and response through a wide range of plant sterol intake occur. In some studies 18 g of beta-sitosterol was used, but 9 to 12 g seemed to have similar effects. Given that these doses were at least an order of magnitude greater than that which occurs in natural human diets, studies at lower doses would have been desirable. The major conclusion to be drawn from work done in the 1950s was that the degree of cholesterol lowering achieved was considerable and, if achievable under nonresearch settings, could have a major impact on atherosclerosis and coronary heart disease, then and now still the largest cause of premature death and disability in all western industrialized countries.

Following 1958 through 1993 only two more studies were reported on the cholesterol-lowering action of plant sterols, perhaps reflecting in part that pharmaceutical firms placed their energies on synthesized organic cholesterol-lowering compounds that can receive patents, rather than substances found in nature. Work after 1958 did, however, continue in ways that settled the questions of the mechanisms of phytosterol's effect on cholesterol absorption and on the dose and timing of ingestion needed for a maximum blockage of cholesterol absorption. Lees and co-workers showed that 3 g of tall oil sterols (98% beta-sitosterol) was equivalent to 9 g/d in reaching equivalent decreases in both cholesterol absorption and blood cholesterol reductions.

Additional studies by Grundy and Mok[12] using an intestinal perfusion technique suggested that even smaller amounts of beta-sitosterol would produce maximum blockage of cholesterol absorption, and they and Mattson and colleagues[13] emphasized that timing, mixing, and close physical proximity to dietary cholesterol were important determinants of effectiveness of plant sterols on absorption and blood cholesterol reduction. These findings are quite relevant to the role of plant sterols in human diets in the amounts ordinarily consumed, a topic to be discussed below.

THE CONTENT OF PHYTOSTEROLS IN EDIBLE OILS AND FOODS

The most recent and most thorough review of the types, amounts, and distribution of plant sterols appeared in 1978 and reported on many edible oils, margarines, seeds, nuts, grains, and fruits and vegetables. Higher values were present in a few selected edible oils and in more calorically dense foods such as seeds and nuts. Figures were given in milligrams of sterol per 100 g of oil or edible food portion. Content of three major phytosterols (beta-sitosterol, campesterol, and stigmasterol) was given for all oils and foods. For oils only, the six so-called "minor sterols" were also measured. For most of the 39 edible oils analyzed the sterols were present in amounts of 100 to 500 mg sterol per 100 g oil. The exceptions were oils from alfalfa seed, corn, rice bran, rye germ, sesame seed, and wheat germ oils, which, respectively, contain about 2.1, 1.4, 3.2, 2.4, 2.9, and 2.0 g phytosterol per 100 g. The authors noted that the content of phytosterols in crude oils was 20 to 60% higher than in refined oils and that hydrogenation removed an additional 20 to 40% more sterol from refined oils. Among margarines (80% fat), tub margarines had somewhat higher levels than stick margarines, and corn oil margarines, containing 430 to 586 mg sterol per 100 g margarine, were higher than soy bean oil/margarines (136 to 430 mg/100 g).[1]

The sterol content of vegetables and fruits measured as fresh weight were quite low, as might be expected from their water content, and values over 100 mg/100 g were found only in seedlings. Legumes were generally higher in plant sterol content than in fruits and non-leguminous vegetables, whereas the highest values in plant foods were present in seeds and nuts, with sesame and sunflower seeds having values of 714 and 534 mg/100 g edible portion, about five times greater than the average. Among cereal grains, buckwheat, corn, and sorghum, averaging almost 100 mg/100 g of edible portion, were each about three times higher in plant sterols than were oats or wheat flour, whereas rice bran (at about 1300 mg/100 g) had a much higher context, exceeding that of most vegetable oils.

The main generalizations derived from these extensive data are that great differences occur in plant sterol content in the plant kingdom, edible oils or seeds and nuts (containing high oil contents) have higher amounts than do less calorically dense plant foods, and much remains to be learned about reasons for the great variability in content, even among similar foods. For example, the free sterol, containing a hydroxyl at C7, is less soluble in oils than are the fatty acyl esters of these sterols. Therefore, the proportion of free to esterified sterol in a particular food could explain the variability of phytosterol content of edible oils derived from those foods; however, such analyses have not been done on either the intact foods or on their oils. In their paper, Weihrauch and Gardner present some evidence of variability in sterol content depending on food variety, seasonality, geography, and processing.[1] Ultimately, one can see the need to know more of the biology of the plant. For example, are the free sterols concentrated in cell membranes (as in higher animals) and what is the concentration, function, and location of any esterified phytosterols within a plant? These questions, including the incompleteness of data on many foods commonly consumed, must relate in part to new scientific interest generated by a belief that this new knowledge would be relevant to human nutrition and, perhaps, to disease prevention — topics to be considered below.

DO PHYTOSTEROLS PLAY A ROLE IN DISEASE PREVENTION IN HUMANS?

Blood cholesterol reductions of 10 to 20% are achieved by feeding supplements of beta-sitosterol before meals in total daily doses of three or more grams,[9-11] and these cholesterol reductions, if prolonged in duration, would have large effects in preventing cardiovascular disease. The question arises whether the plant sterol content of natural foods (which usually amounts to less than 600 mg/d in most contemporary dietary patterns) would have an effect sufficient to decrease cardiovascular disease rates. There are three reasons why the answer might be positive. First, Grundy and Mok have evidence that quite small amounts of plant sterols have the ability to block cholesterol absorption, but only if a test meal presents its dietary cholesterol and its dietary plant sterol to the gut at the same time.[12]

Second, as a corollary, the conditions of mixing and presentation to the gut in an appropriate physical state might well occur naturally during the eating process; therefore, normal meal habits may achieve the process and, thus, the optimum conditions discovered in intestinal intubation studies. However, this concept has not yet been tested in humans.

Third, even small differences in cholesterol levels (present for decades) can influence coronary heart disease rates in contemporary society, with estimates that each 1% decrease in blood cholesterol levels would reduce coronary events by 2 to 3%.[14]

If the experiments were done to show an even slight cholesterol-lowering effect on plant sterols present in current dietary intakes, then benefits would accrue in contemporary society, given the fact that cardiovascular diseases are so prevalent. It is also conceivable that some of the known cholesterol-lowering effects of vegetarian diets could be related to plant sterols, as well as lowered content of saturated fat and cholesterol, which probably accounts for most of the effects of a vegetarian diet on blood cholesterol.

Following this reasoning, given the previously described wide variation among food and edible oils of plant sterol mentioned, selection of foods particularly high in plant sterol content would have an even greater effect on reducing blood cholesterol levels. Experiments on these issues have not yet been done, but many such studies would be relevant. Only after considerably more research would it be possible to speculate on whether plant sterols have played a role in human evolution. Given our status as omnivores, could our foods of plant origin have acted as partial counterbalances to the excessive cholesterol content of many foods of animal origin?

Since the recent, rapid increase in knowledge of the antioxidant properties of many plant micronutrients, it now seems feasible to surmise that these properties of plants may have been positive influences on human survival, especially given our long life span. In this same vein, recent findings have shown a correlation of plasma vitamin E levels and longevity among a large number of mammalian species.[15] At this point it would be highly speculative to assume a beneficial effect of plant sterols on human evolution, except to note that exposure to them has occurred over many millenia.

It is, however, germaine to think of how many positive properties there are in foods of plant origin, beyond micronutrients such as antioxidants (which are thought to have benefits in both cancer and coronary heart disease prevention). One such substance much studied in the past two decades is dietary fibers; these soluble fibers particularly have been shown to have cholesterol-lowering properties, probably through decreasing bile acid reabsorption in the terminal ileum.[16] Perhaps certain fibers and plant sterols could interact synergistically in cholesterol-lowering properties, but only further research could answer this and other questions.

SUMMARY

Clearly, more research is needed to determine the extent of benefit of plant sterols in disease prevention, especially in studies that attempt to avoid the confounding that a reductionist approach can bring. Since both dietary fibers and plant sterols may be altered, and in the case of plant sterols, decreased in amount following extraction from plants that we consume, it seems wise to study effects on blood cholesterol of nonsupplemented diets that vary amounts of fiber, micronutrients, and plant sterols through choosing foods that vary in their content of these materials. Various useful topics for such research are suggested in this chapter. The view was also presented that plant sterols in smaller amounts than have been used experimentally may well have important benefits in disease prevention, due to their effects on reducing blood cholesterol levels.

REFERENCES

1. **Weihrauch, J.L. and Gardner, J.M.,** Sterol content of foods of plant origin, *J. Am. Diet Assoc.,* 73, 39, 1978.
2. **Cerqueira, M.T., McMurry, M., and Connor, W.E.,** The food and nutrient intakes of the Tarahumara Indians of Mexico, *Am. J. Clin. Nutr.,* 32, 905, 1979.
3. **Subbiah, M.T., Ravi,** Significance of dietary plant sterols in man and experimental animals, *Mayo Clin. Proc.,* 46, 549, 1971.
4. **Cerqueira, M.T., McMurry, M., and Connor, W.E.,** The food and nutrient intakes of the Tarahumara Indians of Mexico, *Am. J. Clin. Nutr.,* 32, 905, 1979.
5. **Salen, G., Ahrens, E.H., and Grundy, S.M.,** Metabolism of B-sitosterol in man, *J. Clin. Invest.,* 49, 952, 1970.

6. **Schonheimer, R.,** New contributions in sterol metabolism, *Science,* 74, 579, 1931.
7. **Peterson, D.W.,** Effect of soybean sterols in the diet on plasma and liver cholesterol in chicks, *Proc. Soc. Exp. Biol. Med.,* 78, 143, 1951.
8. **Pollack, O.J.,** Successful prevention of experimental hypercholesterolemia and cholesterol atherosclerosis in the rabbit, *Circulation,* 7, 696, 1953.
9. **Farquhar, J.W., Smith, R.E., and Dempsey, M. S.,** The effect of beta sitosterol on the serum lipids of young men with artheriosclerotic heart disease, *Circulation,* 14, 77, 1956.
10. **Farquhar, J.W. and Sokolow, M.,** Response of serum lipids and lipoproteins of man to beta-sitosterol and safflower oil, *Circulation,* 17, 890, 1958.
11. **Lees, A.M. et al.,** Plant sterols as cholesterol-lowering agents: clinical trials in patients with hypercholesterolemia and studies of sterol balance, *Atherosclerosis,* 28, 325, 1977.
12. **Grundy, S.M. and Mok, H.Y.I.,** Effects of low dose phytosterols on cholesterol absorption in man, in *Lipoprotein Metabolism,* Greten, H., Ed., Springer-Verlag, Berlin, 1976, 112.
13. **Mattson, F.H., Grundy, S.M., and Crouse, J.R.,** Optimizing the effect of plant sterols on cholesterol absorption in man, *Am. J. Clin. Nutr.,* 35, 697, 1982.
14. Lipid Research Clinic Program, The lipid research clinics coronary primary prevention program. I. Reduction in the incidence of coronary heart disease, *J.A.M.A.,* 251, 351, 1984.
15. **Cutler, R.G.,** Antioxidants and aging, *Am. J. Clin. Nutr.,* 53, 373S, 1991.
16. **Spiller, G.A., Farquhar, J.W., Gates, J.E., and Nichols, S.F.,** Guar gum and plasma cholesterol, *Arterioscler. Thromb.,* 11, 1204, 1991.

Chapter 2.6

SAPONINS IN THE TREATMENT OF HYPERCHOLESTEROLEMIA

David Oakenfull

INTRODUCTION

Dietary saponins have been known for many years to lower plasma cholesterol concentrations in various animal species.[1,2] Subsequently, a number of workers in this area have proposed that saponin-containing foods might usefully be advised for human consumption for the treatment of hypercholesterolemia[3-5] — a proposal that has aroused considerable interest and some controversy, with apparently conflicting results appearing in the literature.[6-8]

WHAT ARE SAPONINS?

Saponins are a structurally diverse group of triterpene or steroid glycosides.[9,10] The molecules are amphiphilic, the triterpene or steroid part being hydrophobic and the sugar part hydrophilic. This gives saponins their characteristic surface activity from which the name is derived. The structure of a typical saponin, one of those present in soybeans, is shown in Figure 1. Saponins have been identified in many hundreds of plant species, but relatively few of these are used as food by humans. These are shown in Table 1.

EFFECTS OF SAPONINS ON PLASMA LIPIDS IN BIRDS AND MAMMALS

Some of the first indications of cholesterol-lowering activity of saponins appeared some 30 years ago in experiments with poultry.[1,2] Feeds containing saponin-rich alfalfa meal or isolated saponins from *Quillaia saponaria* were found to lower plasma and liver cholesterol levels. Since then, similar observations have been made in a number of mammalian species — particularly rats and monkeys.[3,11-13]

Kritchevsky et al.[14] appear to have been the first to suggest that the saponins in alfalfa were the cause of the lower plasma cholesterol levels in rats fed alfalfa-based diets. Malinow et al.[3,11,12] have since provided conclusive evidence that the saponins in alfalfa are indeed responsible by demonstrating the cholesterol-lowering activity of isolated alfalfa saponins.

Saponins from other plant species have also been investigated. Malinow et al.[3] demonstrated the cholesterol-lowering effect in monkeys fed digitonin. Oakenfull et al. have found cholesterol-lowering in rats fed commercial saponin white[13] or isolated saponins from quillaja bark,[15] soybeans,[15] navy beans *(Phaseolus vulgaris)*,[16] or chickpeas *(Cicer arietinum).*[17]

EFFECTS OF SAPONINS ON PLASMA LIPIDS IN HUMANS

Probably because saponins have a reputation for being toxic,[18] there is only one report to date of humans having been given saponins directly. Bingham et al.[19] gave a group of arthritic patients tablets of a saponin-rich extract from *Yucca schidigen* in an experiment that also included diets, exercise, and physiotherapy. Substantial reductions in plasma cholesterol were observed, particularly in those individuals with initially higher levels.

There are also a number of reports in the literature of experiments in which human volunteers were fed various saponin-containing foods. In some cases convincing reductions

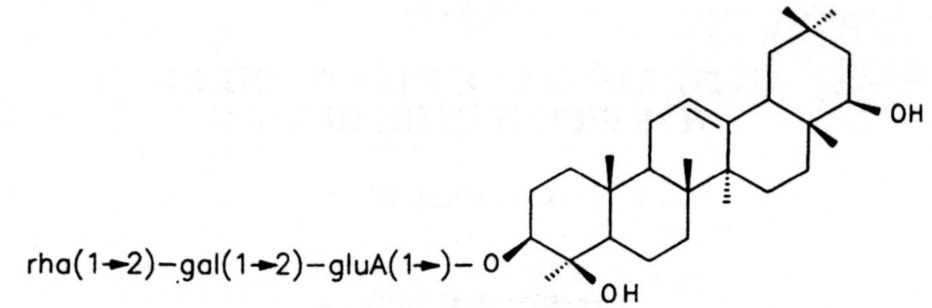

rha(1→2)−gal(1→2)−gluA(1→)−O

OH

OH

FIGURE 1. Structure of one of the saponins from soybeans. (From Oakenfull, D.G. and Sidhu, G.S., Saponins, in *Toxicants of Plant Origin*, Vol. 2, Cheeke, P., Ed., CRC Press, Boca Raton, FL, 1989, chap. 4. With permission.)

TABLE 1
Plant Foods That Contain
Significant Levels of Saponins

Plant	Saponin content[a] (g/kg)
Alfalfa sprouts *(Medicago sativa)*	80
Chickpea *(Cicer arietinum)*	2.3-60
Soybean *(Glycine max)*	5.6-56
Navy bean *(Phaseolus vulgaris)*	4.5-21
Kidney bean *(P. vulgaris)*	2-16
Mung bean *(P. mungo)*	0.5-6
Broad bean *(Vicia faba)*	3.5
Green pea *(Pisum sativum)*	1.8-11
Lentil *(Lens culinaris)*	1.1-5.1
Azuki bean *(Vigna angularis)*	—
Peanut *(Arachis hypoglycaea)*	0.05-16
Onion *(Allium cepa)*	—
Garlic *(A. sativum)*	—
Leek *(A. porrum)*	—
Asparagus *(Asparagus officinalis)*	15
Spinach *(Spinacea oleracia)*	47
Silver beet *(Beta vulgaris)*	58
Egg plant *(Solanum melongena)*	—
Sunflower *(Helianthus annus)*	—
Sesame seed *(Sesamum indicum)*	3
Oats *(Avena sativa)*	1-13
Quinoa *(Chenopodium quinoa)*	—
Blackberry *(Rubus fructiosus)*	—

[a] Range given where literature sources differ.

in plasma cholesterol concentrations were obtained; in others there was no observable effect (Table 2). Here, the picture is complicated by the presence of other dietary components with cholesterol-lowering activity (particularly the soluble dietary fiber in legumes[20]). Considered collectively, though, these results suggest that saponins can have usefully significant cholesterol-lowering properties.[21]

MECHANISMS

There appear to be two mechanisms by which saponins can affect cholesterol metabolism:

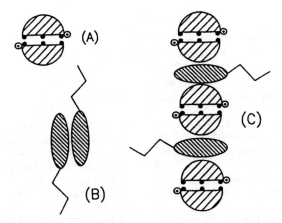

FIGURE 2. Schematic diagram of the formation of large mixed micelles by saponins and bile acids. (A) A bile salt micelle — aggregation of the molecules restricted by electrostatic repulsion between the charged groups; (B) saponin micelles — aggregation of the molecules restricted by the bulky sugar groups; (C) large mixed micelle of saponin and bile acid. Interleaving of the molecules relieves the steric and electrostatic barriers to aggregation of the molecules. (From Oakenfull, D.G. and Sidhu, G.S., Saponins, in *Toxicants of Plant Origin*, Vol. 2, Cheeke, P., Ed., CRC Press, Boca Raton, FL, 1989, chap. 4. With permission.)

TABLE 2
Summary of Observed Effects of Saponin-Containing Diets on Cholesterol Metabolism in Humans

	Approx saponin intake	Plasma cholesterol		Plasma	Fecal sterols		
Saponin source	(mg/d)	Initial (m*M*)	Change (%)	TG	BA	NS	Ref.
Chickpea	a	5.3	-22	N	+	+	28
Soy protein isolate	a	8.6	-23	-	N	N	29
Yucca	a	7.2	-21	-	N	N	30
Alfalfa seeds	400	6.4	-24	N	+	+	31
Soy flour	500	4.2	0	0	+	+	32
	500	6.9	0	0	0	0	33
Bean meal	300	7.3	-17	0	N	N	34
Ginseng	a	>5.7	0	-	N	N	35
Alfalfa seeds	400	9.6	-16	+	N	N	36

Note: + indicates *increase*, - indicates *decrease*, and 0 indicates *no significant change;* N indicates *not stated.* Abbreviations: TG = triglycerides; NS = neutral sterols; BA = bile acids.

a Insufficient information given for a reliable estimate.

1. Some saponins, with particular defined structural characteristics, form insoluble complexes with cholesterol (as, for example, in the well-known precipitation of cholesterol by digitonin). Complexation in the gut then inhibits cholesterol absorption.[22,23]

2. Saponins can also affect cholesterol metabolism indirectly by interfering with the enterohepatic circulation of bile acids. Some saponins form large mixed micelles with bile acids (Figure 2).[24] These can have molecular weights of several millions and the reabsorption of bile acids from the terminal ileum is effectively blocked.[25] Bile acids are thus diverted from the enterohepatic cycle and lost by fecal excretion. This loss is then offset by increased synthesis from endogenous cholesterol, resulting in lower plasma and liver levels.[26]

TABLE 3
Mean Daily Intake of Saponins (Mg per Person) in Different
Population Groups in the U.K.[26]

Population studied	Intake (mg/d)	Main sources
All U.K. households (7193[a])	14.6	Baked beans, lentils, peas
All U.K., omnivore (17)	13.3	Baked beans, lentils, chickpeas
All U.K., vegetarian (12)	109.9	Soya, baked beans, lentils
All U.K., children	12.6	Baked beans, lentils, kidney beans
Male Caucasian (2)	9.6	Baked beans, peas, soya
Male West Indian (11)	46.9	Kidney beans, baked beans, blackeye beans
Male Asian (15)	167.7	Kidney beans, guar beans, chickpeas
Male Asian vegetarian (10)	213.4	Kidney beans, guar beans, chickpeas

[a] Number of subjects.

WHAT DAILY INTAKE WOULD BE REQUIRED TO ACHIEVE A USEFUL REDUCTION IN PLASMA CHOLESTEROL?

An indication of what might be a therapeutically useful daily intake of saponins is provided by comparing amounts that were demonstratively effective in Table 2 with some estimates of daily saponin intakes of different population groups in the U.K.[27] These are given in Table 3. The mean daily intake is small, only 14.6 mg, but in vegetarian and Asian households the daily intake per person is 100 to 200 mg. These latter levels are comparable with the daily intakes that were experimentally effective in lowering plasma cholesterol concentrations. These data suggest that daily intakes of 100 to 200 mg are likely to be effective and are also attainable. It should be borne in mind, though, that saponins differ in their ability to interact with cholesterol and bile acids.[9,10] Consequently, saponins from different plant sources will differ in their ability to lower plasma cholesterol.

CONCLUSIONS

There is now a substantial body of evidence that dietary saponins can lower plasma cholesterol concentrations.[21] Saponin-containing foods thus could usefully contribute to cholesterol-lowering diets. Plant foods that could best be recommended to hypercholesterolemic patients are chickpeas, the different varieties of *Phaseolus vulgaris* such as navy beans ("baked beans"), lentils, soybeans, and alfalfa or fenugreek sprouts. Increasing intake of these foods would have the additional benefit of displacing meat and thus decreasing consumption of saturated fat.

REFERENCES

1. **Newman, H.A.I., Kummerov, F.A., and Scott, H.M.,** Dietary saponin, a factor which may influence liver and serum cholesterol levels, *Poultry Sci.,* 38, 42, 1958.
2. **Griminger, P. and Fisher, H.,** Dietary saponin and plasma cholesterol in the chicken, *Proc. Soc. Exp. Biol. Med.,* 99, 424, 1958.
3. **Malinow, M.R., McLaughlin, P., and Stafford, C.,** Prevention of hypercholesterolemia in monkeys *(Macaca fascicularis)* by digitonin, *Am. J. Clin. Nutr.,* 31, 814, 1978.

4. **Potter, J.D., Topping, D.L., and Oakenfull, D.G.,** Soya products, saponins and plasma cholesterol, *Lancet,* 1, 223, 1979.
5. **Rao, A.V. and Kendall, C.W.,** Dietary saponins and plasma lipids, *Food Chem. Toxicol.,* 24, 441.
6. **Calvert, G.D., Blight, L., Illman, R.J., Topping, D.L., and Potter, J.D.,** A trial of the effects of soya-bean flour and soya-bean saponins on plasma lipids, faecal bile acids and neutral sterols in hypercholesterolaemic men, *Br. J. Nutr.,* 45, 277, 1981.
7. **Gibney, M.J., Pathirana, C., and Smith, L.,** Saponins and fibre: lack of interactive effects on serum and liver cholesterol in rats and hamsters, *Atherosclerosis,* 45, 365, 1982.
8. **Oakenfull, D.G. and Topping, D.L.,** Saponins and plasma cholesterol: a reply to the letter of Gibney, Pathirana and Smith, *Atherosclerosis,* 48, 301, 1983.
9. **Price, K.R., Johnson, I.T., and Fenwick, G.R.,** The chemistry and biological significance of saponins in foods and feedingstuffs, *CRC Crit. Rev. Food Sci. Nutr.,* 26, 27, 1987.
10. **Oakenfull, D.G. and Sidhu, G.S.,** Saponins, in *Toxicants of Plant Origin,* Vol. 2, Cheeke, P., Ed., CRC Press, Boca Raton, FL, 1989, chap. 4.
11. **Malinow, M.R., McLaughlin, P., Papworth, L., Stafford, C., Livingston, A.L., and Cheeke, P.R.,** Effect of alfalfa saponins on intestinal cholesterol absorption in rats, *Am. J. Clin. Nutr.,* 30, 2061, 1977.
12. **Malinow, M.R., Connor, W.E., McLaughlin, P., Stafford, C., Lin, D.S., Livingston, A.L., Kohler, G.O., and McNulty, W.P.,** Cholesterol and bile acid balance in *Macaca fascicularis.* Effects of alfalfa saponins, *J. Clin. Invest.,* 67, 156, 1977.
13. **Oakenfull, D.G., Fenwick, D.E., Topping, D.L., Illman, R.J., and Storer, G.B.,** Effects of saponins on bile acids and plasma lipids in the rat, *Br. J. Nutr.,* 42, 209, 1979.
14. **Kritchevsky, D., Tepper, S.A., and Story, J.A.,** Isocaloric, isogravic diets in rats. III. Effect of non-nutritive fiber (alfalfa or cellulose) on cholesterol metabolism, *Nutr. Rep. Int.,* 9, 301, 1975.
15. **Oakenfull, D.G., Topping, D.L., Illman, R.J., and Fenwick, D.E.,** Prevention of dietary hypercholesterolaemia in the rat by soya bean and quillaja saponins, *Nutr. Rep. Int.,* 29, 1039, 1984.
16. **Kozuharov, S., Oakenfull, D.G., and Sidhu, G.S.,** Navy beans and navy bean saponins lower plasma cholesterol in rats, *Proc. Nutr. Soc. Aust.,* 11, 162, 1986.
17. **Oakenfull, D.G. and Sidhu, G.S.,** Prevention of dietary hypercholesterolaemia by chickpea saponins and navy beans, *Proc. Nutr. Soc. Aust.,* 9, 104, 1984.
18. **George, A.J.,** Legal status and toxicity of saponins, *Food Cosmet. Toxicol.,* 3, 85, 1965.
19. **Bingham, R., Harris, D.H., and Laga, T.,** Yucca plant saponin in the treatment of hypertension and hypercholesterolemia, *J. Appl. Nutr.,* 30, 127, 1978.
20. **Kingman, S.M.,** The influence of legume seeds on human plasma lipid concentrations, *Nutr. Res. Rev.,* 4, 97, 1991.
21. **Oakenfull, D. and Sidhu, G.S.,** Could saponins be a useful treatment for hypercholesterolaemia?, *Eur. J. Clin. Nutr.,* 44, 79, 1990.
22. **Coulson, C.B. and Evans, R.A.,** Effect of saponin, sterols and linoleic acid on the weight increase of growing rats, *Br. J. Nutr.,* 14, 121, 1960.
23. **Gestetner, B., Birk, Y., and Tencer, Y.,** Soybean saponins: fate of ingested soybean saponins and physiological aspect of their hemolytic activity, *J. Agric. Food Chem.,* 16, 1301, 1968.
24. **Oakenfull, D.,** Aggregation of saponins and bile acids in aqueous solution, *Aust. J. Chem.,* 39, 1671, 1986.
25. **Sidhu, G.S. and Oakenfull, D.G.,** A mechanism for the hypocholesterolaemic activity of saponins, *Br. J. Nutr.,* 55, 643, 1986.
26. **Heaton, K.W.,** *Bile Salts in Health and Disease,* Churchill Livingstone, Edinburgh, 1972.
27. **Ridout, C.L., Wharf, S.G., Price, K.R., Johnson, I.T., and Fenwick, G.R.,** UK mean daily intakes of saponins — intestine-permeabilizing factors in legumes, *Food Sci. Nutr.,* 42F, 111, 1988.
28. **Mathur, K.S., Khan, M.A., and Sharma, R.D.,** Hypocholesterolaemic effect of Bengal gram: a long-term study in man, *Br. Med. J.,* 1, 30, 1968.
29. **Sirtori, C.R., Agradi, E., Conti, F., Mantero, I., and Gatti, E.,** Soybean-protein diet in the treatment of type-II hyperlipoproteinaemia, *Lancet,* 1, 275, 1977.
30. **Bingham, R., Harris, D.H., and Laga, T.,** Yucca plant saponin in the treatment of hypertension and hypercholesterolemia, *J. Appl. Nutr.,* 30, 127, 1978.
31. **Malinow, M.R., McLaughlin, P., and Stafford, C.,** Alfalfa seeds: effects on cholesterol metabolism, *Experientia,* 36, 562, 1980.
32. **Potter, J.D., Illman, R.J., Calvert, G.D., Oakenfull, D.G., and Topping, D.L.,** Soya saponins, plasma lipids, lipoprotein turnover and fecal bile acids: a double-blind cross-over study, *Nutr. Rep. Int.,* 22, 521, 1979.
33. **Calvert, G.D., Blight, L., Illman, R.J., Topping, D.L., and Potter, J.D.,** A trial of the effects of soya-bean flour and soya-bean saponins on plasma lipids, faecal bile acids and neutral sterols in hypercholesterolaemic men, *Br. J. Nutr.,* 45, 277, 1981.
34. **Bingwen, L., Zhaofeng, W., Wanzhen, L., and Rongjue, Z.,** Effects of bean meal on serum cholesterol and triglycerides, *Chin. Med.,* 94, 455, 1981.

35. **Yamamoto, M., Uemura, T., Nakayama, S., Uemiya, M., and Kumagai, A.,** Serum HDL-cholesterol-increasing and fatty acid liver-improving actions of *Panax ginseng* in high cholesterol diet-fed rats with clinical effect on hyperlipidemia in man, *Am. J. Clin. Nutr.,* 11, 96, 1983.

36. **Molgaard, J., von Schenck, H., and Olsson, A.G.,** Alfalfa seeds lower low density lipoprotein cholesterol and apoprotein B concentrations in patients with type II hyperlipoproteinemia, *Atherosclerosis,* 65, 173, 1987.

Chapter 2.7

FOOD LIPIDS AND THERMOGENESIS IN RELATION TO OBESITY

Fabio Armellini, Mauro Zamboni, Tiziana Todesco, and Ottavio Bosello

INTRODUCTION

Obesity, from the quantitative point of view, is experienced when energy intake exceeds energy expenditure, following the unchanging law of thermodynamics: energy intake - energy expenditure = energy excess. This equation, however, does not take qualitative aspects of the question into account. Isocaloric meals can, in fact, generate different energy expenditures depending on their relative carbohydrate and fat contents. The metabolic expense for digestion, absorption, and storage in the form of high energy content bonds is higher for carbohydrates than it is for fats, energy ingestion being equal;[1] and while energy excesses coming from carbohydrates tend to be metabolized by the organism, those coming from fats tend to be deposited as body fat.[2]

This paper reviews those studies which, in the authors' opinion, are most representative of this outlook. It also discusses the thermogenetic properties of alcohol.

DIET COMPOSITION AND BODY WEIGHT

EPIDEMIOLOGICAL DATA

Vegetarian populations are a good study model, permitting us to study the incidence of obesity in subjects with different eating habits living in the same area. Ophir et al.[3] compared members of the Israeli Vegetarian Association with omnivorous subjects living in the Tel Aviv urban area (Table 1). Overweightness was more prevalent in nonvegetarians. Of these subjects, 18% had a relative weight higher than 10%, while 46% of the vegetarians had a relative weight under -10%. Snowdon[4] gives results on alimentary studies (Table 2) performed on Seventh-Day Adventists living in California[5,6] and a comparable group of non-Adventists from the American Cancer Society study.[7] Seventh-Day Adventists are prohibited from consuming alcohol and pork and are discouraged from consuming other foods such as meat, fish, and eggs. Milk consumption is encouraged as a source of protein. In non-Adventists meat consumption was substantially higher, egg consumption was higher, and cheese consumption was slightly higher than in Adventists, who consumed more milk. Nutrient intake was quite different and is well illustrated in Table 3, where a subsample of the population of Adventists was compared with a control group from New York.[8] There was striking evidence for higher fat and lower carbohydrate intake in omnivorous subjects. The association between low incidence of obesity and low fat diet is even clearer when we look at the strictest vegetarians, vegans, who do not consume any foods of animal origin, not even milk and eggs.[9] Vegetarians may be leaner than non-vegetarians because of their lower energy consumption, which is primarily the result of their lower fat intake. It has been demonstrated that caloric intake is lower in a low-fat diet both with obese and normal-weight subjects when this is compared with a high-fat diet.[10-12]

Table 4 compares the values of several anthropometric parameters belonging to the highest and lowest quartiles of percentage energy intake in fats in a representative sample of the patients of a general practitioner in Castel D'Azzano, a town in northern Italy.[13] Higher weight indices are associated with higher energy intake percentages in the form of fat. Similar observations were performed by Tremblay et al.[14] on 244 men and by Romieu et al.[15] on 141

TABLE 1
Relative Weight Distribution in 98 Vegetarians and 98 Controls Living in the Tel Aviv Urban Area

Relative weight (%)	Controls (%)	Vegetarians (%)
<20	1	11
-20 to -10	14.5	35
-10 to +10	66.5	44
+10 to +20	11	7
>20	7	3

Modified from Ophir, O., et al., *Am. J. Clin. Nutr.*, 37, 755, 1983.

TABLE 2
Percent Distribution of Animal Product Consumption in 22,940 Seventh-Day Adventists and 112,726 Non-Adventists

	Days/week	Glasses/day	Seventh-Day Adventists (%)	Non-Adventists (%)
Meat	<1		55	2
	1-3		20	9
	4-6		13	45
	7		6	44
Eggs	<1		17	7
	1-3		53	42
	4-6		20	26
	7		10	25
Milk		<1	25	41
		1-2	55	51
		3+	20	3
Cheese	<1		27	15
	1-3		52	58
	4-6		17	20
	7		5	7

Modified from Snowdon, D.A., *Am. J. Clin. Nutr.*, 48, 739, 1988.

TABLE 3
Nutrient Intake (Percent of Total Energy) of a Sample of 145 Seventh-Day Adventists and 433 Non-Adventists

	Seventh-Day Adventists	Non-Adventists
Total carbohydrate %	55	46
Total protein %	15	13
Total fat %	30	41
Saturated fat %	11	18
Polyunsaturated fat %	10	6
Monounsaturated fat %	9	17

Modified from Snowdon, D.A., *Am. J. Clin. Nutr.*, 48, 739, 1988.

TABLE 4
Anthropometric Measurements in 294 Adult Males of the Lower and Upper Quartiles for Percentage of Energy Intake from Lipids[13]

Variable	Lower quartile	Upper quartile	Statistical significance
Lipids (% energy intake)	28 ± 5	49 ± 6	—
Age (years)	36.5 ± 11.2	38.8 ± 8.5	n.s.
Height (cm)	173.6 ± 6.3	171.9 ± 7	n.s.
Body weight (kg)	74.1 ± 9.9	78.3 ± 12	$p < 0.05$
Relative body weight (%)	112 ± 16	120 ± 14	$p < 0.01$
Body mass index (kg/mm²)	24.7 ± 3.4	26.4 ± 3.1	$p < 0.005$
Tricepital skinfold (mm)	12.1 ± 5.8	13.2 ± 5.4	n.s.
Subscapular skinfold (mm)	19.3 ± 8.4	21.9 ± 9	n.s.
Hypomesogastric skinfold (mm)	15.4 ± 8.6	18.2 ± 7.9	$p < 0.05$
Epimesogastric skinfold (mm)	23.2 ± 9.7	27.6 ± 8.7	$p < 0.01$
Suprailiac skinfold (mm)	19.8 ± 8.2	23 ± 8.2	$p < 0.05$
Epitrochanteric skinfold (mm)	11.6 ± 5.8	13.2 ± 4.6	n.s.

women. Their results also emphasize the importance of considering factors that could confound the relationship between energy and body weight, suggesting that fat intake could play a role in obesity that is independent from total energy intake.

EXPERIMENTAL DATA

Resting metabolic rate, physical activity, and the thermic effect of food are fundamental components of daily energy expenditure. The thermic effect of food appears, as a body weight determinant in function of diet nutritional content, to be involved in a different manner.[16-18]

It has been shown that alimentary carbohydrates induce the depositing of glycogen and that the share that cannot be stored in the form of glycogen is oxidized. Carbohydrates also suppress oxidation of fats. Transformation of glucose into fats, even though possible in theory, is in fact irrelevant.[16,19,20] Fats, on the other hand, do not seem to have any effect on fat oxidation itself.[19] These biochemical phenomena have an important impact on the thermic effect of food, which seems to be closely correlated to carbohydrate intake, while the energy cost for transforming alimentary fats into triglyceride deposits seems to be insignificant.

The law of thermodynamics indicates two pathways that lead to obesity: either increase energy intake or reduce energy expenditure. A third path can also be maintained to exist: change the type of food that is ingested, without modifying energy intake, and do this by incrementing the share that can be stored with least expense — the lipids.[16]

Insignificance of Diet Carbohydrates as a Source of Body Fat

Studies on vegetarians led to the hypothesis that chronic adaptation to a low fat/high carbohydrate diet could promote higher resting and postprandial energy expenditure. Table 5 gives the results of research by Poehlman et al.[21] on vegetarians and non-vegetarians. They paradoxically encountered slightly lower post-prandial thermogenesis in vegetarians with respect to controls. Results from this study seem to suggest that a hyperglucidic diet demonstrates its high thermogenetic properties in an acute manner, and not through chronic adaptation to a diet rich in carbohydrates and poor in fats.

Acheson et al.,[22] in order to simulate the effects of a large but still physiological intake of carbohydrates, gave volunteers a large meal composed of bread, jam, and fruit juice, supplying the equivalent of 479 g of polysaccharides (93% of the energy was in the form of carbohydrates, 5% as proteins, and 2% as fat). Resting metabolism was measured using an open-circuit indirect calorimeter and a ventilated hood. Fat synthesis, as Table 6 shows, did not exceed fat oxidation after this load. This means that when a mixed diet is consumed, the rate of *de novo*

TABLE 5
Characteristics, Nutrient Intake, Resting Metabolic Rate, and Thermic Effect of a Meal in 12 Male Vegetarians and 11 Male Nonvegetarians

	Vegetarians	Nonvegetarians	
Age (years)	27 ± 1.9	22.5 ± 0.9	$p < 0.05$
Height (cm)	1.8 ± 0.6	1.8 ± 2	n.s.
Body mass index (kg/m²)	22.7 ± 0.4	24.4 ± 0.8	n.s.
Energy intake (MJ/d)	13.5 ± 0.8	15.9 ± 1.6	n.s.
Protein % energy	13 ± 1.1	15.8 ± 0.5	$p < 0.05$
Carbohydrates % energy	61.1 ± 3.6	45 ± 3	$p < 0.01$
Fat % energy	26 ± 2.8	39.2 ± 3.2	$p < 0.02$
RMR[a] (kJ/min)	4.77 ± 1.67	4.94 ± 0.17	n.s.
RMR (kJ/kg FFW[b]/h)	4.6 ± 1.26	4.39 ± 0.13	n.s.
TEM[c] (kJ/180 min)	233 ± 14	320 ± 15	$p < 0.01$
TEM (kJ/kg FFW/h)	3.77 ± 2.51	4.77 ± 0.25	$p < 0.01$

[a] Resting metabolic rate.
[b] Fat-free weight.
[c] Thermic effect of a meal.

Modified from Poehlman, E.T., et al., *Am. J. Clin. Nutr.*, 48, 209, 1988.

TABLE 6
Energy and Substrate Balance 10 h after Ingestion of the Test Meal in Six Healthy Male Volunteers

	Intake	Oxidation	Balance
Energy (MJ)	8.99 ± 0.5	3.44 ± 0.14	5.54 ± 0.44
Carbohydrates (g)	479 ± 23	133 ± 12	346 ± 12
Fat (g)	8 ± 4	17 ± 4	-9 ± 6

Modified from Acheson, K.J., et al., *Metabolism*, 31, 1234, 1982.

lipogenesis is not very likely to exceed the concomitant rate of fatty acid oxidation. The conclusion was that dietary carbohydrates do not increase individual fat content by *de novo* lipogenesis.

Acheson et al.[23] also performed an elegant study on healthy volunteers in a respiration chamber to identify the upper limit for glycogen storage. Volunteers went on a 7-d carbohydrate overfeeding protocol after 3 d of low energy, high fat diet plus an exercise program to induce depletion of glycogen storage. The composition of the restricted diet (6.69 MJ) was 15% protein, 75% fat, and 10% carbohydrate. The composition of the diet during the overfeeding period (15.06 MJ) was 11% protein, 3% fat, and 86% carbohydrate. Energy intake was increased day by day to add 6.28 MJ to the previous day's energy expenditure. Energy intake was 20.92 MJ on Day 10. Table 7 gives energy balance results encountered during this study. There was a dramatic increase in carbohydrate oxidation (from 74 to 398 g/d) and in glycogen storage (from -46 to 339 g/d) with the onset of carbohydrate overfeeding (Day 4). Oxidation and storage were no longer sufficient after 2 d of carbohydrate overfeeding (Day 5) and the excess began being converted into fat.

This study shows that human glycogen stores are far from full under normal conditions and that they need to be filled by about 500 g before significant *de novo* lipogenesis begins. Fat synthesis can start when glycogen stores are almost saturated. It was necessary to reach a

TABLE 7
Daily Substrate Balance (g) of Three Volunteers after 3 d (Day 3) of the
Experimental Low Energy, Low Carbohydrate Diet, and 1 (Day 4), 2 (Day
5), 4 (Day 7), and 7 (Day 10) d of the High Energy, High Carbohydrate Diet

	Intake		Oxidation		Balance	
Day	Fat	Carbohydrate	Fat	Carbohydrate	Fat	Carbohydrate
3	117 ± 6	28 ± 16	164 ± 46	74 ± 140	-47 ± 41	-46 ± 48
4	56 ± 19	737 ± 12	49 ± 62	398 ± 87	7 ± 43	339 ± 82
5	97 ± 13	813 ± 78	-30 ± 38	622 ± 96	127 ± 25	192 ± 53
7	78 ± 14	868 ± 58	-81 ± 17	792 ± 83	160 ± 12	76 ± 31
10	69 ± 23	981 ± 43	-149 ± 14	1010 ± 37	218 ± 20	-29 ± 6

Modified from Acheson, K.J., et al., *Am. J. Clin. Nutr.*, 48, 240, 1988.

TABLE 8
Energy Balance of Three Different Types of Breakfast in Seven Healthy Male Volunteers

		TEF[b]			Carbohydrate		Fat	
Meal	REE[a] (kJ/min)	kJ/9 h above baseline	% energy intake	Balance (kJ)	Oxidation[c] (kJ)	Balance (kJ)	Oxidation[c] (kJ)	Balance (kJ)
Low fat, low energy	5.56 ± 0.29	318 ± 42	15.8 ± 2.1	-1297 ± 159	1347	-92 ± 134	1423	-1201 ± 251
High fat	5.52 ± 0.29	402 ± 46	11.2 ± 1.3	205 ± 151	1351	-92 ± 155	1536	251 ± 138
High MCT[d]	5.56 ± 0.25	439 ± 54	12.3 ± 1.5	134 ± 117	1356	-100 ± 92	1548	238 ± 105

[a] Resting energy expenditure.
[b] Thermic effect of food.
[c] The original paper only reported mean values and standard errors as bars in bar diagram.
[d] Medium chain triglycerides.

Modified from Flatt, J.P., et al., *J. Clin. Invest.*, 76, 1019, 1985.

carbohydrate intake of 1 kg/d to achieve significant *de novo* lipogenesis from carbohydrates. This is an experimental load that is never consumed in everyday life. The conclusion was that *de novo* lipogenesis from carbohydrates is an insignificant pathway to lipogenesis in man.

The Easy Path from Dietary Fat to Body Fat

Flatt et al.[2] studied the effects of three meals, with different energy and fat contents, on postprandial oxidation of substrata. Their study was on young students and used a ventilated hood system. The three meals were (1) low energy (2.02 MJ), low fat (11% of energy); (2) normal energy (3.59 MJ), high fat with long chain triglycerides (50% of energy: 50 g of margarine containing long chain triglycerides); (3) normal energy (3.58 MJ), high fat with medium chain triglycerides (50% of energy: 50 g of margarine containing 9 g of long chain and 41 g of medium chain triglycerides). The quantity of carbohydrates (75 g) and proteins (32 g) in the three meals was always the same. Each test was separated from the others by a period of at least 1 week. Addition of diverse types and quantities of fats to a fixed quantity of proteins and carbohydrates did not, during the 9-h calorimetric study (Table 8), lead to any

TABLE 9
Energy Balance Before and After 1 Week of
Overfeeding with MCT[a] or LCT[b] in Ten Male
Volunteers

	Thermic effect of food			
	Day one		Day six	
	Above baseline (kJ)	Energy intake (%)	Above baseline (kJ)	Energy intake (%)
MCT[a]	335 ± 33	8	502 ± 54	12
LCT[b]	243 ± 33	5.8	276 ± 42	6.6

[a] Medium chain triglycerides.
[b] Long chain triglycerides.

Modified from Hill, J.O., et al., *Metabolism*, 38, 641, 1989.

increase in lipid oxidation. When fat was provided the fat balance was positive (251 ± 13 and 238 ± 105 kJ after long and medium chain fat meals, respectively). The balance was negative (-1201 ± 251 kJ) when the breakfast contained very few fatty ingredients. It is interesting to note that energy balances were essentially the same as fat balances. Results demonstrate that fat and carbohydrate oxidation rates are not influenced by the fat content of the meal.

Hill et al.[24] studied the effects of overfeeding using a ventilated face mask to compare medium with long chain triglycerides. Their study involved a group of volunteers that were overfed (150% of estimated energy requirement) for 1 week using a synthetic liquid diet containing 15% proteins, 45% carbohydrates, and 40% medium or long chain triglycerides. No differences in resting metabolic rates were noted between the two groups, neither before nor after 1 week of overfeeding (Table 9). However, the thermic effect of food resulting from a 4.2-MJ mixed meal with the same composition as the basic diet was significantly higher in the group that consumed medium chain triglycerides both before (8 vs. 5.8%) and even more after the week of overfeeding (12 vs. 6.6%). This study shows that the energy cost of conversion of dietary medium chain triglycerides into body fat is higher than the conversion cost for long chain triglycerides. The authors themselves concluded that the difference between the two types of fat only became evident in cases of overfeeding with medium chain triglycerides and that this effect could lose practical significance in normal feeding or restricted diet conditions.

Schutz et al.[25] demonstrated the inability to increase energy expenditure by adding an extra dose of about 4.2 MJ in the form of fats to the normal diet. During the first 24 h, nutrients were provided in proportions similar to those usually consumed by the individuals. During the following day the subjects received four meals containing the same amount of protein and carbohydrates as during the first day of the study, but twice the amount of fat: caloric intakes were increased by 4130 ± 230 kJ/24 h. Table 10 shows study results performed for 2 d in a respiration chamber. The fat supplement was entirely stored (energy balance = +3966 ± 908 kJ/d) without altering 24 h energy expenditure. Results from this and the preceding study further point out the lack of any metabolic response to increase fat oxidation in the presence of an increase in lipid intake.

Particular Aspects in Obese Subjects

Up to this point the differences in use of nutrients can be explained by the greater energy cost in depositing body fat when this is synthesized from carbohydrates rather than when fat

TABLE 10
24-h Energy Balance in Seven Healthy Male Volunteers

Intake (MJ/d)	Expenditure (MJ/d)	Activity count (%)	Energy balance (MJ/d)
First 24-h Period (Mixed Maintenance Diet)			
11.78 ± 0.70	11.64 ± 0.97	16.6 ± 0.7	+0.14 ± 0.75
Second 24-h Period (Fat-Supplemented Diet)			
15.92 ± 0.94	11.8 ± 1.19	16.7 ± 0.7	+4.12 ± 0.91

Modified from Schutz, Y., et al., *Am. J. Clin. Nutr.*, 50, 307, 1989.

TABLE 11
24-h Energy Expenditure at the End of Each Dietary Phase in Six Obese and Five Lean Women

	Basal metabolic rate			
	Baseline	Overfeeding	VLED[a]	Refeeding
Obese (kJ/d)	9,683 ± 1,223	10,184 ± 1,175	8,844 ± 1,180	9,003 ± 1,139
% above the preceding diet		5.2[b]		1.8[b]
% of the energy of the supplement		7.8 ± 4.5		3.6 ± 4.9
Lean (kJ/d)	7,472 ± 703	8,123 ± 570	6,948 ± 548	7,303 ± 620
% above the preceding diet		8.7[b]		5.1[b]
% of the energy of the supplement		14.2 ± 3.4		8.1 ± 3.8

[a] Very low energy diet.
[b] Calculated from the mean values.

Modified from Zed, C.A. and James, W.P.T., *Int. J. Obesity*, 10, 375, 1986.

is simply transferred from foods to fat stores. It is not surprising that we encounter greater obesity in populations that have adopted a fat-rich diet. The susceptibility of individuals to become obese may also be explained by the selective propensity of the pre-obese to store rather than oxidize fat.[26]

Zed and James[27] studied thermogenic responses to overfeeding with fats during a weight-maintenance diet and during a very low energy diet in both normal weight and in obese subjects. Table 11 gives the results of this study measured by whole body calorimetry during a 28-d period: 6 d of weight maintenance diet (16.7% proteins, 41.7% carbohydrates), 6 d of overfeeding (weight maintenance diet + 4.32 MJ: 93.6% fat, 1.3% proteins, 5.1% carbohydrates), 10 d of very low energy diet (5 MJ/d with a composition like that of the weight maintenance diet), and 6 d of refeeding (very low energy diet + the same supplement as with the second period). Measurements were performed at the end of each period. In obese subjects

TABLE 12
Differences in Energy Balance (%) in Five Control and Five
Postobese Subjects Overfed from Low- and High-Fat Energy-
Balance Diets over 24 h

% Change in 24-h energy expenditure			
From low fat (fat: +37%; carbohydrate: +13%)		From high fat (fat: no change; carbohydrate: +50%)	
Control (%)	Postobese (%)	Control (%)	Postobese (%)
-0.1 ± 1.6	1.19 ± 1.26	1.09 ± 1.30	5.79 ± 1.19[a]

[a] Postobese vs. controls: $p < 0.05$.

Modified from Lean, M.E.J., et al., "Obesity Without Overeating? Reduced Diet-Induced Thermogenesis in Post-Obese Women, Dependent on Carbohydrate and Not Fat Intake," in *Obesity in Europe 88*, Björntorp, P. and Rössner, S., Eds., John Libbey, London, 1989.

the increase in energy expenditure in response to the hyperlipidic supplement averaged less than that of controls, especially if it is expressed as percentage of energy supplement (7.8 ± 4.5 vs. 14.2 ± 3.4%). The same phenomena can be observed at the end of the refeeding period subsequent to the very low energy diet (3.6 ± 4.9 vs. 8.1 ± 3.8%). The authors conclude that fat overfeeding does not lead to thermogenesis in excess of the minimum thermodynamic cost of fat storage and that individuals with familial obesity display subnormal thermogenetic response to dietary fat.

Lean et al.[28] used whole body calorimetry to study the effects of overfeeding with carbo-hydrates in normal weight and post-obese weight-stable subjects. Subjects were fed at differ-ent periods for 3 d with two isoenergetic diets: low fat (3% fat, 82% carbohydrate) and high fat (40% fat, 45% carbohydrate). On the fourth day the diet was identical to the high fat diet but with an addition of an extra 50% of energy fed as carbohydrates. Table 12 gives changes to the energy balance that were observed during this last day compared to the preceding days. A significant increase in energy expenditure was only seen in post-obese subjects when carbohydrates were added to the hyperlipidic diet. Overfeeding with carbohydrates conse-quently seems only able to increase energy expenditure if it is added to a hyperlipidic basal diet. Post-obese subjects undergoing a hyperlipidic diet seem more sensitive than do controls to the carbohydrate supplement.

Swaminathan et al.[29] studied the thermic effect of food using the ventilated hood technique and isocaloric meals (1.67 MJ) composed of carbohydrates, fats, proteins, and mixed compo-nents in normal and obese subjects. The thermogenetic response to the mixed meal and the lipid meal was significantly lower in obese subjects than it was in controls (Table 13). The authors also noted that the change in metabolic rate observed in response to the mixed meal was related to that seen in response to fat ($r = 0.62$; $p < 0.05$). This supports the hypothesis that the lower thermic effect in obese subjects in response to a mixed meal is due to its fat content. The authors also observed a significant negative relationship ($r = -0.50$; $p < 0.05$) between body mass index and thermogenetic response to a fat meal.

ALCOHOL AND ENERGY BALANCE

Alcohol is a common dietary component and may cover more than 20% of daily energy intake even in non-alcoholic subjects.[30] The thermogenetic response to carbohydrate, fat, and protein has received intense study. Much less research has been dedicated to the effects of

TABLE 13
Changes in Metabolic Rate during the 30- to
120-Min Period in Response to 1.67 MJ of
Carbohydrate, Fat, Protein, and Mixed Meal
in 11 Lean and 11 Obese Subjects

	Percentage changes (%)		
	Lean subjects	Obese subjects	
Carbohydrate	13.9 ± 2.4	11.5 ± 3.4	n.s.
Fat	14.4 ± 3.4	-0.9 ± 2	$p < 0.05$
Protein	22.5 ± 5	18.7 ± 3.8	n.s.
Mixed meal	25 ± 4.8	12.9 ± 2.3	$p < 0.05$

Note: Metabolic rate after a test meal was compared to metabolic rate after a low energy drink.

Modified from Swaminathan, R., et al., *Am. J. Clin. Nutr.*, 42, 177, 1985.

TABLE 14
Mean Values, Adjusted by Analysis of Covariance for Age, Body
Mass Index, Smoking Habits, Physical Activity, and Menopause, of
Energy and Lipid Intake in a Sample of 294 Males and 307 Females
with Different Alcohol Intake

		Males				Females			
Alcohol intake	(g/d)	0	<16	16-20	>20	0	<10	10-15	>15
Subjects	(n)	26	62	61	80	79	69	57	68
Total energy	(MJ)	9.7	10.1	11.3	14.9[a]	7.6	7.9	8.9	10[a]
Lipids	(g/d)	71	90	106	152[a]	64	72	87	102[a]

[a] $p < 0.001$.

Modified from Armellini, F., et al., *Eur. J. Clin. Nutr.*, 47, 52, 1993.

alcohol on energy metabolism and substrate oxidation. Alcohol ingested in a quantity of 0.75 g/kg in half an hour is totally metabolized in a 4-h time period,[31] but has a much longer lasting effect as far as energy metabolism and substrate oxidation are concerned.[32,33] The quantity of alcohol habitually ingested, the composition of the diet, the duration of studies, and the dose of alcohol employed in these metabolic studies: these appear to be the main factors explaining differences observed in several different studies.

POPULATION STUDIES

Armellini et al.[34] studied a 50% random sample of a general practitioner's patients. Energy and lipid intake increased alongside increasing alcohol intake (Table 14) both in males and females after taking into account several possible confounding factors, principally the body mass index. No associations were observed between carbohydrates and alcohol intake. These results, and those regarding energy intake in particular, agree with the results of other authors[35-40] and show that subjects with greater alcohol intake are in some way protected against weight increase even in the presence of high energy intakes. The qualitative character of the diet is especially interesting: the diet, being hyperlipidic, should have been more fattening.[1]

TABLE 15
Body Weight and Energy Expenditure Before and at the
End of the Experiment Diets With or Without Alcohol in
Eight Healthy Males

	Body weight (kg)	Resting metabolic rate (kJ/min)	Postprandial energy expenditure (kJ/4 h)
Basal	74.6 ± 4.2	4.887 ± 0.56	—
Without alcohol	74.3 ± 4.6	4.544 ± 0.47	221.8 ± 45.9
With alcohol	74.3 ± 4.1	4.544 ± 0.47	210.1 ± 76.1

Modified from Contaldo, F., et al., *Metabolism,* 38, 166, 1989.

CLINICAL STUDIES

Contaldo et al.[41] studied eight healthy middle-aged men. Four subjects underwent an isocaloric diet (20% protein, 50% carbohydrate, 30% fat) without alcohol and four subjects an isocaloric diet with wine (12% protein, 29% carbohydrate, 25% fat, 75 g of alcohol served as 750 ml of red wine). The groups were inverted after 2 weeks. Resting energy expenditure was measured using a ventilated hood system both during fasting and after a 3.77-MJ mixed meal (15% protein, 55% carbohydrate, 30% fat). The measurement was protracted for 4 h and calculated as the incremental area above the resting value. Table 15 shows that body weight, resting metabolic rate, and postprandial thermogenesis were similar after both diets. The authors concluded that short-term moderate alcohol intake does not seem to have any adverse effects on energy metabolism. The failure of this study to demonstrate any differences between drinkers and non-drinkers was probably mainly due to the fact that the test meal was performed more than 10 h after the last meal with alcohol.

Prentice et al.[33] studied a group of moderate male drinkers in a whole body calorimeter on three occasions. Three different test meals were ingested at 12 o'clock noon after an overnight fast. These provided half of the maintenance energy expenditure. Fat, carbohydrate, protein, and alcohol compositions were as follows: controls, 40:46:14:0; isoenergetic replacement of 50% of carbohydrate with alcohol, 40:23:14:23; addition of alcohol, 34:36:12:18. Endogenous and exogenous fat oxidation was considered by a radioisotopic technique that collected breath samples. Breath alcohol levels were also measured. Subjects received a supper providing one third of the maintenance energy expenditure with the composition of the control meal at 7:00 P.M. and measurements continued until 9:00 A.M. the following day. Table 16 gives the effects of alcohol (38 g as Calvados in each of the alcohol meals) on postprandial thermogenesis and the oxidation rate of substrates. The alcohol was completely oxidized in the first 6 h. During this period fat oxidation was suppressed and lipogenesis was induced after the meal in which the alcohol was added. Carbohydrate oxidation was related to carbohydrate intake since it was blunted only when carbohydrates were replaced with alcohol. In the replacement study, alcohol reduced fat oxidation during the first 6 h. Afterwards fat oxidation increased probably as a consequence of carbohydrate deficiency. The overall consequence was that fat oxidation was unaffected. On the contrary, fat oxidation remained significantly reduced in the alcohol addition study. Postprandial thermogenesis was similar between the control group and the group with isoenergetic substitution of carbohydrates with alcohol in the first 6 h. A delayed positive effect of alcohol on thermogenesis was evident even when alcohol was completely oxidized. The net effect at the end of the study, despite documented prolonged inhibition of lipolysis in the subjects in which alcohol was added to the diet, was an increased energy expenditure. These results agree with those of other studies,[42-44] demonstrating that addition of 2.1 to 2.7 MJ of excess daily energy as alcohol over 3 to 4 weeks was not associated with

TABLE 16
Effects of Alcohol on Postprandial Thermogenesis and on Substrate Oxidation in Five Males on Three Different Types of Diet

	Test meals		
	Control	50% carbohydrate replacement by alcohol	Alcohol addition
Postprandial Thermogenesis			
0-6 h (kJ)	2301 ± 74	2333 ± 163	2460 ± 84[a]
6-20.5 h (kJ)	4426 ± 385	4774 ± 350[a]	4729 ± 213[a]
0-20.5 h (kJ)	6727 ± 443	7107 ± 509[a]	7189 ± 309[a]
Alcohol Oxidation			
0-6 h (kJ)	0	1142[a]	1135[a]
6-20.5 h (kJ)	0	0	0
0-20.5 h (kJ)	0	1142[a]	1135[a]
Carbohydrate Oxidation			
0-6 h (kJ)	1326	817[a,b]	1112
6-20.5 h (kJ)	2372	2087	2636
0-20.5 h (kJ)	3698	2904[a]	3748
Fat Oxidation			
0-6 h (kJ)	619	80[a]	-125[a]
6-20.5 h (kJ)	1400	2010[a,b]	1345
0-20.5 h (kJ)	2019	2090	1220[b,c]

[a] $p < 0.05$ vs. control.
[b] $p < 0.05$ vs. alcohol addition.
[c] $p < 0.05$ vs. carbohydrate replacement.

Modified from Prentice, A.M., et al., "Overview: Energy Requirements and Energy Storage," in *Energy Metabolism. Tissue Determinants and Cellular Corollaries,* Kinney, J.M. and Tucker, H.N., Eds., Raven Press, New York, 1992, 211.

weight gain. There does seem to be an interaction between alcohol and food with a prolongation and amplification of the thermic response from about 1 h postprandially onward.[45,46]

Suter et al.[32] studied the effect of alcohol on 24-h substrate oxidation in a group of males during two 48-h sessions in an indirect-calorimetry chamber. The first 24 h served as control. On the second day of one session, an additional 25% of the total energy requirement was added as alcohol (96 ± 4 g/d). During the other session, 25% of the total energy requirement was replaced by alcohol, which was isocalorically substituted for lipids and carbohydrates. Both addition and substitution of alcohol (Table 17) caused increased energy expenditure and reduced lipid oxidation. Substitution of alcohol alone caused reduction in carbohydrate oxidation. Addition of alcohol to the diet resulted in a positive energy balance as compared with the control day. This was not the case as far as substitution of alcohol is concerned — here the trend was negative. The authors conclude that long-term ingestion of alcohol in place of other foods can lead to loss of body weight. On the contrary, the ingestion of alcohol as additional energy above nutritional requirements decreases lipid oxidation and favors lipid storage and, as a consequence, there is weight gain.

TABLE 17

**Daily Energy Expenditure and Substrate-Oxidation Rate in Eight
Men Before and During the Ingestion of Ethanol**

	Session 1		Session 2	
	Control	Ethanol added	Control	Ethanol substituted
Energy intake MJ/24 h	11.4 ± 0.2	14.3 ± 0.2	11.4 ± 0.2	11.4 ± 0.2
Energy expenditure MJ/24 h	11.8 ± 0.3	12.6 ± 0.2[a]	11.9 ± 0.2	12.4 ± 0.3[b]
Balance kJ/24 h	-381 ± 166	1688 ± 120	498[c]	841[c]
Fat oxidation kJ/24 h	5439 ± 443	3487 ± 276[a]	5401 ± 321	3660 ± 322[d]
Carbohydrate oxidation kJ/24 h	4308 ± 360	4150 ± 261	4482 ± 201	3798 ± 191[e]
Protein oxidation kJ/24 h	2081 ± 69	2134 ± 80	2056 ± 48	2136 ± 106
Ethanol oxidation kJ/24 h	0	2843 ± 44	0	2847 ± 42

[a] $p < 0.001$ vs. control.
[b] $p < 0.025$ vs. control.
[c] Difference between mean values of intake and expenditure.
[d] $p < 0.0025$ vs. control.
[e] $p < 0.05$ vs. control.

Modified from Suter, P.M., et al., *N. Engl. J. Med.*, 326, 983, 1992.

CONCLUSIONS

FAT

Appetizing high-calorie foods are easily available in the developed nations. It has been a common opinion that the availability of these foods is an important cause of obesity.[47] These eating habits are characterized by a low complex carbohydrate content, and consequently are poor in fibers and rich in fats and sugars. This type of diet has a high energy concentration, low volume, and by its very nature can facilitate onset of obesity even when energy consumption is not excessive.[4] Carbohydrates are stored after transformation into fats at a cost, in energy, equal to 25% of their energy contents. Fats, on the other hand, are deposited at a cost of only 3%.[1] In reality carbohydrates, in physiological conditions, are not transformed into fats. Proof of this is that the composition of the fat in the adipose tissue reflects that of the diet.[1,19]

Studies that aim at underlining the practical utility of low-fat diets to cure obesity are extremely difficult to enact. Kendall et al.[11] studied 13 healthy unrestrained-eater women to measure the effect of varying the macronutrient composition of the diet on intake and body weight. Subjects were randomly assigned to either a low fat (20 to 25% energy as fat) or a control diet (35 to 40% fat) with the same palatability for 11 weeks. After a 7-week washout each subject received the alternate treatment. Subjects and food intake were strictly controlled during the entire study. Table 18 shows study results. Pooling the data from both periods,

TABLE 18
Initial Weight and Weight Loss in 13
Women on Two Different Diets

	Control diet	Low fat diet
Period 1		
Subjects (n)	8	5
Initial weight (kg)	61.4 ± 7.2	65.4 ± 11.6
Weight loss (kg)	1.2 ± 0.5	3.7 ± 0.8
Period 2		
Subjects (n)	5	8
Initial weight (kg)	63.2 ± 9.7	60.9 ± 8.6
Weight loss (kg)	1.4 ± 0.6	1.8 ± 0.5

Modified from Kendall, A., et al., *Am. J. Clin. Nutr.*, 53, 1124, 1991.

weight loss was significantly greater on the low-fat diet than on the control diet ($p = 0.04$). Strong linear relationships between initial weight and amount of weight lost was found with the low-fat diet ($p = 0.01$) but not with the control diet. These results suggest that the use of a palatable low-fat diet may be an effective means of achieving weight reduction even when no limitations are placed on the quantity consumed.

Obesity, consequently, could be prevented or treated with greater success by adopting a cereal- and vegetable-rich diet, generally, according to the Mediterranean model.[48] This is especially so when considering studies that testify to the greater "vulnerability" toward alimentary fats of obese or ex-obese subjects.[26-29]

ALCOHOL

Alcohol combustion in a bomb calorimeter releases 29.3 kJ/g. Alcohol use by the human body, however, seems to be quite different. Clinical studies[32,33] agree in affirming that alcohol intake inhibits fat oxidation and that this phenomenon could favor storage of triglycerides and consequently lead to weight gain. On the other hand, these same studies evidence a total increase of energy expenditure over a 24-h period in subjects who ingest alcohol compared to controls. This increase cannot be simply explained by studying the combustion of fundamental nutrients. The phenomenon becomes dramatically evident in population studies[34] where greater alcohol consumption demonstrates a wasting effect, associated not only with higher energy intake, but, paradoxically, with a greater lipid intake. Lieber recently confronted this problem. In a lucid review[40] he suggested that the phenomenon is closely related to the drinking habits of the subjects being studied. Social drinkers, meaning women who do not ingest more than one drink and men who do not ingest more than two drinks per day, metabolize alcohol by means of alcohol dehydrogenase. This path leads to efficient use of the energy that comes from alcohol. Alcohol dehydrogenase, in fact, transforms alcohol into acetaldehyde by synthesis of high energy compounds. At higher doses the microsomal alcohol oxidizing system is activated, transforming alcohol into acetaldehyde without forming high energy compounds but only generating heat. It is clear that these mechanisms are also activated by light enological aliments, as is the case in studies described by Suter et al.[32] and by Prentice et al.[33] Many other energy-wasting phenomena have also been called into the picture to explain the energy deficit, which becomes increasingly great as the alcohol intake increases.[40] Autoptic studies by Hutten and Kortelainen[50] even suggest that chronic alcohol intake can induce white perivascular fat to transform into brown fat. The experimental observation that an alcohol-induced energy deficit cannot be observed with a low-fat diet[51] is of great interest, suggesting that chronic alcohol consumption may be associated with a decrease in energy derived from fat oxidation in the liver. As a consequence the energy deficit cannot be elicited by a low-fat diet. In fact, those who have greater alcohol and energy intakes also have diets that are qualitatively richer in fats.[34] The mood-altering effects of alcohol could be extended by a fat-rich diet. It has, in fact, been demonstrated that the liver damage caused by alcohol is greater when alcohol is associated with a fat-rich diet.[52,53] However, the potential that the energy-wasting power of alcohol may have to control body weight must be considered in the framework of the ethical, social, and legal aspects of alcohol consumption.[54]

REFERENCES

1. Danforth, E., Jr., Obesity and thermogenesis, in *Diabetes 1988,* Larkins, R., Zimmet, P., and Chisholm, D., Eds., Elsevier, New York, 1989, 301.
2. Flatt, J.P., Ravussin, E., Acheson, K.J., and Jéquier, E., Effects of dietary fat on postprandial substrate/ oxidation and on carbohydrate and fat balances, *J. Clin. Invest.,* 76, 1019, 1985.
3. Ophir, O., Peer, G., Gilad, J., Blum, M., and Aviram, A., Low blood pressure in vegetarians: the possible role of potassium, *Am. J. Clin. Nutr.,* 37, 755, 1983.
4. Snowdon, D.A., Animal product consumption and mortality because of all causes combined, coronary heart disease, stroke, diabetes, and cancer in Seventh-Day Adventists, *Am. J. Clin. Nutr.,* 48, 739, 1988.
5. Snowdon, D.A., Phillips, R.L., and Choi, W., Diet, obesity, and risk of fatal prostate cancer, *Am. J. Epidemiol.,* 120, 244, 1984.
6. Phillips, R.L., Kuzma, J.W., Beeson, W.L., and Lotz, T., Influence of selection versus lifestyle on risk of fatal cancer and cardiovascular disease among Seventh-Day Adventists, *Am. J. Epidemiol.,* 112, 296, 1980.
7. Hammond, E.C., Smoking in relation to the death rates of one million men and women, *Natl. Cancer Inst. Monogr.,* 19, 127, 1966.
8. Walden, R.T., Schaefer, L.S., Lemon, F.R., Sunshine, A., and Wynder, E.L., Effect of environment on the serum cholesterol-triglycerides distribution among Seventh-Day Adventists, *Am. J. Med.,* 36, 269, 1964.
9. Dwyer, J., Health aspects of vegetarian diets, *Am. J. Clin. Nutr.,* 48, 712, 1988.
10. Duncan, K.H., Bacon, J.A., and Weinsier, R.L., The effects of high and low energy density diets on satiety, energy intake and eating time of obese and nonobese subjects, *Am. J. Clin. Nutr.,* 37, 763, 1983.
11. Kendall, A., Levitsky, D.A., Strupp, B.J., and Lissner, L., Weight loss on a low fat diet: consequence of the imprecision of the control of food intake in humans, *Am. J. Clin. Nutr.,* 53, 1124, 1991.
12. Lissner, L., Levitsky, D.A., Strupp, B.J., Kalkwarf, H.J., and Roe, D.A., Dietary fat and the regulation of energy intake in human subjects, *Am. J. Clin. Nutr.,* 46, 886, 1987.
13. Armellini, F., unpublished data, 1990.
14. Tremblay, A., Plourde, G., Despres, J.-P., and Bouchard, C., Impact of dietary fat content and fat oxidation on energy intake in humans, *Am. J. Clin. Nutr.,* 49, 799, 1989.
15. Romieu, I., Willett, W.C., Stampfer, M.J., Colditz, G.A., Sampson, L., Rosner, B., Hennekens, C.H., and Speizer, F.E., Energy intake and other determinants of relative weight, *Am. J. Clin. Nutr.,* 47, 406, 1988.
16. Ravussin, E. and Swinburn, B.A., Energy metabolism, in *Obesity: Theory and Therapy,* Stunkard, A.J. and Wadden, T.A., Eds., Raven Press, New York, 1993, chap. 6.
17. Acheson, K.J., Schutz, Y., Bessard, T., Ravussin, E., Jéquier, E., and Flatt, J. P., Nutritional influences on lipogenesis and thermogenesis after a carbohydrate meal, *Am. J. Physiol.,* 246, E62, 1984.
18. Jéquier, E., Calorie balance versus nutrient balance, in *Energy Metabolism: Tissue Determinants and Cellular Corollaries,* Kinney, J.M. and Ticker, H.N., Eds., Raven Press, New York, 1992, 123.
19. Abbott, W.G.H., Howard, B.V., Christin, L., Freymond, D., Lillioja, S., Boyce, V.L., Anderson, T.E., Bogardus, C., and Ravussin, E., Short term energy balance: relationship with protein, carbohydrate, and fat balances, *Am. J. Physiol.,* 255, E332, 1988.
20. Felber, J.P., Ferranini, E., Golay, A., Meyer, H.U., Theibaud, D., Curchod, B., Maeder, E., Jéquier, E., and De Fronzo, R.A., Role of lipid oxidation in pathogenesis of insulin resistance of obesity and type II diabetes, *Diabetes,* 36, 1341, 1987.
21. Poehlman, E.T., Arciero, P.J., Melby, C.L., and Badylak, S.F., Resting metabolic rate and postprandial thermogenesis in vegetarians and nonvegetarians, *Am. J. Clin. Nutr.,* 48, 209, 1988.
22. Acheson, K.J., Flatt, J.P., and Jéquier, E., Glycogen synthesis versus lipogenesis after 500 gram carbohydrate meal in man, *Metabolism,* 31, 1234, 1982.

23. **Acheson, K.J., Schutz, Y., Bessard, T., Anantharaman, K., Flatt, J.-P., and Jéquier, E.,** Glycogen storage capacity and de novo lipogenesis during massive carbohydrate overfeeding in man, *Am. J. Clin. Nutr.,* 48, 240, 1988.

24. **Hill, J.O., Peters, J.C., Yang, D., Sharp, T., Kaler, M., Abumrad, N.N., and Greene, H.L.,** Thermogenesis in humans during overfeeding with medium-chain triglycerides, *Metabolism,* 38, 641, 1989.

25. **Schutz, Y., Flatt, J.P., and Jéquier, E.,** Failure of dietary fat intake to promote fat oxidation: a factor favoring the development of obesity, *Am. J. Clin. Nutr.,* 50, 307, 1989.

26. **James, W.P.T., McNeill, G., and Ralph, A.,** Metabolism and nutritional adaptation to altered intakes of energy substrates, *Am. J. Clin. Nutr.,* 51, 264, 1990.

27. **Zed, C.A. and James, W.P.T.,** Dietary thermogenesis in obesity: fat feeding at different energy intakes, *Int. J. Obesity,* 10, 375, 1986.

28. **Lean, M.E.J., James, W.P.T., and Garthwaite, P.H.,** Obesity without overeating? Reduced diet-induced thermogenesis in post-obese women, dependent on carbohydrate and not fat intake, in *Obesity in Europe 88,* Björntorp, P. and Rössner, S., Eds., John Libbey, London, 1989, chap. 43.

29. **Swaminathan, R., King, R.F.G.J., Holmfield, J., Siwek, R.A., Baker, M., and Wales, J.K.,** Thermic effect of feeding carbohydrate, fat, protein and mixed meal in lean and obese subjects, *Am. J. Clin. Nutr.,* 42, 177, 1985.

30. **Mitchell, M.C. and Herlong, H.F.,** Alcohol and nutrition: caloric value, bioenergetics, and relationship to liver damage, *Annu. Rev. Nutr.,* 6, 457, 1986.

31. **Shelmet, J.J., Reichard, G.A., Skutches, C.L., Hoeldtke, R.D., Owen, O.E., and Bosen, G.,** Ethanol causes acute inhibition of carbohydrate, fat and protein oxidation and insulin resistance, *J. Clin. Invest.,* 81, 1137, 1988.

32. **Suter, P.M., Schutz, Y., and Jéquier, E.,** The effect of ethanol on fat storage in healthy subjects, *N. Engl. J. Med.,* 326, 983, 1992.

33. **Prentice, A.M., Stubbs, R.J., Sonko, B.J., Diaz, E., Goldberg, G.R., Murgatroyd, P.R., and Black, A.E.,** Overview: energy requirements and energy storage, in *Energy Metabolism. Tissue Determinants and Cellular Corollaries,* Kinney, J.M. and Tucker, H.N., Eds., Raven Press, New York, 1992, 211.

34. **Armellini, F., Zamboni, M., Frigo, L., Mandragona, R., Robbi, R., Micciolo, R., and Bosello, O.,** Alcohol consumption, smoking habits and body fat distribution in Italian men and women aged 20-60 years, *Eur. J. Clin. Nutr.,* 47, 52, 1993.

35. **McDonald, J.T. and Margen, S.,** Wine versus ethanol in human nutrition. I. Nitrogen and calorie balance, *Am. J. Clin. Nutr.,* 29, 1093, 1976.

36. **Gruchow, H.W., Sobocinskii, K.A., Barboriak, J.J., and Scheller, J.G.,** Alcohol consumption, nutrient intake and relative body weight among US adults, *Am. J. Clin. Nutr.,* 42, 289, 1985.

37. **Camargo, C.A., Vrazinan, K.M., Dreon, D.M., Frey-Hewitt, B., and Wood, P.D.,** Alcohol, caloric intake, and adiposity in overweight men, *J. Am. Coll. Nutr.,* 6, 271, 1987.

38. **Williamson, D.F., Forman, M.R., Binkin, N.J., Gentry, E.M., Remington, P.L., and Trowbridge, F.L.,** Alcohol and body weight in United States adults, *Am. J. Public Health,* 77, 1324, 1987.

39. **Colditz, G.A., Giovannucci, E., Rimm, E.B., Stampfer, M.J., Rosner, B., Speizer, F.E., Gordis, E., and Willet, W.C.,** Alcohol intake in relation to diet and obesity in women and men, *Am. J. Clin. Nutr.,* 54, 49, 1991.

40. **Lieber, C.S.,** Perspectives: do alcohol calories count?, *Am. J. Clin. Nutr.,* 54, 976, 1991.

41. **Contaldo, F., D'Arrigo, E., Carandente, V., Cortese, C., Coltorti, A., Mancini, M., Taskinen, M., and Nikkilä, E.A.,** Short-term effects of moderate alcohol consumption on lipid metabolism and energy balance in normal men, *Metabolism,* 38, 166, 1989.

42. **Bertiere, M.C., Betoulle, D., Apfelbaum, M., and Girard-Globa, A.,** Time-course magnitude and nature of the changes induced in HDL by moderate alcohol intake in young non-drinking males, *Atherosclerosis,* 61, 7, 1986.

43. **Crouse, J.R. and Grundy, S.M.,** Effects of alcohol on plasma lipoproteins and cholesterol and triglycerides metabolism in man, *J. Lipid Res.,* 25, 486, 1984.

44. **Välimäki, M., Taskinen, M.-R., Ylikahri, R., Roine, R., Kuusi, T., and Nikkilä, E.A.,** Comparison of the effects of two different doses of alcohol on serum lipoproteins, HDL-subfractions and apolipoproteins A-I and A-II: a controlled study, *Eur. J. Clin. Invest.,* 18, 472, 1988.

45. **James, W.P.T.,** From SDA to DIT to TEF, in *Energy Metabolism. Tissue Determinants and Cellular Corollaries,* Kinney, J.M. and Tucker, H.N., Eds., Raven Press, New York, 1992, 163.

46. **Rosemberg, K. and Durnin, J.V.G.A.,** The effect of alcohol on resting metabolic rate, *Br. J. Nutr.,* 40, 293, 1978.

47. **U.S. Department of Agriculture,** Consumption of food in the US 1909-1952, Agricultural Handbook 62 (Suppl.), U.S. Government Printing Office, Washington, D.C., 1962.

48. **Bosello, O., Armellini, F., and Zamboni, M.,** Obesity, in *The Mediterranean Diet in Health and Disease,* Spiller, G. A., Ed., Van Nostrand Reinhold, New York, 1990, 252.

49. **Oscai, L.B. and Miller, W.C.,** Dietary-induced severe obesity: exercise implications, *Med. Sci. Sports Exercise,* 18, 6, 1986.
50. **Hutten, P. and Kortelainem, M.L.,** Long-term alcohol consumption and brown adipose tissue in man, *Eur. J. Appl. Physiol.,* 60, 418, 1990.
51. **Reinus, J.F., Heymsfield, S.B., Wiskind, R., Casper, K., and Galambos, J.T.,** Ethanol: relative fuel value and metabolic effects in vivo, *Metabolism,* 38, 125, 1989.
52. **Lieber, C.S. and DeCarli, L.M.,** Quantitative relationship between the amount of dietary fat and the severity of the alcoholic fatty liver, *Am. J. Clin. Nutr.,* 23, 474, 1970.
53. **Lieber, C.S. and DeCarli, L.M.,** Hepatotoxicity of ethanol, *J. Hepatol.,* 12, 404, 1991.
54. **Ruderman, A.J.,** Dietary restraint: a theoretical and empirical review, *Psychol. Bull.,* 99, 247, 1986.

Chapter 2.8

DIETARY LIPIDS AND COLON CANCER

Bandaru S. Reddy

INTRODUCTION

During the past three decades, epidemiologic studies have demonstrated the influence of the environment and life styles on the development of certain forms of cancer.[1] These encouraging observations have led several investigators to identify the ways in which the environmental factors increase the risk of cancer and to manipulate the environment and reduce the risk to a minimum. The observation that the causes of several types of cancer may be mainly environmental in origin has led to a great deal of interest in the role of diet in cancer etiology. Although the exact proportion of diet involvement in cancer causation is unknown, Wynder and Gori[1] estimated that as many as 60% of cancers in women and more than 40% of cancers in men are related to nutritional factors. Doll and Peto[2] also estimated that about 35% (range 10 to 70%) of all cancer mortality in the U.S. may be attributable to dietary factors. Because few relationships between specific dietary components and cancer risk are well established, it is not possible to quantify precisely the contribution of diet to individual cancers and thus to total cancer rates.[3] Nevertheless, these estimates help to emphasize the importance of nutritional factors in the etiology and prevention of several types of cancer in the U.S. and other countries.[3]

The literature relating dietary fat to cancer has been reviewed several times in recent years.[4,5] The purpose of this review is to provide an overview of the results thus far generated in laboratory animal model studies on the relationship between dietary fat and the development of colon cancer. Although the primary emphasis of this review article will be on laboratory animal model studies, reference will be made to human studies where appropriate.

DIETARY FAT AND CANCER OF COLON

A variety of compounds, namely, 1,2-dimethylhydrazine (DMH), azoxymethane (AOM), methylazoxymethanol acetate (MAM acetate), 3,2'-dimethyl-4-aminobiphenyl (DMAB), methylnitrosourea (MNU), and N-methyl-N'-nitro-N-nitrosoguanidine (MNNG) that are carcinogenic for the colon have been used in a number of animal models to study the effect of dietary fat on tumorigenesis at the site.[6,7] Additionally, these animal models have been used as unique tools for systematic studies of risk factors observed in human setting and for determining whether or not suspected etiologic factors can be reproduced under controlled laboratory conditions. Thus, as is described here, a number of major elements observed in humans could not have been established without careful, deliberate investigations carried out in laboratory animals.

Dietary fat has received considerable attention as a possible risk factor in the etiology of colon cancer. Based on Japanese data and case-control studies, Wynder et al.[8] proposed in the late 1960s that colon cancer incidence was mainly associated with dietary fat. They further suggested that dietary fat influenced the composition of the gut flora and thus may be involved in the pathogenesis of cancer of the colon. These pioneering studies led to several correlation and case-control studies on the relationship of dietary fat and colon cancer. Since then, a substantial amount of progress has been made in understanding the relationship between dietary factors and the development of colon cancer in humans. Recent case-control studies have demonstrated that the colon cancer risk enhances with increased intake of dietary fat.[10-12]

0-8493-4248-1/96/$0.00+$.50
© 1996 by CRC Press Inc.

Laboratory animal model studies provided evidence that not only the amount but also the types of fat (differing in fatty acid composition) are important factors in determining the enhancing effect of this nutrient in colon tumor development.[7] In addition, the stage of carcinogenesis at which the effect of dietary fat is exerted appears to depend on the fatty acid composition.[5] In several earlier animal model studies on dietary fat and colon cancer, interpretation of results between high-fat and low-fat diets was complicated by the use of diets of varying caloric density and confounded by different intakes of other nutrients. For example, in experiments comparing a relatively low-fat diet with a high-fat diet, there will be major changes in the consumption of other dietary components unless the content has been changed isocalorically. Generally, laboratory animals adjust their food intake so that similar energy intake is maintained even with diets containing substantially different energy density. Therefore, a diet with a low energy value per unit weight (low fat) will be consumed at a greater rate than a diet with more highly concentrated energy (high fat). Accordingly, in addition to the changes in fat and carbohydrate intake, the intakes of protein, minerals, vitamins, and fiber will be lower in animals fed the high-fat diet. Therefore, it is necessary to formulate high-fat diets that would ensure an intake of protein, vitamins, minerals, and fiber comparable to low-fat diets so that the effect of high fat on tumorigenesis can be measured.

With the above limitations, numerous experiments in animal models have shown that certain dietary lipids influence tumorigenesis in the colon. This promoting effect can be modified, however, by the type of dietary fat. In general, the overall evidence from the animal model studies is consistent with the epidemiological data.

EFFECT OF TYPES AND AMOUNT OF FAT

Several early experiments indicated a strong relationship between the amount of dietary fat and colon cancer in animal models.[12-15] Investigations were also carried out to test the effect of diets comprising 20 and 5% beef fat on colon carcinogenesis by a variety of carcinogens — DMH, MAM acetate, DMAB, or MNU which differ in metabolic activation.[16,17] Combined results of these two studies indicate that, irrespective of colon carcinogens used, animals fed the diet containing a high amount of beef fat had a greater incidence of colon tumors than did rats fed a low beef-fat diet.

Nigro et al.[12] studied the effect of diets containing 5 and 35% beef fat on azoxymethane (AOM)-induced intestinal tumors in Sprague-Dawley rats. Animals fed the high beef-fat diet developed more intestinal tumors and more metastases in the abdominal cavity, lungs, and liver than did rats fed the low-fat diet. Howarth and Pihl[14] demonstrated that DMH-induced colon tumors were increased in male D/A rats fed a high-fat diet compared with those in such rats fed a low-fat diet. Studies conducted in our laboratory indicate that animals fed 20% lard or 20% corn oil diets were more susceptible to DMH-induced colon tumors compared with those fed 5% lard or 5% corn oil diets.[18] The type of fat appears to be immaterial at the 20% level; however, at the 5% fat level, unsaturated fat (corn oil) induced more colon tumors than did saturated fat (lard). Studies by Nutter et al.[19] demonstrated that Sprague-Dawley rats fed the beef-fat diet had a higher colon tumor incidence than those fed the corn oil diet. Sakaguchi et al.[20] demonstrated a significantly higher incidence of colon tumors in rats fed 5% linoleic acid than in those fed 4.7% stearic acid plus 0.3 linoleic acid. Another study by Pence and Buddingh[21] indicated that DMH-induced colon carcinogenesis was increased in F344 rats fed a high-fat (corn oil) diet. However, Nauss et al.[22] showed that a high-fat diet containing 24% corn oil, 24% beef fat, or 24% Crisco had no promoting effect on colon tumors induced by DMH in rats.

The role of type and amount of dietary fat in colon carcinogenesis has been studied in several laboratories.[23-25] In these studies, the effect of corn oil, coconut oil, olive oil, safflower oil, and fish oil was investigated. The animals fed the high corn oil or high safflower oil (23.5%) diets had a higher incidence of AOM-induced colon tumors than did those fed the diets low in fat (5%). By contrast, diets high in coconut oil, olive oil, or menhaden fish oil had no colon tumor-promoting effect compared with diets high in corn oil or safflower oil. The varied effects of different types of fat on colon carcinogenesis suggest that the fatty acid composition is one of the determining factors in colon tumor promotion. Corn oil and safflower oil are very high in omega-6 fatty acids (linoleic acid), olive oil is rich in monounsaturated fatty acid (oleic acid), coconut oil is high in medium chain fatty acids (lauric acid), and fish oil is rich in omega-3 fatty acids (docosahexanoic acid and eicosapentanoic acid).

Although high dietary fish oil (22.5%) inhibited colon carcinogenesis in the above study,[24] large amounts of fish oil may induce a variety of physiopathological conditions and harmful side effects. Therefore, the author has conducted another study to investigate the efficacy of varying levels of fish oil and corn oil on colon carcinogenesis to determine the optimum dietary levels at which the combination of two sources of fat elicit maximum inhibition.[25] In this study, feeding of high-fat diets containing 17.6% corn oil + 5.9% fish oil, 11.8% corn oil +11.8% fish oil, or 5.9% corn oil +17.6% fish oil significantly inhibited colon carcinogenesis as compared to 23.5% corn oil diet. High-fat diets containing high levels of fish oil and low levels of corn oil induced fewer colon adenocarcinomas than did the diet containing high corn oil alone.

There was an increase in per capita consumption of fat for the last 50 years in the U.S. and this increase can be partly accounted for by a trend toward greater use of vegetable oils, which include partially hydrogenated fats and processed vegetable oils that contain *trans* fatty acids. The most common isomers that occur in processed fats are 9-*trans*-18:1 (elaidic acid) and *cis-trans*-octadecadienoic acid. Experiments conducted in a colon cancer model failed to support the hypothesis that *trans* fatty acids increase tumorigenesis.[26] High-fat diets (23.5%) containing low *trans* fat (5.8% corn oil + 5.9% *trans* fat + 11.8% oleate), intermediate level of *trans* fat (5.8% corn oil + 11.8% *trans* fat + 5.9% oleate), high level of *trans* fat (5.8% corn oil + 17.6% *trans* fat), and no *trans* fat (23.5% corn oil) were tested for colon tumor-promoting effect.[26] Animals fed the high-fat diets containing different levels of *trans* fat developed significantly fewer colon tumors than did the rats fed the high corn oil diet. Thus, high *trans* fat had little or no promoting effect in colon carcinogenesis when used in a diet that contained about 5.9% corn oil.

In summary, laboratory animal studies have provided useful data for evaluating the role of dietary fat in the development of colon carcinogenesis. The majority of studies in colon cancer models in which the intakes of all nutrients and total calories were controlled in both high- and low-fat diet groups clearly suggest that not only the amount but also the type of fat (differing in fat acid composition) are important factors in determining the modulating effect of dietary fats in colon tumor development. The stage of carcinogenesis at which the effect of dietary fat is exerted appears to be mostly during the promotional phase of carcinogenesis; however, certain dietary fats such as lard or fish oil also act during the initiation phase of colon carcinogenesis. The strength of the association between dietary fat and colon cancer risk, the experimental evidence, as well as the biological plausibility indicate that these associations are real.

ACKNOWLEDGMENTS

The experiments discussed here were supported by USPHS grants CA-17613 and CA-37663 from The National Cancer Institute.

TABLE 1
Studies on the Colon Tumor-Enhancing (Promoting) Effect of High Dietary Fats in Experimental Animals

Carcinogen and dosage	Dietary fat source and amount in diet	Animal model and gender (M or F)	Effect on carcinogenesis as compared to low (5%) fat diet	Ref.
DMH, s.c., 150 mg/kg body wt, one dose	Beef fat, 20%	F344 rats, M	Enhancement	16
MAM acetate, i.p., 35 mg/kg body wt, one dose	Beef fat, 20%	F344 rats, M	Enhancement	16
MNU, i.r., 2.5 mg/rat, once weekly for 2 weeks	Beef fat, 20%	F344 rats, M	Enhancement	16
DMAB, s.c., 50 mg/kg body wt, once weekly for 20 weeks	Beef fat, 20%	F344 rats, M	Enhancement	27
AOM, s.c., 8 mg/kg body wt, once weekly for 24 weeks	Beef fat, 35%	Sprague-Dawley rats, M	Enhancement	12
AOM, s.c., 8 mg/kg body wt, once weekly for 8 weeks	Beef fat, 30%	Sprague-Dawley rats, M	Enhancement	23
DMH, s.c., 7.5 mg/kg body wt, once weekly for 5 weeks	Beef fat, 25%	Sprague-Dawley rats, M	No effect	22
MNU, i.r., 1.5 mg/rat, 4 times in 2 weeks	Beef fat, 24%	Sprague-Dawley rats, M	No effect	28
DMH, s.c., 20 mg/kg body wt, once weekly for 20 weeks	Beef fat + soybean oil, 33.5%	D/A rats, M	Enhancement	29
DMH, s.c., 15 mg/kg body wt, twice weekly for 20 weeks	Lard, 30%	W/Fu rats, M	Enhancement	14
DMH, i.m., 10 mg/kg body wt, once weekly for 20 weeks	Safflower oil, 20%	Sprague-Dawley rats, M	Enhancement	13
AOM, s.c., 20 mg/kg body wt, once	Safflower oil, 23.5%	F344 rats, F	Enhancement	30
AOM, s.c., 20 mg/kg body wt, once	Corn oil, 23.5%	F344 rats, F	Enhancement	30
AOM, s.c., 15 mg/kg body wt, once weekly for 2 weeks	Corn oil, 23.5%	F344 rats, M	Enhancement	23, 24
AOM, s.c., 15 mg/kg body wt, once weekly for 2 weeks	Corn oil, 23.5%	F344 rats, M	Enhancement	23, 24
AOM, s.c., 15 mg/kg body wt, once weekly for 2 weeks	Corn oil, 23.5%	F344 rats, M	Enhancement	23, 24
DMH, i.p., 30 mg/kg body wt, once weekly for 20 weeks	Corn oil, 20%	F344 rats, M	Enhancement	21
MNU, i.r., 1.5 mg/rat, 4 times in 2 weeks	Corn oil, 24%	Sprague-Dawley	No effect	22
DMH, s.c., 15 mg/kg body wt, once weekly for 5 weeks	Corn oil, 24%	Sprague-Dawley rats, M	No effect	28
AOM, s.c., 20 mg/kg body wt, once	Olive oil, 23.5%	F344 rats, F	No effect	30
AOM, s.c., 20 mg/kg body wt, once	Coconut oil, 23.5%	F344 rats, F	No effect	30
AOM, s.c., 15 mg/kg body wt, once weekly for 2 weeks	Fish oil, 22.5% + corn oil, 1%	F344 rats, M	No effect	23, 24
AOM, s.c., 15 mg/kg body wt, once weekly for 2 weeks	Fish oil, 23.5% + corn oil, 1%	F344 rats, M	No effect	23, 24
AOM, s.c., 15 mg/kg body wt, once weekly for 2 weeks	*Trans* fat, 17.6% + corn oil, 5%	F344 rats, M	No effect	26
DMH, s.c., 7.5 mg/kg body wt, once weekly for 5 weeks	Crisco, 24%	Sprague-Dawley rats, M	No effect	22

REFERENCES

1. **Wynder, E.L. and Gori, G.B.,** Contribution of the environment to cancer incidence: an epidemiologic exercise, *J. Natl. Cancer Inst.,* 58, 825, 1975.
2. **Doll, R. and Peto, R.,** The causes of cancer: quantitative estimates of avoidable risks of cancer in the United States today, *J. Natl. Cancer Inst.,* 66, 1191, 1981.
3. **Committee on Diet and Health, Food and Nutrition Board,** *Diet and Health,* National Research Council, National Academy of Sciences, Washington, D.C., 1989, 593.
4. **Welsch, C.,** *Carcinogenesis and Dietary Fat,* Abraham, S., Ed., Khuwer Academic Publishers, Boston, 1989, 115.
5. **Reddy, B.S. and Cohen, L.A.,** Eds., *Diet, Nutrition, and Cancer: A Critical Evaluation,* Vol. 1, CRC Press, Boca Raton, FL, 1986.
6. **Shamsuddin, A.K.,** *Experimental Colon Carcinogenesis,* Autrup, H. and Williams, G.M., Eds., CRC Press, Boca Raton, FL, 1983, 51.
7. **Reddy, B.S.,** Diet and colon cancer: evidence from human and animal model studies, in *Diet, Nutrition, and Cancer: A Critical Evaluation,* Vol. 1, Reddy, B.S. and Cohen, L.A., Eds., CRC Press, Boca Raton, FL, 1986, 47.
8. **Wynder, E.L., Kajitani, T., Ishikawa, S., Dodo, H., and Takano, A.,** Environmental factors of cancer of the colon and rectum. II. Japanese epidemiological data, *Cancer,* 23, 1210, 1969.
9. **Jain, M., Cook, G.M., Davis, F.G., Grace, M.G., Howe, G.R., and Miller, A.B.,** A case-control study of diet and colorectal cancer, *Int. J. Cancer,* 26, 757, 1980.
10. **Miller, A.B., Howe, G.R., and Jain, M.,** Food items and food groups as risk factors in a case-control study of diet and colorectal cancer, *Int. J. Cancer,* 32, 155, 1983.
11. **Graham, S., Marshall, J., Hanghey, B., Mittleman, A., Swanson, M., Zielexny, M., Byers, T., Wilkinson, G., and West, D.,** Dietary epidemiology of cancer of the colon in western New York, *Am. J. Epidemiol.,* 128, 490, 1988.
12. **Nigro, N.D., Singh, D.V., Campbell, R.L., and Pak, M.S.,** Effect of dietary beef fat on intestinal tumor formation by azoxymethane in rats, *J. Natl. Cancer Inst.,* 54, 439, 1975.
13. **Broitman, S.A., Vitale, J.J., Vavrousek-Jakuba, E., and Gottlieb, L.S.,** Polyunsaturated fat, cholesterol and large bowel tumorigenesis, *Cancer,* 40, 2455, 1977.
14. **Howarth, A.E. and Pihl, E.,** High-fat diet promotes and causes distal shift of experimental rat colonic cancer — beer and alcohol do not, *Nutr. Cancer,* 6, 229, 1985.
15. **Reddy, B.S., Mangat, S., Weisburger, J.H., and Wynder, E.L.,** Effect of high-risk diets for colon carcinogenesis on intestinal mucosal and bacterial β-glucuronidase activity in F344 rats, *Cancer Res.,* 37, 3533, 1977.
16. **Reddy, B.S., Watanabe, K., and Weisburger, J.H.,** Effect of high-fat diet on colon carcinogenesis in F344 rats treated with 1,2-dimethylhydrazine, methylazoxymethanol acetate and methylnitrosourea, *Cancer Res.,* 37, 4156, 1977.
17. **Reddy, B.S. and Ohmori, T.,** Effect of intestinal microflora and dietary fat on 3,2′-dimethyl-4-aminobiphenyl-induced colon carcinogenesis in F344 rats, *Cancer Res.,* 451, 1363, 1981.
18. **Reddy, B.S., Narisawa, T., Vukusich, D., Weisburger, J.H., and Wynder, E.L.,** Effect of quality and quantity of dietary fat and dimethylhydrazine in colon carcinogenesis in rats, *Proc. Soc. Exp. Biol. Med.,* 151, 237, 1976.
19. **Nutter, R.L., Kettering, J.D., Apprecio, R.M., Weeks, D.A., and Gridley, D.S.,** Effects of dietary fat and protein on DMH-induced tumor development and immune responses, *Nutr. Cancer,* 13, 141, 1990.
20. **Sakaguchi, M., Hiramatsu, Y., Takada, H., Yamamura, M., Hioki, K., Sato, K., and Yamamoto, M.,** Effect of dietary unsaturated and saturated fats on azoxymethane-induced colon carcinogenesis in rats, *Cancer Res.,* 44, 1472, 1984.
21. **Pence, B.C. and Buddingh, F.,** Inhibition of dietary fat-promoted colon carcinogenesis in rats by supplemental calcium or vitamin D3, *Carcinogenesis,* 9, 187, 1988.
22. **Nauss, K.M., Locniscar, M., and Newberne, P.M.,** Effect of alterations in the quality and quantity of dietary fat on 1,2-dimethylhydrazine-induced colon tumorigenesis in rats, *Cancer Res.,* 43, 4083, 1983.
23. **Bull, A.W., Soullier, B.K., Wilson, P.S., Haydon, M.T., and Nigro, N.D.,** Promotion of azoxymethane-induced intestinal cancer by high-fat diet in rats, *Cancer Res.,* 39, 4956, 1979.
24. **Reddy, B.S. and Maruyama, H.,** Effect of different levels of dietary corn oil and lard during the initiation phase of colon carcinogenesis in F344 rats, *J. Natl. Cancer Inst.,* 77, 815, 1986.
25. **Reddy, B.S. and Maruyama, H.,** Effect of dietary fish oil on azoxymethane-induced colon carcinogenesis in male F344 rats, *Cancer Res.,* 46, 3367, 1986.
26. **Reddy, B.S., Tanaka, T., and Simi, B.,** Effect of different levels of dietary trans fat or corn oil on azoxymethane-induced colon carcinogenesis in F344 rats, *J. Natl. Cancer Inst.,* 75, 791, 1985.
27. **Reddy, B.S. and Ohmori, T.,** Effect of intestinal microflora and dietary fat on 3,2′-dimethyl-4-aminobiphenyl-induced colon carcinogenesis in F344 rats, *Cancer Res.,* 41, 1363, 1981.

28. **Nauss, K.M., Locniskar, M., Sondergaard, D., and Newberne, P.M.,** Lack of effect of dietary fat on N-nitrosomethyl urea (NMU)-induced colon tumorigenesis in rats, *Carcinogenesis*, 5, 255, 1984.
29. **Bansal, B.R., Rhoads, J.E., Jr., and Bansal, S.C.,** Effects of diet on colon carcinogenesis and the immune system in rats treated with 1,2-dimethylhydrazine, *Cancer Res.*, 38, 3293, 1978.
30. **Reddy, B.S. and Maeura, Y.,** Tumor promotion by dietary fat in azoxymethane induced colon carcinogenesis in female F344 rats: influence of amount and source of dietary fat, *J. Natl. Cancer Inst.*, 72, 745, 1984.

Section 3: Effects of Selected Whole Foods and Interaction of Antioxidants and Fiber on the Health Effects of Dietary Fatty Acids and Cholesterol

INTRODUCTION

Gene A. Spiller

This section of the Handbook focuses on the physiological and health effects of some whole foods such as nuts and olive oil and some important factors in foods such as fiber and antioxidants, factors that have recently found an important niche in the study of the effects of fatty acids and cholesterol in human nutrition and medicine.

For example, short chain fatty acids may play an important role in the metabolism of food lipids in humans. Their production in the human colon and absorption is probably one of the mechanisms by which fiber favorably affects plasma lipoproteins.

The complex interactions of these various components of foods — earlier considered of minor importance — are one of the great challenges for the student of disease prevention at the dawn of the 21st century.

Chapter 3.1

ALMONDS, WALNUTS, AND SERUM LIPIDS

Joan Sabaté and Debra Geary Hook

Tree nuts have existed as staples of Mediterranean diets since antiquity.[1] Originally brought to the western U.S. by Spanish priests, the majority of the world supply of almonds and walnuts are currently grown in California. Despite this wide availability, the nut consumption of most Americans is relatively low.[2] One reason may be due to the nut's reputation as a significant source of fat and calories. Almost two thirds of a nut's weight is fat, which is why nuts have traditionally been considered inappropriate for individuals with elevated blood lipids.[3] However, it is now widely accepted that the fatty acid composition, rather than the total amount of fat consumed, is the greater predictor of cholesterolemic effects. With the exception of chestnuts, which are high in carbohydrate, most commonly eaten nuts are high in fat, consisting largely of unsaturated fatty acids (see Table 1, Chapter 3.2). Many nuts, including almonds, are high in monounsaturated fatty acids (MUFAs) (oleic acid), while walnuts are high in two polyunsaturated fatty acids (PUFAs): linoleic and linolenic. The beneficial effects of polyunsaturated fat on cholesterol levels have long been known, but until recently, MUFAs were thought to have a neutral effect.[4,5] Recent studies challenge this notion by suggesting that there is an inverse relationship between dietary MUFAs and serum cholesterol levels and they beneficially alter other coronary heart disease risk factors.[4,6]

Despite their high fat content, nuts, when consumed as a whole food, are more than just a source of fat. They contain several nutrients postulated to provide a protective effect against coronary heart disease (CHD), such as fiber, arginine, and antioxidants.[3,7] Knowing the beneficial effects of dietary fiber[8] and fat as well as recent work on antioxidants, it is surprising that nuts as a whole food had not been considered earlier for CHD research. This oversight may be due to earlier animal studies with peanut oil, which implied that this nut oil was atherogenic.[9] Contradictory findings from recent research plus epidemiological data from Mediterranean countries and the Adventist Health Study have stimulated interest in research using whole nuts.[10] This chapter will review human nutrition studies conducted on the effects of whole almonds and walnuts on serum lipid levels, since the majority of work has utilized these two nuts. Preliminary work on hazelnuts and macadamias has been communicated but not yet published; hence it will not be discussed here. A brief summary of the studies referenced in this chapter is provided in Table 1.

ALMOND STUDIES

Inadvertently, Berry et al.,[11,12] in the Jerusalem Nutrition Study, provided some of the first data on the effects of whole nut consumption on serum lipids. The purpose of this project was to investigate the effects of MUFAs on lipoprotein structure and function. During the first trial,[11] a diet high in PUFA was compared to a diet high in MUFA. Due to their fatty acid composition, almonds and walnuts were included in the MUFA and PUFA diets, respectively. Instead of having a neutral effect upon serum lipids, the MUFA-rich diet lowered low-density-lipoprotein cholesterol (LDL-C) and total cholesterol (TC) levels, without significantly altering high-density-lipoprotein cholesterol (HDL-C) levels.

To further verify whether MUFAs have a hypocholesterolemic effect, the investigators[12] conducted a second feeding trial. Here, a diet high in MUFA was compared with a carbohydrate-rich diet. Whole almonds were used in the MUFA diet to provide a portion of the fat. This diet lowered TC concentrations and LDL-C levels with respect to baseline values, while

0-8493-4248-1/96/$0.00+$.50
© 1996 by CRC Press Inc.

TABLE 1
Summary of Published Almonds and/or Walnut Studies on Serum Lipids

Type of nut	Ref.	Where conducted	Subjects	Study design	Length of study	Diet descriptions	Baseline mean serum cholesterol	Results from dietary intervention Total cholesterol	LDL	HDL	LDL:HDL
Almonds and walnuts	11	Israel	18 young men	Randomized crossover feeding trial	Two 12-week dietary periods	MUFA diet: 34% total fat with almonds, olive oil, and avocado PUFA diet: 34% total fat with walnuts, safflower, and soy oils	147 mg/dl (3.81 mmol/l)	vs. baseline values MUFA: ↓ 10% PUFA: ↓ 16%	MUFA: ↓ 14% PUFA: ↓ 20%	MUFA: NS PUFA: NS	
Almonds	12	Israel	17 young men	Randomized crossover feeding trial	Two 12-week dietary periods	MUFA diet: 33% total fat with almonds (≈100 g/d), olive oil and avocado CHO diet: 18% total fat	156 mg/dl (4.03 mmol/l)	vs. CHO diet ≈↓ 8%	≈ ↓14%	NS	
Almonds	13	California	26 adult men and women	Uncontrolled supplemental field study	9 weeks	Modified Mediterranean-type diet: 37% total fat with 100 g/d almonds and almond oil Usual diet	235 mg/dl (6.08 mmol/l)	vs. baseline values ↓9%	↓ 12%	N/S	↓ 12%
Almonds	14	California	30 adult men and	Parallel groups supplemental	4 weeks	Almond group: low-fat base	251 mg/dl (6.49 mmol/l)	vs. control group ↓ 15%		NS	

					Diet	Baseline				
Walnuts	9	California	women	field study	diet, plus 100 g/d almonds					↓ 12%
					Control group: low-fat base diet, plus 48g of fat from cheese and butter					
	18 young and adult men			Tightly controlled randomized, crossover feeding trial	Walnut diet: 31% total fat including 85g/d walnuts	198 mg/dl (5.11 mmol/dl)	vs. control diet ↓ 12.4%	↓ 16.3%	↓ 4.9%	
					Control diet: 30% total fat (10% sat, 10% mono, 10% poly)					
Almonds and walnuts	15	Australia	16 adult men	Consecutive supplemental field study	Almond diet: 36% total fat with 84 g/d almonds	199 mg/dl (5.15 mmol/l)	vs. control diet Almond: ↓ 7% Walnut: ↓ 5% (↓ 7%)	Almond: ↓ 10% Walnut: ↓ 9% (↓ 12%)	Almond: NS Walnut: ↑ 3% (NS)	Almond: ↓ 10% Walnut: ↓ 12% (↓ 12%)
				Three 3-week dietary periods	Walnut diet: 37% total fat with 68 g/d walnuts					
					Control diet: 36% total fat with 50 g/d peanuts, 40 g/d coconut, and 50 g/d confectionary					

Note: Numerical results reported were easily available. NS denotes not statistically significant. Unless so indicated results are statistically significant ($p < 0.05$).

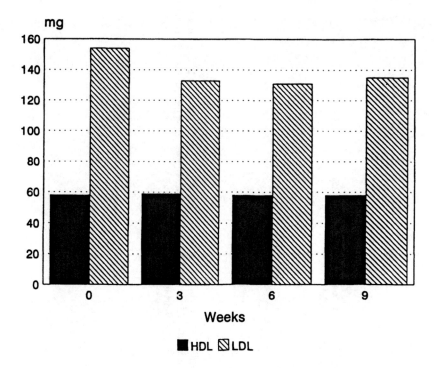

FIGURE 1.   Mean serum concentrations of HDL and LDL cholesterol for 26 subjects consuming a Mediterranean-style diet supplemented with almonds. Measurements were taken at 3-week intervals.

not changing HDL-C levels significantly. In comparison, no significant changes were observed with the carbohydrate diet (see Table 1).

To examine the effects of whole almonds on plasma cholesterol independent of total dietary fat content, Spiller et al.[13] conducted a 9-week study using 26 men and women in a free-living environment. These moderately hypercholesterolemic (total serum cholesterol >220 mg/dl) subjects were already consuming a fairly low-fat diet containing an average of 67 g of fat per day. After baseline measurements, subjects were instructed to consume a Mediterranean-style diet in which the major portion of fat was supplied by 100 g/d of raw almonds with the use of almond oil as the only allowable free fat. Compared to the habitual diet, the almond-supplemented diet provided about 23 g/d more fat. This 9% increase in fat was almost entirely the result of an increase in MUFA content (70% of total fatty acids), not a replacement of saturated fatty acids with MUFA. Within 3 weeks, mean TC levels were reduced by 20 ± 5 mg/dl with the almond diet, and a sustained reduction in LDL-C and TC was achieved throughout the remainder of the study despite the higher fat intake (see Figure 1). HDL-C levels remained unchanged; thus the TC/HDL-C ratio showed a significant decrease of about 0.4. There were no significant changes in very low density lipoprotein cholesterol or triglyceride levels from baseline values, or between measurements taken during the study's diet period. The lack of a control group and study design prevents assessment as to whether lipid changes were due to the almond diet or to other factors such as the well-documented seasonal variations in serum lipids or regression to the mean effect.

In another study, Spiller et al.[14] investigated whether the plasma lipid lowering effects ascribed to consumption of olive oil could be achieved with whole almonds, both of which are rich in oleic acid. Using three diet groups, they compared diets supplemented daily with 100 g of almonds, 48 g of olive oil, or 85 g of cheddar cheese. The latter two diets contained a combination of cheese, butter, and rye crackers, hence the total fat, protein, and fiber content of the olive oil and cheddar cheese diets was similar to that of the almond diet.

Forty-five hypercholesterolemic men and women were allocated to either the almond, olive oil, or cheese diet. After 2 weeks, a significant difference in TC and LDL-C levels was seen between the three groups. The almond-supplemented diet showed a significant reduction from baseline values in TC and LDL-C levels. After 4 weeks, the almond-supplemented diet produced 12, 11, and 12% reductions in TC and LDL-C levels and TC/HDL-C ratio, respectively. Although the olive oil group produced significant differences from the control group, it was not significantly lower from its own baseline values after 4 weeks.

Both the olive oil and almond diets were hypocholesterolemic; however, greater lipid reductions were observed in the almond-supplemented diet. The differences in plasma lipid lowering effects between the olive oil and almond diets may not be due to their MUFA content, but rather to other nutritional constituents of the almonds such as vegetable protein and fiber. The olive oil diet contained casein protein from cottage cheese, and there is an indication from animal studies that almond protein may have beneficial effects on blood lipids compared to casein.[16] Besides MUFA, vegetable protein, and fiber, other components in almonds could possibly behave as hypocholesterolemic agents.[8] The fine skin that covers the almond is rich in substances such as saponins, while the whole nut contains phytosterols.[14]

WALNUT STUDY

Recently, Sabaté et al.[9] conducted a controlled, randomized, crossover study comparing the effects on serum lipids of a diet rich in walnuts with a control diet that followed the current recommendation for lowering cholesterol. Both diets conformed to the National Cholesterol Education Program Step 1 diet, with 30% of total calories from fat and less than 10% from saturated fat. Each was comprised of identical foods and macronutrients, except that 20% of the calories from the walnut diet were derived from walnuts. To maintain a similar quantity of fat, the walnut diet contained lesser amounts of visible fats and fat-rich foods. The fatty acid composition of the walnut diet more closely reflected that of the walnut itself, having contained a larger proportion of fat calories from PUFA, especially alpha-linolenic acid. During the investigation, 18 normocholesterolemic (mean total serum cholesterol 198 mg/dl) men were assigned to one of the two study groups. Each group was fed in alternating order the two mixed natural diets, each with a duration of 4 weeks.

As expected, the changes in lipid levels for both diets were lower in relation to baseline values. However, the walnut diet, when compared to the control diet, further decreased TC and LDL-C by 12 and 16%, respectively. In comparison with the control diet, the walnut diet produced TC and LDL-C levels 22.4 and 18.2 mg/dl lower, respectively. Despite a 2.3-mg/dl reduction in HDL-C level with the walnut diet, a significant decrease in the LDL-C/HDL-C ratio occurred. As shown in Figure 2, a clear crossover pattern was observed for TC and LDL-C values with no evidence of a carryover effect between the diet periods. The investigators concluded that replacing a portion of the fat in a cholesterol-lowering diet with walnuts further reduces TC levels in men and may be more beneficial than current dietary recommendations. It is remarkable that these serum lipid reductions were observed in all subjects despite relatively low baseline cholesterol levels.

ALMOND AND WALNUT STUDIES

Although Berry et al.[11] used almonds and walnuts to increase the dietary MUFA and PUFA content of their experimental diets, the first study to directly compare nuts of different fatty compositions on plasma lipids was conducted by Abbey et al.[15] In a 9-week consecutive supplemental field study, 16 normolipidemic males consumed a background diet which provided 18% of energy as fat. A supplementation was added to this diet contributing additional fat, raising the total calories from fat to 36%. During the first 3-week period, the supplementation consisted of peanuts, coconut, and a confectionery bar, creating a fatty acid

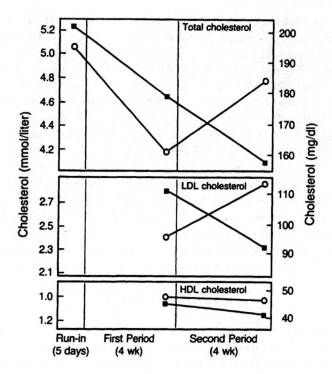

FIGURE 2.   Mean serum concentrations of total, LDL, and HDL cholesterol during each diet period. All 18 subjects followed each diet. Ten followed the walnut diet first (O) and eight followed the reference diet first (■).

profile reflecting that of the typical Western diet. This first supplemental diet served as the reference diet. During the second and third 3-week periods, the background diet was supplemented daily with 86 g of raw almonds and finally 68 g of raw walnuts.

There was no change in plasma cholesterol between the start of the study and the end of the first 3-week supplementation period. After supplementation with almonds, 7 and 10% reductions in TC and LDL-C levels were noticed, respectively. Following supplementation with walnuts, there was a 5 and 9% reduction in TC and LDL-C levels, respectively. These results were statistically significant. HDL-C levels remained unchanged. However, due to the study design, it is difficult to disentangle dietary effects from secular trends.

DIETARY IMPACT OF OXIDATION ON SERUM LIPIDS

The role of oxidized LDL-C in the atherogenic process is well documented. Modification of LDL-C by oxidation creates a LDL particle that is more rapidly taken up into macrophages via specific receptors, which do not recognize unoxidized LDL-C. Bonanome et al.[17] demonstrated in a crossover randomized dietary trial an inverse relationship between oxidation rate of plasma LDL-C and the ratio of oleic acid to linoleic acid of the LDL particle. For this reason, it has been suggested that although almonds and walnuts are both hypocholesterolemic agents, it may be possible that almonds are more beneficial in reducing the risk of CHD.[15] The MUFA content of the almonds could be more resistant to oxidation. In the trial conducted by Berry et al.[11] as part of the Jerusalem nutrition studies, the susceptibility of plasma lipids to oxidation was reduced in subjects on the almond/MUFA diet as compared to the walnut/PUFA diet. In light of recent research with MUFA and PUFA, this is theoretically plausible because MUFAs have fewer double bonds than PUFAs, making them less vulnerable to oxidation.

However, in a recent analysis of a subsample of participants of the walnut study,[9] Sabaté and Fish[18] failed to demonstrate that PUFAs supplied by walnuts increase the susceptibility of LDL particles to oxidation. The fat content of both diets was approximately 30%, with the proportion of PUFA at 17% in the walnut diet vs. 10% in the control diet. Even though subjects were fed unusually high amounts of walnuts in this study (three daily servings, contributing more than 50% of the total fat in the diet), a greater difference in PUFA content may be required to assess an increased propensity toward oxidation. The investigators speculated that antioxidants present in the walnuts may prevent increased LDL-C oxidation.

Due to the high alpha-tocopherol content of almonds, Berry et al.[12] hypothesized that the degree of unsaturation of dietary fats may not be the sole determinant for the almond/MUFA diet's resistance to oxidation. Although this is definitely plausible, the study conducted by Bonanome et al.[17] demonstrated that the LDL oxidation rate was independent of the antioxidant (vitamin E) content within the LDL particles. Scaccini et al.,[19] on the other hand, argue that other natural endogenous antioxidants besides vitamin E could be responsible for the prevention of oxidative modification.

If the differences in LDL oxidation rates seen between diets supplemented with almonds and walnuts are truly independent of antioxidant content, then the 14% difference in plasma linoleic acid levels in the study conducted by Abbey et al.[15] may be sufficient to influence LDL oxidation rates in such a manner that almonds are conceivably more protective against CHD. However, it remains to be proven whether it is the fatty acid composition of the nuts, the alpha-tocopherol content, or some other natural endogenous antioxidant which is responsible for reducing LDL's susceptibility to oxidation. More research is needed to identify which nutrients or combination of nutrients in the whole food is responsible for the reduction in LDL oxidation, and whether there is a difference in oxidation susceptibility between the two nuts.

CONCLUSION

Although each study differs in methodological rigor and dietary control, there seems to be a consensus among the human nutritional studies published thus far that walnuts, almonds, and possibly other nuts with a similar nutrient composition have a cholesterol lowering effect. Thus, the protective effect of nuts on CHD may in part be mediated through reduction of serum lipids by the fatty acid composition of the nut as well as through other intrinsic components. Future research may prove that protection against LDL oxidation is likewise an important factor in the nut's ability to reduce risk for CHD. The effects of whole nut consumption on cardiovascular risk factors and the atherogenic process deserve further exploration.

REFERENCES

1. **Braun, T.,** Ancient Mediterranean food, in *The Mediterranean Diets in Health and Disease,* Spiller, G.A., Eds., Van Nostrand Reinhold, New York, 1991, chap. 2.
2. **Sabaté, J.,** Does nut consumption protect against ischaemic heart disease?, *Eur. J. Clin. Nutr.,* 47(Suppl 1), S71, 1993.
3. **Sabaté, J. and Fraser, G.,** The probable role of nuts in preventing coronary heart disease, *Primary Cardiol.,* 19, 65, 1993.
4. **Riccardi, G. and Rivellese, A.,** An update on monounsaturated fatty acids, *Curr. Opinion Lipidol.,* 4, 13, 1993.
5. **Grundy, S.,** Monounsaturated fatty acids and cholesterol metabolism: implications for dietary recommendations, *J. Nutr.,* 119, 529, 1989.
6. **Trevisan, M., Krough, V., Freudenheim, J., Blake, A., Muti, P., Panico, S., et al.,** Consumption of olive oil, butter, and vegetable oils and coronary heart disease risk factors, *J.A.M.A.,* 263, 688, 1990.
7. **Sabaté, J. and Fraser, G.,** Nuts: a new protective food against coronary heart disease, *Curr. Sci.,* 5, 11, 1994.

8. **Anderson, J. and Akanji, A.,** Treatment of diabetes with high fiber diet, in *The Mediterranean Diets in Health and Disease,* Spiller, G.A., Eds., Van Nostrand Reinhold, New York, 1991, chap. 7.3.

9. **Sabaté, J., Fraser, G., Burke, K., Knutsen, S., Bennett, H., and Lindsted, K.,** Effects of walnuts on serum lipid levels and blood pressure in normal men, *N. Engl. J. Med.,* 328, 603, 1993.

10. **Fraser, G., Sabaté, J., Beeson, W., and Strahan, M.,** A possible protective effect of nut consumption on risk of coronary heart disease, *Arch. Intern. Med.,* 152, 1416, 1992.

11. **Berry, E., Eisenberg, S., Haratz, D., Friedlander, Y., Norman, Y., Kaufmann, N., et al.,** Effects of diets rich in monounsaturated fatty acids on plasma lipoproteins — the Jerusalem Nutrition Study: high MUFAs vs. high PUFAs, *Am. J. Clin. Nutr.,* 53, 899, 1991.

12. **Berry, E., Eisenberg, S., Friedlander, Y., Harats, D., Kaufmann, N., Norman, Y., et al.,** Effects of diets rich in monounsaturated fatty acids on plasma lipoproteins — the Jerusalem Nutrition Study. II. Monounsaturated fatty acids vs. carbohydrates, *Am. J. Clin. Nutr.,* 56, 394, 1992.

13. **Spiller, G. et al.,** Effect of a diet high in monounsaturated fat from almonds on plasma cholesterol and lipoproteins, *J. Am. Coll. Nutr.,* 11, 126, 1992.

14. **Spiller, G., Gates, J., Jerkins, D., Bosello, O., Nichols, S., and Cragen, L.,** Effect of two foods high in monounsaturated fat on plasma cholesterol and lipoproteins in adult humans, *Am. J. Clin. Nutr.,* 51 (Abstr.), 524, 1990.

15. **Abbey, M., Noakes, M., Belling, G., and Nestel, P.,** Partial replacement of saturated fatty acids with almonds or walnuts lowers total plasma cholesterol and low-density-lipoprotein cholesterol, *Am. J. Clin. Nutr.,* 59, 995, 1994.

16. **Sanchez, A., Rubano, D., Shavlik, G., Hubbard, R., and Horning, M.,** Cholesterolemic effects of the lysine/arginine ratio in rabbits after initial early growth, *Arch. Latinoam. Nutr.,* 38, 229, 1998.

17. **Bonanome, A., Pagnan, A., Biffanti, S., Opportuno, A., Sorgato, F., Dorella, M., et al.,** Effect of dietary monounsaturated and polyunsaturated fatty acids on the susceptibility of plasma low density lipoproteins to oxidative modification, *Arterioscler. Thromb.,* 12, 529, 1992.

18. **Sabaté, J. and Fish, D.,** Consumption of a High Polyunsaturated Fat Food and the Oxidative Tendency of Low Density Lipoproteins in Normocholesterolemic Males, paper presented at the 10th Int. Symp. on Atherosclerosis, Montreal, 1994.

19. **Scaccini, C., Nardini, M., D'Aquino, M., Gentili, V., Di Felice, M., and Tomassi, G.,** Effect of dietary oils on lipid peroxidation and on antioxidant parameters of rat plasma and lipoprotein fractions, *J. Lipid Res.,* 33, 627, 1992.

Chapter 3.2

NUT CONSUMPTION AND CORONARY HEART DISEASE RISK

Joan Sabaté, Heather E. T. Bell, and Gary E. Fraser

INTRODUCTION

Although the majority of the literature connecting diet to coronary heart disease (CHD) and serum lipids has focused on nutrient intake, recent studies have emphasized the effect of individual foods on serum lipids and the risk of CHD. Because foods are unique combinations of nutrient and non-nutrient chemicals with many unknown biological actions, it is important to study the effects of natural foods on risk factors and the risk of chronic disease. Epidemiological research on nutrients is often translated into dietary recommendations for the general population. It is important for such research to evaluate foods as well as nutrients, however, since most individuals and institutions will change their nutrient intake largely by their choice of foods.[1]

This chapter will summarize the epidemiological and mechanistic evidence to date on the relationship between the consumption of a food — nuts — and CHD. The majority of epidemiological evidence relating nuts and risk of CHD was provided by the Adventist Health Study.[2] Recently, however, additional support for the protective role of nuts was provided by the Iowa Women's Health Study,[3] which will be discussed later.

THE ADVENTIST HEALTH STUDY

The Adventist Health Study examined the association between the intake of particular foods and the risk of CHD in 34,198 Californian members of the Seventh-Day Adventist Church. Seventh-Day Adventists, members of a conservative Christian denomination, possess health habits different from the general population. A large majority of Adventists neither smoke nor drink, and about 50% of Californian Adventists follow a lacto-ovo vegetarian diet.

As a result of their health habits, the Adventist population is considered uniquely suited to the prospective study of the effects of diet on disease risk. First, this population is homogeneous with respect to cigarette and alcohol use, which greatly diminishes the confounding power of these risk factors. Second, Adventists are usually well educated and interested in health issues, which contributes to a higher response rate and validity with self-administered questionnaires, the most widely used information-gathering tool in large epidemiological studies. Last, Adventists display a greater variety of dietary practices than the general population (including a varied frequency of meat and nut consumption), which lends a greater statistical power to the results.

Subjects in the Adventist Health Study completed a life style questionnaire at baseline which determined the habitual frequency of consumption of 55 foods, in addition to other variables related to CHD. The foods were evaluated on a scale of 1 to 8, from foods never consumed to foods consumed more than once a day.[2] The Adventist subjects tended to have a greater routine consumption of nuts than the general population. Of the participants, 24% consumed nuts five or more times per week, 42% ate nuts one to four times per week, and the remainder of the population ate nuts less than once a week. During the 6-year follow-up, new cases of CHD and death were identified.

The study found that the frequency of nut consumption was inversely related to the risk of myocardial infarction or dying of CHD. This relationship was found to be independent of coronary risk factors including age, gender, smoking habits, hypertension, relative weight, exercise habits, and other common foods including meats. Compared to those who ate nuts less than once a week, those who consumed nuts one to four times per week had a 27% reduced risk of CHD death. Adventist subjects who consumed nuts more often than four times per week experienced a 38% decrease in risk of CHD death. The risk reduction for myocardial infarction of those consuming nuts one to four times per week was 26%, and with an intake of five or more times per week, the risk reduction was 48% compared to subjects whose nut consumption was infrequent.

One important aspect of the study results was that the protective effect of nut intake was borne out in several population subgroups. Figure 1 shows the age- and sex-stratified analyses of associations between the consumption of nuts and definite CHD events in different subgroups of the Adventist Health Study population. The frequency of nut consumption was inversely associated with CHD risk in both males and females, and similar results were seen for normotensive and hypertensive subjects regardless of their relative weight. Nuts were also shown to be protective for both vegetarians and those consuming a mixed diet.

No increase in cancer risk was found among frequent consumers of nuts, and, in fact, a clear decrease in risk for total mortality was demonstrated. For all-cause mortality, the corresponding relative risks for the three consumption categories of nuts (<1/week, 1 to 4/week, >5/week) were, respectively, 1.0, 0.72, and 0.67 (*p* <0.00001). It is important to note that the decrease in CHD mortality was not compensated for by an increase in any other cause of mortality.[4]

THE IOWA WOMEN'S STUDY

A second epidemiological study that provides further support for the protective role of nuts on CHD is the Iowa Women's Health Study. In this prospective cohort study, approximately 34,000 women were asked about the frequency of their nut consumption at baseline. Data collected showed that 41% never ate nuts, 35% consumed nuts one to three times per month, 15% ate nuts once a week, and 9% ate nuts more than twice a week.[5] Because only a small percentage consumed nuts frequently, the statistical power of the results is somewhat limited. Nevertheless, the 5-year follow-up showed that coronary mortality was inversely associated with nut intake in this population. After adjustment for traditional coronary risk factors and energy intake, those women with the highest nut intake had only 40% the risk of CHD as those women who never ate nuts. Thus, the Iowa Women's Study independently duplicated in women the earlier findings of the Adventist Health Study. Results from both epidemiological studies would seem to suggest that frequent nut consumption is protective against CHD. In what way, however, does nut consumption affect cardiovascular physiology?

SUGGESTED MECHANISMS OF ACTION

Several mechanisms have been suggested to explain the protective effects of nut consumption for CHD. One obvious factor is the well-documented effect of dietary fatty acids on serum cholesterol. Table 1 presents the fat composition of several nuts.[6] The total fat content of nuts is high, ranging from 48 to 74% by weight, but this fat is largely unsaturated. Most nuts are high in oleic acid (monounsaturated), and there are significant amounts of linoleic and linolenic acid (polyunsaturated) in walnuts, in particular. Recent studies have recognized a beneficial effect on serum lipids with a high intake of monounsaturated fatty acids,[7,8] and the cholesterol lowering effect of polyunsaturated fats has long been known.[9] However, previous studies with peanut oil in laboratory animals suggested an atherogenic role for the oil.[10-13] These studies, however, were high in dietary cholesterol, and corrected studies produced results which were much less consistent.[14-16] More recent studies have contradicted the earlier results.[17]

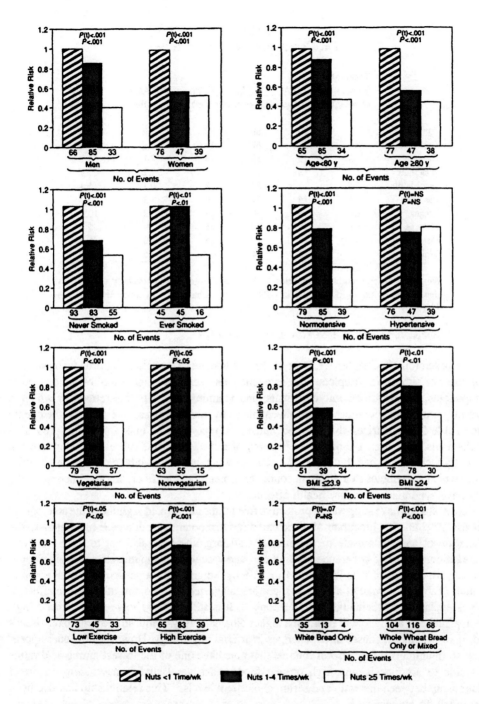

FIGURE 1. The age- and sex-stratified breakdown of associations between the consumption of nuts and heart attacks in different subgroups of the Adventist Health Study population. P(t) is the *p* value for a test of trend; P is the *p* value for the overall test of difference between categories. NS indicates not significant; BMI = body mass index.

The differences seen relative to atherogenesis and CHD risk when nut oils are compared to nuts raise the issue of the role of whole nuts as a nutrient source. Traditionally, nuts have been regarded chiefly as a source of fat in the diet. In fact, the mono- and polyunsaturated oils found in nuts and their effects on serum lipids have been researched in several studies[18-20] and

TABLE 1
Fat Composition of Several Nuts

	Total fat (percentage of weight)	SFA	MUFA (percentage of total fat)	PUFA	PUFA/SFA ratio	UFA/SFA ratio
Almonds	52	10	68	22	2.2	9.0
Cashews	46	20	62	18	0.9	3.9
Chestnuts	2.3	19	38	43	2.1	3.9
Hazelnuts	63	8	82	10	1.2	11.9
Macadamias	74	16	82	2	0.1	5.4
Peanuts	49	15	51	34	2.3	5.7
Pecans	68	8	66	26	3.2	10.9
Pinyons	61	15	40	45	2.9	5.4
Pistachios	48	13	72	15	1.2	6.6
Walnuts	62	10	24	66	6.5	9.0

Note: SFA denotes saturated fatty acids; MUFA denotes monounsaturated fatty acids; PUFA denotes polyunsaturated fatty acids. UFA denotes unsaturated fatty acids (MUFA plus PUFA).

Elaborated from data in the USDA Handbook No. 8-12.[6]

are discussed fully in Chapter 3.1. Despite the fact that nuts are about two thirds fat by weight, nuts are not simply fat droplets. The remaining one third of a nut supplies, among other components, protein, fiber, micronutrients, and vitamins, and it is this remainder which is currently providing new ideas for the mechanisms by which nuts assume an anti-atherogenic role. Table 2 summarizes the selected nutrient components of nuts for which protective mechanisms have been hypothesized.[21] Some of the mechanisms discussed below are not currently well established, but the evidence is growing that a high intake of nuts on a regular basis, which contributes to a different source of protein, fat, and fiber, among other nutrients, may have synergistic coronary health effects.

Recent studies have suggested a protective role for the amino acid arginine in the atherogenic process. Arginine is a precursor of nitric oxide (NO), a potent vasodilator.[22] In addition to the relaxation of smooth muscle of the arterial wall, arginine may also act to inhibit platelet aggregation, monocyte adherence, chemotaxis, and vascular smooth muscle cell proliferation. The protein content of nuts is high, 14 to 26% by weight, with a particularly high arginine content (Table 2). Cooke et al.[23] have therefore suggested that the anti-atherogenic effect of nuts is explained by their high arginine content. It is still unclear, however, whether a high dietary intake of arginine will result in a higher bioavailability of this amino acid to the tissues and, if so, whether it causes direct changes in arterial physiology. However, this does appear to be so in animals.[24] It is also of interest that nuts have one of the lowest amino acid ratios of lysine to arginine in any high protein food, and several studies have shown a direct relationship between this ratio and serum cholesterol levels.[25] This relationship has also been postulated for humans.[26]

Nuts contain more vitamin E than any other source except vegetable oils. Almonds and hazelnuts are especially rich in α-tocopherol (24 mg/100 g), while walnuts and pecans are high in γ-tocopherol (18 mg/100 g). The U.S. Recommended Dietary Allowances for vitamin E are 10 and 8 mg/d of tocopherol for men and women, respectively, and a single serving of these nuts easily exceeds this requirement. In two reports from large prospective studies, an independent protective effect of dietary vitamin E and vitamin E supplements relative to CHD

TABLE 2
Nonfat Components of Nuts with Hypothesized Antiatherogenic Properties[a]

	α-Tocopherol (mg)	Dietary fiber[b] (g)	Protein (g)	Lysine (g)	Arginine (g)	Lys/Arg ratio	Mg (mg)	Cu (mg)
Almonds	24.0	11.2	20	0.7	2.5	0.4	296	0.9
Cashews	0.6	6.0	15	0.8	1.7	0.5	260	2.2
Chestnuts	—	19.6	2	0.1	0.2	0.8	32	0.5
Hazelnuts	23.9	6.4	13	0.4	2.2	0.2	285	1.5
Macadamia	—	5.2	8	0.3	0.9	0.4	116	0.3
Peanuts	9.1	8.8	26	1.0	3.5	0.3	180	1
Pecans	3.1[c]	6.5	8	0.3	1.1	0.3	128	1.2
Pinyon	—	—	12	0.4	2.3	0.2	234	1
Pistachios	5.2	10.8	21	1.3	2.2	0.6	158	1.2
Walnuts	2.6[c]	4.8	14	0.4	2.1	0.2	169	1

[a] Values per 100 g of edible food. Data were taken from U.S. Department of Agriculture Handbook No. 8-12 unless otherwise indicated.
[b] Data on dietary fiber elaborated from Spiller G. A., Appendix I, in *Handbook of Dietary Fiber in Human Nutrition*, CRC Press, Boca Raton, FL 1993, 567.
[c] Pecans contain 19.1 mg of γ-tocopherol; walnuts contain 17.2 mg.

was seen in men and women.[27,28] The protective effects of vitamin E are thought to derive from its role as anti-oxidant and preventor of lipid peroxidation. Vitamin E traps free radicals, preventing the oxidation of unsaturated fatty acids. Recent evidence suggests that vitamin E levels may thus determine the oxidative resistance of low density lipoprotein (LDL) particles,[29,30] as vitamin E is transported with LDL in the plasma.

Last, nuts are also good dietary sources of magnesium and copper, nutrients often low in the Western diet. Although the evidence is as yet weak, there is some indication that copper and magnesium may protect against CHD death.[31-34] Studies done in some populations have shown an inverse relationship for magnesium and CHD mortality, and the association between hypomagnemia and ventricular arrythmias is well documented.[35] In addition, magnesium supplementation of acute myocardial infarction survivors has been found to be protective against a second event.[36] Copper has also been shown to protect against a variety of disorders related to atherogenesis. Persons with a low intake of this metal have presented with increased serum cholesterol levels, hypertension, abnormal EKGs, and impaired glucose tolerance.[37-39]

CONCLUSION

Independent research seems to suggest that nut consumption has a protective effect on CHD. The only two epidemiological studies published to date looking at the association between consumption of nuts and risk of CHD found a graded and inverse relationship (decreasing coronary risk with increasing nut intake). In the Adventist Health Study, this relationship was consistent among several population subgroups and different clinical manifestations of CHD. In addition, the protective effect was not compromised by an increase in mortality from other causes. As potential mechanisms are explored, it appears increasingly likely that nuts possess a unique nutrient composition which affects both serum lipids and other risk indicators for atherogenesis, making more biologically coherent the connection between nuts and CHD.

REFERENCES

1. **Willet, W.,** Foods and nutrients, in *Nutritional Epidemiology,* Oxford University Press, New York, 1990, chap. 2.
2. **Fraser, G., Sabaté, J., Beeson, W., and Strahan, M.,** A possible protective effect of nut comsumption on risk of coronary heart disease, *Arch. Intern. Med.,* 152, 1416, 1992.
3. **Munger, R.G., Folsom, A.R., Kushi, L.H., Kaye, S.A., and Sellers, T.A.,** Dietary assessment of older Iowa women with a food frequency questionnaire: nutrient intake, reproducibility, and comparison with 24-hour dietary recall interviews, *Am. J. Epidemiol.,* 136, 192, 1992.
4. **Fraser, G., Sabaté, J., and Beeson, W.L.,** Nuts, nuts good for your heart?, *Arch. Intern. Med.,* 152(letter), 2511, 1992.
5. **Prineas, R.J., Kushi, L.H., Folsom, A.R., Bostick, R.M., and Wu, Y.,** Walnuts and serum lipids, *N. Engl. J. Med.,* 329(letter), 359, 1993.
6. U.S. Department of Agriculture, Composition of foods: nuts and seed products, in Agricultural Handbook No. 8-12, U.S. Government Printing Office, Washington, D.C., 1984.
7. **Mattson, F.H. and Grundy, S.M.,** Comparison of effects of saturated, monounsaturated and polyunsaturated fatty acids on plasma lipids and lipoproteins in man, *J. Lipid Res.,* 26, 192, 1985.
8. **Grundy, S.M.,** Comparison of monounsaturated fatty acids and carbohydrates for lowering plasma cholesterol, *N. Engl. J. Med.,* 314, 745, 1986.
9. **Keys, A., Anderson, J.T., and Grande, F.,** Prediction of serum cholesterol responses of men to changes in fats in the diet, *Lancet,* 2, 959, 1957.
10. **Kritchevsky, D., Tepper, S.A., Vesselino-Vitch, D., and Wessler, R.W.,** Cholesterol vehicle in experimental atherosclerosis. II. Peanut oil, *Atherosclerosis,* 14, 53, 1971.
11. **Vesselinovich, D., Getz, G.S., Hughes, R.H., and Wessler, R.W.,** Atherosclerosis in the rhesus monkey fed three food fats, *Atherosclerosis,* 20, 303, 1974.
12. **Kritchevsky, D., Tepper, S.A., Scott, D.A., Klurfeld, D.M., Vesselinovich, D., and Wessler, R.W.,** Cholesterol vehicle in experimental atherosclerosis. XVIII. Comparison of North American, African and South American peanut oils, *Atherosclerosis,* 38, 291, 1981.
13. **Kritchevsky, D., Tepper, S.A., Kim, H.K., Story, J.A., Vesselinovich, D., and Wessler, R.W.,** Experimental atherosclerosis in rabbits fed cholesterol-free diets. XVI. Comparison of peanut oil on pre-established lesions, *Atherosclerosis,* 31, 365, 1978.
14. **Kritchevsky, D., Tepper, S.A., Klurfeld, D.M., Vesselinovich, D., and Wessler, R.W.,** Experimental atherosclerosis in rabbits fed cholesterol-free diets. V. Comparison of corn, butter and coconut oils, *Exp. Mol. Pathol.,* 24, 375, 1976.
15. **Kritchevsky, D., Tepper, S.A., Klurfeld, D.M., Vesselinovitch, D., and Wessler, R.W.,** Experimental atherosclerosis in rabbits fed cholesterol-free diets. XII. Comparison of peanut and olive oils, *Atherogenesis,* 50, 253, 1984.
16. **Funch, J.P., Krogh, B., and Dam, H.,** Effects of butter, some margarines and arachis oil in purified diets on serum lipids and atherosclerosis in rabbits, *Br. J. Nutr.,* 14, 355, 1960.
17. **Alderson, L.J., Hayes, K.C., and Nicolasi, R.J.,** Peanut oil reduces diet-induced atherosclerosis in cynomolgus monkeys, *Arteriosclerosis,* 6, 465, 1986.
18. **Spiller, G., Jenkins, D., Cragen, L., Gates, J., Bosello, O., Berra, K., Rudd, C., Stevenson, M., and Superko, R.,** Effect of a diet high in monounsaturated fat from almonds on plasma cholesterol and lipoproteins, *Am. Coll. Nutr.,* 11, 126, 1992.
19. **Spiller, G., Gates, J., Jenkins, D., Bosello, O., Nichols, S., and Cragen, L.,** Effect of two foods high in monounsaturated fatty acids on plasma cholesterol and lipoproteins in adult humans, *Am. J. Clin. Nutr.,* 51 (Abstr.), 524, 1990.
20. **Sabaté, J., Fraser, G., Burke, K., Knutsen, S., Bennett, H., and Lindsted, K.,** Effects of walnuts on serum lipid levels and blood pressure in normal men, *N. Engl. J. Med.,* 328, 603, 1993.
21. **Sabaté, J. and Fraser, G.,** The probable role of nuts in preventing coronary heart disease, *Primary Cardiol.,* 19, 65, 1993.
22. **Moncada, S. and Higgs, A.,** The L-arginine-nitric-oxide pathway, *N. Engl. J. Med.,* 239, 2002, 1993.
23. **Cooke, J.P., Tsao, P., Singer, A., Wang, B., Koesk, J., and Drexler, H.,** Anti-atherogenic effect of nuts: is the answer NO?, *Arch. Intern. Med.,* 153(letter), 896, 1993.
24. **Cooke, J.P., Singer, A.H., Tsao, P., Zera, P., Rowan, R.A., and Billingham, M.E.,** Antiatherogenic effects of L-arginine in the hypercholesterolemic rabbit, *J. Clin. Invest.,* 90, 1168, 1992.
25. **Kritchevsky, D., Tepper, S.A., Czarnecki, S.K., and Klurfield, D.M.,** Artherogenicity of animal and vegetable protein-influence of the lysine to arginine ratio, *Atherosclerosis,* 41, 429, 1982.
26. **Sanchez, A. and Hubbard, R.W.,** Plasma amino acids and the insulin/glucagon ratio as an explanation for the dietary protein modulation of atherosclerosis, *Med. Hypotheses,* 35, 324, 1991.

27. Stampfer, M.J., Hannekens, C.H., Manson, J.E., Colditz, G.A., Rosner, B., and Willet, W.C., Vitamin E consumption and the risk of coronary heart disease in women, *N. Engl. J. Med.*, 328, 1444, 1993.

28. Rimm, E.B., Stampfer, M.J., Acherio, A., Giovannucci, E., Colditz, G., and Willet, W.C., Vitamin E consumption and the risk of coronary heart disease in women, *N. Engl. J. Med.*, 328, 1450, 1993.

29. Esterbauer, H., Dieber-Rotheneder, M., Striegl, G., and Waeg, G., Role of vitamin E in preventing the oxidation of low density lipoprotein, *Am. J. Clin. Nutr.*, 53, 314S, 1991.

30. Sato, K., Niki, E., and Shimisaki, H., Free radical-mediated chain oxidation of low density lipoprotein and its synergistic inhibition by vitamin E and vitamin C, *Arch. Biochem. Biophys.*, 279, 402, 1990.

31. Klevay, L., Copper in nuts may lower heart disease risk, *Arch. Intern. Med.*, 153(letter), 401, 1993.

32. Elin, R.J. and Hosseini, J.M., Is the magnesium content of nuts a factor for coronary heart disease?, *Arch. Intern. Med.*, 153(letter), 779, 1993.

33. Allen, K.G.D. and Klevay, L., Copper: an antioxidant nutrient for cardiovascular health, *Curr. Opinion Lipidol.*, 5, 22, 1994.

34. Marier, J.F., Cardio-protective contribution of hard waters to magnesium intake, *Rev. Can. Biol.*, 37, 115, 1978.

35. Eisenberg, M.J., Magnesium deficiency and cardiac arrythmias, *N.Y. State J. Med.*, March, 133, 1986.

36. Schechter, M., Had, H., Marks, N., et al., Beneficial effect of magnesium sulfate in acute myocardial infarction, *Am. J. Cardiol.*, 66, 271, 1990.

37. Klevay, L.M., Inman, L., Johnson, L.K., et al., Increased cholesterol in plasma in a young man during experimental copper depletion, *Metabolism*, 33, 1112, 1984.

38. Reiser, S., Smith, J.C., Jr., Mertz, W., et al., Indices of copper status in humans consuming a typical American diet containing either fructose or starch, *Am. J. Clin. Nutr.*, 49, 870, 1989.

39. Turnlund, J.R., Keyes, W.R., Anderson, H.L., and Acord, L.L., Copper absorption and retention in young men at three levels of dietary copper using the stable isotope 65-Cu, *Am. J. Clin. Nutr.*, 49, 870, 1989.

Chapter 3.3

OLIVE OIL AND HEALTH*

Apostolos (Paul) Kiritsakis, Antony Kafatos, and Maria Hassapidou

INTRODUCTION

Olive oil, which is extracted from the fruits of the tree *Olea europaea,* is of very old origin. The olive tree is prehistoric and its real home can be considered the Mediterranean area. Many ancient physicians such as Hippocrates and Galen recognized olive oil's therapeutic benefit and extolled it as a combined food and medicine. In recent years, doctors and nutritionists have realized that olive oil is a source of significant nutritional value in human health.

OLIVE OIL AND THE CHANGING DIET OF CRETE

Keys, in the "Seven Countries Study",[1] has shown that the risk of coronary disease is significantly lower in populations living in certain Mediterranean areas, where olive oil constitutes the main fat in the diet.

The study of the changing diet of Crete is helpful in understanding the role of olive oil in heart disease. In the mid-1970s, the risk factors for myocardial infarction had risen rapidly in the rural and urban populations of Crete.[2] The total serum cholesterol levels of middle-aged adult Cretans increased 25% over cholesterol levels in the early 1960s. This change can be attributed to the progressive shift in the traditional diet, a diet which had remained uniform for nearly 4000 years.[3] Indeed, olive oil intake accounted for 32.6% of total daily energy intake in 1960 and only 21.7% in 1979.[4] This level of consumption is reflected in the adipose fatty acid composition.

As shown in Table 1, saturated fatty acids in the adipose tissue increased by 24.5% from 1962 to 1988. The monounsaturated fatty acids decreased by 4%, while the polyunsaturated fatty acids decreased by 30%. This decrease in monounsaturated and polyunsaturated fatty acids in the adipose tissue of Cretan males corresponds almost to an analogous increase in death rates from circulatory system diseases in the Greek population from 1960-1964 to 1980-1984.[5]

CHANGES OF OLIVE OIL DURING HEATING

Olive oil under controlled heating conditions does not incur substantial chemical changes.[6] The formation of trans fatty acids during heating is lower in olive oil than in other oils[7] (Table 2).

---

* Editor's Note: Olive oil has played a special role in the diets of Mediterranean countries for millennia. A book on lipids in nutrition would not be complete without a short chapter on this oil. The olive oil discussed in this chapter is referred to as *extra virgin olive oil* and is high in many antioxidants. As other olive oils after refining may be different, the classification of the International Olive Oil Council can be found in Appendix V.

0-8493-4248-1/96/$0.00+$.50
© 1996 by CRC Press Inc.

TABLE 1
Fatty Acid Composition of Adipose Tissue of Cretan Adult
Males in 1962 and 1988

	Year (%)		
	---	---	---
	1962	1988	
Fatty acids	(200 men)	(250 men)	Change (%)
Saturated fatty acids	18.80	23.40	+24.5
Monounsaturated fatty acids	68.70	66.10	-04.0
Polyunsaturated fatty acids	12.30	08.60	-30.0
P/S ratio	00.70	00.36	-44.6
M/S ratio	03.70	02.82	-22.7

From Kafatos, A., et al., Adipose Tissue Fatty Acids Composition of Obese and Non-Obese Adult Cretans, 14th International Congress of Nutrition, Seoul, Korea, 1989. With permission.

TABLE 2
Trans Fatty Acid Content of Heated Vegetable Oils
at 200°C for 3.5 and 7 h (as Elaidic Acid, % by wt)

| | Trans fatty acids after heating for | | |
Oil	0 h	3.5 h	7.0 h
Greek olive oil I	0.5	2.0	5.5
Greek olive oil II	0.6	2.1	6.0
Italian olive oil	0.6	4.5	7.0
Corn oil	0.6	6.4	12.5
Sunflower oil (polyunsaturated)	0.5	5.5	12.0
Safflower oil (polyunsaturated)	0.6	5.5	13.5
Partially hydrogenated vegetable oil	5.5	6.5	12.0

From Kiritsakis, A., Aspris, P., and Markakis, P., Trans Isomerization of Certain Vegetable Oils During Frying, Proc. 6th Int. Flavor Conf., Rethymnon, Crete, Greece, 1989. With permission.

REFERENCES

1. **Keys, A.,** Coronary heart disease, in Seven Countries Study, *Circulation,* 44(Suppl. 1), 1970.
2. **Kafatos, A., Kouroumallis, I., Facourelis, N., and Vlachonikolis, J.,** Adipose Tissue Fatty Acids Composition of Obese and Non-Obese Adult Cretans, 14th Int. Congr. Nutrition, Seoul, Korea, 1989.
3. **Christakis, G., Fordyce, M.K., and Kurtz, C.S.,** The Biological and Medical Aspects of Olive Oil, Int. Olive Oil Council, Madrid, Spain, 1982.
4. **Aravanis, C.,** The Greek Islands Heart Study, 3rd Int. Cong. Biological Value of Olive Oil, Chanea, Crete, Greece, 1980.
5. **WHO,** Annual Statistical Yearbook, 1988.
6. **Kiritsakis, A.,** A book, American Oil Chemists' Society, Champaign, IL, 1991.
7. **Kiritsakis, A., Aspris, P., and Markakis, P.,** Trans Isomerization of Certain Vegetable Oils During Frying, Proc. 6th Int. Flavor Conf., Rethymnon, Crete, Greece, 1989.

Chapter 3.4

MODIFIED LDL IN THE PATHOGENESIS OF ATHEROSCLEROSIS: ROLE OF NUTRITION

L. Cominacini, U. Garbin, A. Fratta Pasini, A. Davoli, M. Campagnola, A. De Santis, A. M. Pastorino, and V. Lo Cascio

LDL AND ATHEROGENESIS

Today coronary heart disease (CHD) is the major cause of death and disability in most developed countries: in the U.S. alone a half million persons die each year of myocardial infarctions or strokes caused by narrowing of the arteries by atherosclerotic plaque.[1]

Until the 1980s only the clinical manifestations of the disease were widely studied, whereas the pathogenesis was largely unknown; in the last 10 years many studies were carried out in order to explain the biochemical mechanisms of atherogenesis.

Hypercholesterolemia is important in the pathogenesis of the atherosclerotic plaque, but it is not the only causal factor, because at any given level of hypercholesterolemia there is important variation in the clinical manifestations of the disease.[2]

The cholesterol that accumulates in the atherosclerotic plaque originates from plasma lipoproteins, especially low-density lipoproteins (LDL),[3] because LDL is the main carrier of cholesterol in plasma. It is now well established that high levels of LDL are associated with an increased risk of atherogenesis[4-6] and that lowering LDL levels decreases the risk of coronary heart disease. In recent years many clinical and experimental studies were performed to make clear the role of LDL in plaque formation.

The earliest macroscopically evident atherosclerotic lesion is the fatty streak, characterized by an accumulation of cells loaded with esterified and free cholesterol (foam cells) in the subendothelial space. These cells are macrophages derived from circulating monocytes that have entered the intima[7-9] and smooth muscle cells proliferating in the region of the plaque.[9,10] Many *in vitro* observations showed that macrophages do not develop into foam cells even if they are exposed for a long period to high concentrations of LDL.[11,12] The internalization of native LDL is mediated by LDL receptor (apo B/E receptor) which is downregulated by the intracellular cholesterol;[13] macrophages indeed possess only few B/E receptors.[14]

Since it is well known that foam cells develop in the presence of high concentrations of LDL, it was thought that native LDL ought to be first "modified" and then taken up to form the foam cell. The treatment of LDL with acetic anhydride, for example, leads to a form of LDL which is rapidly taken up by macrophages. This uptake is due to another receptor (scavenger-acetyl receptor) that is not downregulated by the intracellular cholesterol content;[13] this scavenger receptor mediates the uptake of some kinds of chemically or biologically modified LDLs but not of native LDL. Chemical modification blocks the epsilon amino group of the lysine residues of the apolipoprotein B and increases the negative charge of the LDL.[15]

In 1981 Henriksen showed that the incubation of LDL with endothelial cells induced physical and chemical modifications of LDL that allowed a more rapid uptake by macrophages than native LDL.[16,17] The same workers[17] and others[18,19] demonstrated that smooth muscle cells and macrophages were able to induce similar modifications, thus introducing the evidence that LDL could also be modified biologically by the cells present in the intima wall.

STRUCTURE AND COMPOSITION OF HUMAN LDL

Human LDLs are large, spherical particles with a diameter of 19 to 25 nm, a mean molecular weight of 2.5 million, and a density ranging from 1.019 to 1.063. Using equilibrium density-gradient ultracentrifugation and other techniques, LDL can be separated into two or more subfractions differing somewhat in density, size, and molecular weight.[20]

Each LDL holds cholesteryl ester and triglycerides in the lipophilic core that is surrounded by a monolayer of phospholipid and free cholesterol. Embedded in the outer layer is a large protein (apolipoprotein B) embracing the whole surface of the LDL. Apolipoprotein B is glycosylated and has a molecular weight of 550 kDa.[21]

The native LDL contains 22.3 ± 3.9% of phospholipids (62 to 66% of phosphatidylcoline, 24 to 28% of sphingomyelin, 7% of lysophosphatidylcholine, and 2 to 3% of others), 5.9 ± 2.7% of triglycerides, 9.6 ± 0.7% of free cholesterol, 42.2 ± 3.8% of cholesteryl esters, and 22.0 ± 1.9% of proteins.[21] The different lipid classes of LDL bind many fatty acid molecules; of these nearly half are polyunsaturated fatty acids (PUFAs), 88% of which are linoleic acid and 12% arachidonic acid, with traces of [docosohexaenoic] acid. The PUFAs are not equally distributed between the different lipid classes: linoleic acid is mainly (65%) contained in the cholesteryl esters, whereas arachidonic acid is mainly (68%) present in the phospholipid. [Docosohexaenoic] acid is present only in the phospholipids.[15] In each subject, probably due to different dietary habits, the kind of fatty acids varies greatly.

Moreover, LDL has a number of lipophilic antioxidants which protect the PUFAs from free radical attack and oxidation. The major antioxidant (on molar base) is alpha-tocopherol; a LDL particle holds about six molecules of alpha-tocopherol. All other antioxidants (gamma-tocopherol, carotenoids, oxycarotenoids, and ubiquinol-10) are present in much smaller amounts: for instance, only about one third of the LDL molecules contains beta-carotene.[15]

It has been proposed that alpha-tocopherol is the only important antioxidant, whereas carotenoids and ubiquinol-10 play only a very minor role in protecting LDL from oxidation.[22] Recently more than 20 different carotenoids have been reported,[23] and it seems reasonable that LDL contain several other lipid- and/or water-soluble antioxidants.

As for the PUFAs, there is a large variability in the antioxidant content of LDL between different donors. The reason for this high variability is not completely known, but it seems to reflect different dietary conditions.

MODIFICATION OF LDL

Oxidation of LDL is a lipid peroxidation process in which the PUFAs contained in the phospholipids are transformed into lipid hydroperoxides and then to some unsaturated aldehydes. These products can neutralize some positive charges from lysine, arginine, and histidine residues and increase the apo B affinity to the scavenger receptor.[24]

Endothelial cells,[16] smooth muscle cells,[17] and macrophages[25] have all been shown to be capable of oxidizing LDL *in vitro*. LDL can also be oxidized in a cell-free medium with a sufficiently high concentration of copper and iron.[26] The hypothesis of the mechanism of copper-stimulated oxidation of LDL is the following: traces of transition metal ions (copper or iron) can decompose small amounts of preformed lipid hydroperoxides (LOOH) contained in the LDL, in reactive alkoxy (LO·), and peroxy radicals (LOO·). These radicals begin the lipid peroxidation chain reaction in which the PUFAs of the LDL (mainly linoleic and arachidonic acid) are oxidized to lipid hydroperoxides. These are transformed, in the presence of traces of metal ions, into new alkoxy and peroxy radicals which initiate new chain reactions and amplify the phenomenon. Lipid alkoxy radicals decompose to a great variety of reactive aldehydes which react with the epsilon amino groups of lysine of the apo B. This transformed apo B has an increased negative charge on the surface, is recognized by the scavenger receptor on macrophages, and leads to foam cell formation.

The process starts only when all the LDL antioxidants (alpha-tocopherol, retinoids, carotenoids) are consumed; the antioxidants therefore protect the LDL from the oxidation.[27]

At present the chemical processes involved in the cell-induced LDL modification are not fully identified. It is known that oxidation of LDL is accelerated by metal ions and inhibited by chelating agents. It seems that LDL oxidation by cells is induced by free radicals: their source in the cells, however, is not clear and we do not know whether cells can oxidize LDL by releasing lipid oxidizing agents (superoxide or hydrogen peroxide) which then attack LDL, or by releasing lipid peroxidation products formed during oxidation of the cellular lipids (through the action of the lipoxygenase). These peroxides can initiate the oxidation of the polyunsaturated acid of the LDL to lipid hydroperoxides. These decompose to aldehydes which link with lysine of the apo B.[14,28]

The LDL oxidation can be divided in three consecutive phases: lag phase, propagation phase, and decomposition phase. During the lag phase the LDL antioxidants are consumed and no significant oxidation of PUFAs takes place. When LDL is depleted of its antioxidants, PUFAs are rapidly oxidized to lipid hydroperoxides, which are then converted to a variety of other products including reactive aldehydes. The covalent binding of aldehydes to apo B is probably involved in the characteristic epitopes which are recognized in modified LDL by scavenger receptors. In fact, many of the chemical (loss of NH_2 groups), physicochemical (increased fluorescence and relative electrophoretic mobility), and biological (increased uptake by macrophages, cytotoxicity) characteristics of cell- or copper-oxidized LDL can also be obtained by incubating native LDL with aldehydes (such as malondialdehyde, hydroxynonenal).[29]

In conclusion, whichever mechanism is involved in initiating the modification of LDL, it seems clear that the subsequent processes are always the same: loss of antioxidants, lipid peroxidation, and decomposition of lipid peroxides to aldehydes and other products. At the end of the decomposition phase, LDL has more or less similar biological and chemical properties, independently of how the oxidation is initiated.

POSSIBLE ROLE OF THE MODIFIED LDL IN THE PATHOGENESIS OF THE ATHEROSCLEROTIC PLAQUE

Recent studies make it possible to propose a hypothesis for the development of fatty-streak lesions that is based on the presence of elevated plasma LDL levels plus the oxidative modification of LDL within the artery wall. LDL may enter the intima (where there is a low concentration of antioxidants) and can be modified into mildly oxidized LDL by free radicals or oxidized lipids released from endothelial cells and smooth muscle cells.[16,17,25,30,31] This mildly oxidized LDL may induce endothelial cells to produce chemotactic factors which are able to recruit circulating monocytes in the subendothelial space and to become resident macrophages;[32] these cells can release large amounts of free radicals and lipid peroxides, which transform mildly LDL into highly oxidized LDL. The resident macrophages express the acetyl receptor and can rapidly take up highly oxidized LDL to form foam cells.

Because of its content in lysophosphatidylcholine the highly oxidized LDL can be directly chemotactic for monocytes, which enter the intima faster and lead to the rapid progression of the oxidation process.[32]

Furthermore, oxidized LDL inhibits the mobility of the macrophages which cannot leave the intima and help to rapidly increase the oxidative process of LDL.[32] There is some evidence that these events also occur *in vivo:* LDL extracted from human lesions are oxidized and taken up more rapidly by macrophages and recognized by their scavenger receptors.[33]

SUSCEPTIBILITY TO LDL OXIDATION

As reported above, the modification of LDL can be a pivotal step in the pathogenesis of the atherosclerotic plaque.

Many studies[34,35] have reported that even in normal subjects, the LDL predisposition to copper or cell-induced oxidation varies greatly. Until today the reasons for this marked variability in the extent of LDL oxidation under the same oxidative stress are not known. It has been proposed that some compositional differences in the lipoproteins (i.e., antioxidant content or variability in fatty acid composition and physicochemical properties) may give rise to different susceptibilities of LDL.[15]

Since the oxidative modification of LDL comes from a lipid peroxidation, it was proposed that water- and lipid-soluble antioxidants should be important in delaying this modification; in fact, a great amount of studies, using different antioxidants, have confirmed this hypothesis.

There are at least four nutrients which, at least in theory, could have antioxidant effects: alpha-tocopherol,[19,36] beta-carotene,[37] ascorbic acid (vitamin C),[19,35,36] and flavonoids.[38] Moreover, butylated hydroxytoluene, a common food additive and antioxidant, seems to prevent cholesterol oxidation and atherosclerosis in rabbits.[39]

Alpha-tocopherol content in the LDL is about 30 to 50% of the total plasma concentration and it is generally believed that alpha-tocopherol is one of the principal biological defenses against lipid peroxidation.[40,41] Therefore, an LDL with a high alpha-tocopherol content was supposed to be more resistant to oxidation (expressed as duration of lag phase and rate of the propagation phase) compared to one with a low antioxidant content. Surprisingly, however, it has been repeatedly demonstrated[21,34,42] that there was no correlation between the concentration of alpha-tocopherol and the length of the lag phase.[15,42] The protective effect of alpha-tocopherol and beta-carotene was confirmed in cultures of endothelial cells,[19,43] smooth muscle cells,[19] and monocyte-macrophages.[36,42]

Vitamin C is a water-soluble antioxidant and it has been recently reported that it is capable of inhibiting metal ion and cell-mediated oxidation of LDL.[44] Its major role seems to be the sparing effect on alpha-tocopherol and carotenoids: during copper oxidation it can reactivate alpha-tocopherol and carotenoid radicals in LDL and prolong the lag phase in a concentration-dependent fashion.[45,46]

An important contribution to the susceptibility to LDL oxidation could come from the amount of PUFAs and monounsaturated fatty acids (MUFAs) in the LDL. As described above, the oxidative modification of the LDL involves principally the PUFAs, so LDL with a low PUFA content are potentially less susceptible to oxidation. Nevertheless, it is not known whether LDL fatty acid composition is an important determinant of the susceptibility toward oxidation.

LDL is a heterogeneous class of macromolecules differing in physicochemical composition.[47] Each individual shows a specific LDL subfraction pattern. De Graaf et al.[48] separated human LDL into three subfractions differing in density: LDL_1 (very light), LDL_2 (light), and LDL_3 (dense), and assessed their oxidation resistance by measuring the diene vs. time profile in copper-stimulated oxidation. It has been shown that dense LDL fractions (LDL_2, LDL_3) have a reduced lag phase and contain more conjugated dienes after 4 h oxidation, probably because of their higher content in PUFAs.

EFFECT OF ANTIOXIDANT SUPPLEMENTATION ON ATHEROSCLEROSIS

Many studies performed in the last few decades have tried to assess if an elevated plasma antioxidant content and/or an antioxidant treatment could inhibit or decrease the progression of atherosclerosis.[49]

While in animals the earliest report showed no effect,[50] many recent studies have demonstrated that antioxidant supplementation can reduce experimental atherosclerosis. Wilson et al.,[51] in fact, reported that alpha-tocopherol can prevent the endothelial damages induced by a triglyceride-rich diet in rabbits. Moreover, a recent study in primates showed that supplementation with alpha-tocopherol reduced the severity and rate of progression of atherosclerosis.[52]

Several studies with probucol, a hypocholesterolemic drug with antioxidant properties, showed that it can reduce atherosclerotic lesions in animals.[53-55] Furthermore, in the Watanabe heritable hyperlipidemic rabbit probucol was effective in preventing atherosclerosis in long-term studies at both early and late stages.[56] It is not known if these observations can be extended to humans. The results of the first controlled trial in hypercholesterolemic patients (Probucol Quantitative Regression Swedish Trial — PQRST) are still awaited; the treatment period of the study is 3 years and the progression of atherosclerosis is assessed angiographically.[57]

Controlled clinical trials on the effects of alpha-tocopherol and other antioxidants on the progression of atherosclerosis in humans have not been published yet.

An important contribution, however, derives from epidemiological studies: in the WHO/Monica project,[58,59] for instance, the plasma antioxidant content (alpha-tocopherol, vitamin C, vitamin A, carotenoids, and selenium) was measured in 16 European populations and correlated with the incidence of coronary heart disease. It has been shown that there was a strong inverse correlation between the incidence of coronary heart disease and the level of alpha-tocopherol. The same conclusions were found in the Nurses Health Survey,[60] which was a prospective study on the relationship between dietary alpha-tocopherol intake and cardiovascular risk, and in the Edinburgh study.[61]

EFFECT OF DIETARY FATTY ACIDS ON ATHEROSCLEROSIS

The LDL of people consuming Western diets are rich in linoleic, palmitic, and oleic acid; arachidonic acid (20:4) is the only long-chain polyunsaturated acid present, whereas there are only traces of [eicosapentaenoic] and [docosohexaenoic] acids.[46]

It is well known that the fatty acid composition of the diet influences LDL cholesterol levels: numerous studies, in fact, show that the replacement of dietary saturated fatty acids with both MUFAs or PUFAs significantly reduces cholesterol levels.[62-65] Since the more unsaturated the fatty acids, the more susceptible they are to oxidation, it is usually believed (but not established) that a diet rich in PUFAs may increase the susceptibility of LDL to oxidation. Support for the hypothesis was recently given in a study by Parthasarathi et al. in animals.[66] These authors demonstrated that a diet rich in PUFAs increased the LDL susceptibility to oxidation when compared to a diet rich in MUFAs.

These data were confirmed by Berry[67] and Bonanome[68] in humans; they demonstrated that the degree of LDL oxidation was much higher in the subjects given a diet rich in PUFAs than in MUFAs. The results of these authors support the idea that oleic acid-rich diets render LDL more resistant to oxidation and that this effect is independent of the LDL antioxidant content.[68] Another study,[69] on the contrary, showed that dietary supplementation with rapeseed oil, an oil rich in oleic acid, prolonged the lag phase as measured by the formation of conjugated dienes, but that the augmented resistance of LDL to oxidation was, at least in part, due to the increased alpha-tocopherol content of the oil. Epidemiological and animal studies[70-72] and evidence from human clinical trials[73,74] suggest that dietary supplementation with omega-3 fatty acids contained in fish oil may prevent the progression of atherosclerosis. Unfortunately there are few data on the effect of these PUFAs on LDL oxidation and the only two studies published until now showed contrasting results.[75,76] Like rapeseed oil and vegetable oils, fish oil contains large amounts of antioxidants[77] and so the results of these studies should separate the effect of the PUFAs per se from that of the antioxidants.

CONCLUSIONS

Although the etiology of atherosclerosis is multifactorial, there is increasing evidence that modification of LDL is an important step in the pathogenesis of the atherosclerotic plaque. Several natural antioxidants presumably protect against the initial peroxidation and the subsequent formation of lipid peroxidation breakdown products. This suggests that antioxidant therapy could prove useful in prevention of early atherosclerotic lesions. Epidemiological studies do not, of course, provide evidence of cause and effect: clinical studies are required for this. On the basis of the results of these studies, antioxidants may be included in the recommendations for the prevention of atherosclerosis. For the time being, however, although antioxidant therapy is relatively benign, it should not be viewed as a substitute for lipid-lowering measures but utilized in conjunction with them.

REFERENCES

1. Lipid Research Clinics Program, The Lipid Research Clinics coronary prevention trial results. I. Reduction in incidence of coronary heart disease, *J.A.M.A.*, 251, 358, 1984.
2. Piper, J. and Orrild, L., Essential familial hypercholesterolemia and xanthomatosis: follow-up study of twelve Danish families, *Am. J. Med.*, 21, 34, 1956.
3. Newman, H.A.I. and Zilversmith, D.B., Quantitative aspects of cholesterol flux in rabbit atheromatous lesions, *J. Biol. Chem.*, 237, 2078, 1962.
4. Tyroler, H.A., Lowering plasma cholesterol levels decreases the risk of coronary heart disease: an overview on clinical trials, in *Hypercholesterolemia and Atherosclerosis*, Steinberg, D. and Olefsky, J.M., Eds., Churchill Livingstone, New York, 1987, 99.
5. Goldstein, J.L. and Brown, M.S., The low-density lipoprotein pathway and its relation to atherosclerosis, *Annu. Rev. Biochem.*, 46, 897, 1977.
6. Steinberg, D., Lipoproteins and atherosclerosis: a look back and a look ahead, *Arteriosclerosis*, 3, 283, 1983.
7. Gerrity, R.G., The role of the monocyte in the atherogenesis. I. Transition of blood-borne monocytes into foam cells in fatty lesions, *Am. J. Pathol.*, 103, 181, 1981.
8. Gerrity, R.G., The role of the monocyte in the atherogenesis. II. Migration of foam cells from atherosclerotic lesions, *Am. J. Pathol.*, 103, 191, 1981.
9. Ross, R. and Glomset, J.A., The pathogenesis of atherosclerosis. I, *N. Engl. J. Med.*, 95, 369, 1976.
10. Geer, J.C., McGill, H.C., and Strong, J.P., The fine structure of human atherosclerotic lesions, *Am. J. Pathol.*, 38, 263, 1961.
11. Brown, M.S. and Goldstein, J.L., Lipoprotein metabolism in the macrophage: implications for cholesterol deposition in atherosclerosis, *Annu. Rev. Biochem.*, 52, 223, 1983.
12. Goldstein, J.L., Ho, Y.K., Basu, S.K., and Brown, M.S., Binding site on macrophages that mediates uptake and degradation of acetylated low density lipoprotein, producing massive cholesterol deposition, *Proc. Natl. Acad. Sci. U.S.A.*, 76, 333, 1979.
13. Brown, M.S. and Goldstein, J.L., A receptor-mediated pathway for cholesterol homeostasis, *Science*, 232, 34, 1986.
14. Steinberg, D., Parthasarathy, S., Carew, T.E., et al., Beyond cholesterol. Modifications of low-density lipoprotein that increase its atherogenicity, *N. Engl. J. Med.*, 14, 915, 1989.
15. Esterbauer, H., Rotheneder, M., Striegl, G., et al., Vitamin E and other lipophilic antioxidants protect LDL against oxidation, *Fat. Sci. Technol.*, 8, 316, 1989.
16. Henriksen, T., Mahoney, E.M., and Steinberg, D., Enhanced macrophage degradation of low-density lipoprotein previously incubated with cultured endothelial cells: recognition by receptor for acetylated low density lipoproteins, *Proc. Natl. Acad. Sci. U.S.A.*, 78, 6499, 1981.
17. Henriksen, T., Mahoney, E.M., and Steinberg, D., Enhanced macrophage degradation of biologically modified low-density lipoprotein, *Arteriosclerosis*, 3, 149, 1983.
18. Heinecke, J.W., Rosen, H., and Chait, A., Iron and copper promote modification of low-density lipoprotein by human arterial smooth muscle cells in culture, *J. Clin. Invest.*, 74, 1890, 1984.
19. Morel, D.W., Di Corleto, P.E., and Chisolm, G.M., Endothelial and smooth muscle cells alter low-density lipoprotein in vitro by free radical oxidation, *Arteriosclerosis*, 4, 357, 1984.
20. Myant, N.B., *Cholesterol Metabolism, LDL and the LDL Receptor*, Academic Press, San Diego, 1990, 99.
21. Esterbauer, H., Gebicki, J., Puhl, H., and Jurgens, G., The role of lipid peroxidation and antioxidants in oxidative modification of LDL, *Free Rad. Biol. Med.*, 13, 341, 1992.
22. Halliwell, B., How to characterize a biological antioxidant, *Free Rad. Res. Comm.*, 9, 1, 1990.

23. **Di Mascio, P., Kaiser, S., and Sies, H.,** Lycopene as the most efficient singlet oxygen quencher, *Arch. Biochem. Biophys.,* 274, 532, 1989.
24. **Steinberg, D., Parthasarathy, S., Carew, T.E., et al.,** Modifications of low-density lipoprotein that increase its atherogenicity, *N. Engl. J. Med.,* 14, 915, 1989.
25. **Parthasarathy, S., Printz, D.J., Boyd, D., et al.,** Macrophage oxidation of low-density lipoprotein generates a modified form recognized by the scavenger receptor, *Arteriosclerosis,* 6, 505, 1986.
26. **Steinbrecher, U.P.,** Oxidation of human low density lipoprotein results in derivatization of lysine residues of apoprotein B by lipid peroxide decomposition products, *J. Biol. Chem.,* 262, 3603, 1987.
27. **Esterbauer, H., Jurgens, G., Quehenberger, O., and Koller, E.,** Autoxidation of human low-density lipoprotein: loss of polyunsaturated fatty acids and vitamin E and generation of aldehydes, *J. Lipid Res.,* 28, 495, 1987.
28. **Esterbauer, H., Dieber-Rotheneder, M., Waeg, G., et al.,** Biochemical, structural and functional properties of oxidized low-density lipoprotein, *Chem. Res. Toxicol.,* 3, 77, 1990.
29. **Jurgens, G., Hoff, H.F., Chisolm, G.M., and Esterbauer, H.,** Modification of human serum low-density lipoprotein by oxidation-characterization and pathophysiological implications, *Chem. Phys. Lipids,* 45, 315, 1987.
30. **Leake, D.S. and Rankin, S.M.,** The oxidative modification of low-density lipoproteins by macrophages, *Biochem. J.,* 270, 741, 1990.
31. **Lamb, D.J., Wilkins, G.M., and Leake, D.S.,** The oxidative modification of low-density lipoproteins by human lymphocytes, *Atherosclerosis,* 92, 187, 1992.
32. **Leake, D.S.,** Effects of mildly oxidized low-density lipoprotein on endothelial cell function, *Curr. Opinion Lipidol.,* 2, 301, 1991.
33. **Yla-Herttuala, S., Palinski, W., Rosenfeld, M.E., et al.,** Evidence for the presence of oxidatively modified low-density lipoprotein in atherosclerotic lesions of rabbit and man, *J. Clin. Invest.,* 84, 1086, 1989.
34. **Cominacini, L., Garbin, U., Cenci, B., et al.,** Predisposition to LDL oxidation during copper-catalyzed oxidative modification and its relation to alfatocopherol content in humans, *Clin. Chim. Acta,* 204, 57, 1991.
35. **Jialal, I., Freeman, D.A., and Grundy, S.M.,** Varying susceptibility of different low-density lipoproteins to oxidative modification, *Arterioscler. Thromb.,* 11, 482, 1991.
36. **Cathcart, M.K., Morel, D.W., and Chisolm, G.M.,** Monocytes and neutrophils oxidize low-density lipoprotein making it cytotoxic, *J. Leukocyte Biol.,* 38, 341, 1985.
37. **Jialal, I., Norkus, E.P., Cristol, L., and Grundy, S.M.,** Beta-carotene inhibits the oxidative modification of low-density lipoprotein, *Biochim. Biophys. Acta,* 1086, 134, 1991.
38. **Das, N.P. and Ramanathan, L.,** Studies on flavonoids and related compounds as antioxidants in food, in *Lipid-Soluble Antioxidants: Biochemistry and Clinical Applications,* Ong, A.S.H. and Packer, L., Eds., Birkhauser Verlag, Basel, 1992.
39. **Bjorkhem, I., Henriksson-Freyschuss, A., Breuer, O., et al.,** The antioxidant butylated hydroxytoluene protects against atherosclerosis, *Arterioscler. Thromb.,* 11, 15, 1991.
40. **Fukuzawa, K. and Gebicki, J.M.,** Oxidation of alpha-tocopherol in micelles and liposomes by the hydroxyl, perhydroxyl, and superoxide free radicals, *Arch. Biochem. Biophys.,* 226, 242, 1983.
41. **Burton, G.W., Joice, A., and Ingold, K.U.,** Is vitamin E the only lipid-soluble, chain-breaking antioxidant in human blood plasma erythrocyte membranes?, *Arch. Biochem. Biophys.,* 221, 281, 1983.
42. **Jessup, W., Rankin, S.M., De Whalley, C.V., et al.,** Alpha-tocopherol consumption during low-density-lipoprotein oxidation, *Biochem. J.,* 265, 399, 1990.
43. **Steinbrecher, U.P., Parthasarathy, S., Leake, D.S., et al.,** Modification of low-density lipoprotein by endothelial cells involves lipid peroxidation and degradation of low-density lipoprotein phospholipids, *Proc. Natl. Acad. Sci. U.S.A.,* 81, 3883, 1984.
44. **Jialal, I. and Grundy, S.,** Preservation of the endogenous antioxidants in low-density lipoprotein by ascorbate but not probucol during oxidative modification, *J. Clin. Invest.,* 87, 597, 1991.
45. **Esterbauer, H., Striegl, G., Puhl, H., et al.,** The role of vitamin E and carotenoids in preventing oxidation of low-density lipoproteins, *Ann. N.Y. Acad. Sci.,* 570, 254, 1989.
46. **Esterbauer, H., Dieber-Rotheneder, M., Striegl, G., and Waeg, G.,** Role of vitamin E in preventing the oxidation of low-density lipoprotein, *Am. J. Clin. Nutr.,* 53, 314S, 1991.
47. **La Belle, M. and Krauss, R.M.,** Differences in carbohydrate content of low-density lipoproteins associated with low-density lipoprotein subclass patterns, *J. Lipid Res.,* 31, 1577, 1990.
48. **De Graaf, J., Hak-Lemmers, H., Hectors, M., et al.,** Enhanced susceptibility to in vitro oxidation of the dense low-density lipoprotein subfraction in healthy subjects, *Arteriosclerosis,* 11, 298, 1990.
49. **Shute, W.E., Shute, E.V., and Vogelsang, A.,** The physiological and biochemical basis for the use of vitamin E in cardiovascular disease, *Ann. Intern. Med.,* 30, 1004, 1948.
50. **Dam, H. and Kelman, E.M.,** The effect of vitamin E on the blood plasma lipid of the chick, *Science,* 96, 430, 1942.
51. **Wilson, R.B., Middleton, C.C., and Sun, G.Y.,** Vitamin E, antioxidants and lipid peroxidation in experimental atherosclerosis of rabbits, *J. Nutr.,* 108, 1858, 1978.

52. **Verlangieri, A.J. and Bush, M.J.,** Effects of d-alpha tocopherol supplementation on experimentally induced primate atherosclerosis, *J. Am. Coll. Nutr.,* 11, 131, 1992.

53. **Kita, T., Nagano, Y., Yokode, M., et al.,** Probucol prevents the progression of atherosclerosis in Watanabe heritable hyperlipidemic rabbit, an animal model for familial hypercholesterolemia, *Proc. Natl. Acad. Sci. U.S.A.,* 84, 5928, 1987.

54. **Carew, T.E., Schwenke, D.C., and Steinberg, D.,** Antiatherogenic effect of probucol unrelated to its hypocholesterolemic effect: evidence that antioxidants in vivo can selectively inhibit low density lipoprotein degradation in macrophage-rich fatty streaks and slow the progression of atherosclerosis in the Watanabe heritable hyperlipidemic rabbit, *Proc. Natl. Acad. Sci. U.S.A.,* 84, 7725, 1987.

55. **Mao, S.J.T., Yates, M.T., Rechtin, A.E., et al.,** Antioxidant activity of probucol and its analogs in hypercholesterolemic Watanabe rabbits, *J. Med. Chem.,* 34, 298, 1991.

56. **Nagano, Y., Nakamura, T., Matsuzawa, Y., et al.,** Probucol and atherosclerosis in the Watanabe heritable hyperlipidemic rabbit: long-term antiatherogenic effect and effects on established plaques, *Atherosclerosis,* 92, 131, 1992.

57. **Regnstrom, J., Walldius, G., Carlson, L.A., and Nilsson, J.,** Effect of probucol treatment on the susceptibility of low-density lipoprotein isolated from hypercholesterolemic patients to become oxidatively modified in vitro, *Atherosclerosis,* 82, 43, 1990.

58. **Gey, K.F. and Puska, P.,** Plasma vitamins E and A inversely related to mortality from ischemic heart disease in cross-cultural epidemiology, *Ann. N.Y. Acad. Sci.,* 570, 268, 1989.

59. **Gey, K.F., Puska, P., Jordan, P., and Moser, U.K.,** Inverse correlation between plasma vitamin E and mortality from ischemic heart disease in cross-cultural epidemiology, *Am. J. Clin. Nutr.,* 53, 326B, 1991.

60. **Steinberg, D., Berliner, J.A., Burton, G.W., et al.,** Antioxidants in the prevention of human atherosclerosis. Summary of the proceedings of a National Heart, Lung, and Blood Institute Workshop — September 5-6, 1991, Bethesda, Maryland, *Circulation,* 85, 2338, 1992.

61. **Riemersma, R.A., Wood, D.A., Macintyre, C.C.A., et al.,** Anti-oxidants and pro-oxidants in coronary heart disease, *Lancet,* 337, 677, 1991.

62. **Grundy, S.M. and Denke, M.A.,** Dietary influences on serum lipids and lipoproteins, *J. Lipid Res.,* 31, 1149, 1990.

63. **Mattson, F.H. and Grundy, S.M.,** Comparison of effects of dietary saturated, monounsaturated and polyunsaturated fatty acids on plasma lipids and lipoproteins in man, *J. Lipid Res.,* 26, 194, 1985.

64. **Baggio, G., Pagnan, A., Muraca, M., et al.,** Olive oil-enriched diet: effect on serum lipoprotein levels and biliary cholesterol saturation, *Am. J. Clin. Nutr.,* 47, 960, 1988.

65. **Mensink, R.P. and Katan, M.B.,** Effect of a diet enriched with monounsaturated or polyunsaturated fatty acids on levels of low density and high density lipoprotein cholesterol in healthy men and women, *N. Engl. J. Med.,* 321, 436, 1989.

66. **Parthasarathy, S., Khoo, J.C., Miller, E., et al.,** Low-density lipoprotein rich in oleic acid is protected against oxidative modification: implications for dietary prevention of atherosclerosis, *Proc. Natl. Acad. Sci. U.S.A.,* 87, 3894, 1990.

67. **Berry, E., Eisenberg, S., Haratz, D., et al.,** Effects of diets rich in monounsaturated fatty acids on plasma lipoproteins: the Jerusalem Nutrition Study — High MUFAs vs high PUFAs, *Am. J. Clin. Nutr.,* 53, 899, 1991.

68. **Bonanome, A., Pagnan, A., Biffanti, S., et al.,** Effect of dietary monounsaturated and polyunsaturated fatty acids on the susceptibility of plasma low density lipoproteins to oxidative modification, *Arterioscler. Thromb.,* 12, 529, 1992.

69. **Corboy, J., Sutherland, W.H.F., and Ball, M.J.,** Fatty acid composition and the oxidation of low density lipoproteins, *Biochem. Med. Met. Biol.,* 49, 25, 1993.

70. **Kromhout, D., Bosschieter, E.B., and Coulander, C. de L.,** The inverse relation between fish consumption and 20-year mortality from coronary heart disease, *N. Engl. J. Med.,* 312, 1205, 1985.

71. **Dolecek, T.A. and Grandits, G.,** Dietary polyunsaturated fatty acids and mortality in the Multiple Risk Factor Interventional Trial (MRFIT), *World Rev. Nutr. Diet,* 66, 205, 1991.

72. **Davis, H.R., Bridenstine, R.T., Vesselinovitch, D., and Wissler, R.W.,** Fish oil inhibits development of atherosclerosis in rhesus monkeys, *Arteriosclerosis,* 7, 441, 1987.

73. **Slack, J.D., Pinkerton, C.A., Van Tassel, J., et al.,** Can oral fish oil supplement minimize re-stenosis after percutaneous transluminal coronary angioplasty?, *J. Am. Coll. Cardiol.,* 9(Suppl.), 64A, 1987.

74. **Dehmer, G.J., Popma, J.J., van den Bergh, E.K., et al.,** Reduction in the rate of early re-stenosis after coronary angioplasty by a diet supplemented with n-3 fatty acids, *N. Engl. J. Med.,* 319, 733, 1988.

75. **Harats, D., Dabach, Y., Hollander, G., et al.,** Fish oil ingestion in smokers and nonsmokers enhances peroxidation of plasma lipoproteins, *Atherosclerosis,* 90, 127, 1991.

76. **Nenseter, M.S., Rustan, A.C., Lund-Katz, S., et al.,** Effect of dietary supplementation with n-3 polyunsaturated fatty acids on physical properties and metabolism of low-density lipoprotein in humans, *Arteriosclerosis,* 12, 369, 1992.

77. **Harris, W.S.,** Fish oils and plasma lipid and lipoprotein metabolism in humans: a critical review, *J. Lipid Res.,* 30, 785, 1989.

Chapter 3.5

SHORT CHAIN FATTY ACID PRODUCTION AND POSSIBLE METABOLIC CONSEQUENCES IN HUMANS

Tiziana Todesco, Mauro Zamboni, Fabio Armellini, Luisa Bissoli, Emanuela Turcato, and Ottavio Bosello

The short chain fatty acids (SCFA) — acetic, propionic, and butyric acid — are produced in the colon by fermentation of unabsorbed dietary carbohydrate.[1] It has been demonstrated that SCFA are absorbed rapidly by the human jejunum, colon, and rectum.[2-4] It is estimated that the daily production of SCFA is 200 to 500 mmol, in an acetate:propionate:butyrate molar ratio of approximately 60:25:15.[5,6]

Some butyrate is used by the colonic epithelial cells and the poorly absorbed portion is cleared by the liver as propionate. A small amount of these acids appears in the peripheral circulation. Acetate is absorbed and passes into circulation in larger amounts and can be used as fuel for different tissues.[7-11]

Data regarding the influence of dietary fiber on fecal SCFA in laboratory rodents, pigs, and non-human primates have been reviewed elsewhere.[12] Table 1 summarizes the influence of dietary fiber on fecal SCFA in human subjects.[13-20] The data in the table demonstrate an effect of starch and dietary fiber on colonic fermentation in humans. Most of the studies have shown an increase in SCFA output rather than in fecal concentration, with an excretion closely related to total fecal output.

Different forms of dietary fiber have different effects and there is considerable intersubject variability. More fermentable sources of fiber, such as cabbage, have higher SCFA excretion compared to other less fermentable substrates like cellulose.

Even though the effect of purification and isolation of fiber polymers on their physiological function is not well known, it appears that isolated polymers must be studied as well as natural foods with a view to the possible dietary use of both purified and natural dietary fiber and their clinical implications.

In vitro experiments involving fermentation by fecal incubation of fecal homogenates suggest that the ratio of acetate to propionate produced varies among individuals but is constant over time in the same individual. The fermentation of different pure substrates also yields different acetate-propionate ratios varying from 1.8:1 for guar to >5:1 for pectin, lactose, and glucose.[21-23]

The *in vitro* incubation system has been used because of the very limited possibility of studying SCFA concentrations of relevant size such as in the portal vein in humans. An alternative approach is to study their concentrations in a peripheral vein.

Table 2 summarizes the studies that have been conducted in humans on the effect of different fibers on acetate levels in the blood.[24-32] There are no *in vivo* studies to the best of the authors' knowledge on the effect of fermentation on propionate levels in the blood, possibly because of the very low concentration expected.

The data in Table 2 seem to suggest that it is difficult to assess the total amounts of SCFA produced, but with serial blood determination and areas under the curve, comparisons can be made between carbohydrates. We have to consider that acetate turnover is fast, but that it correlates positively with plasma acetate concentration.[7-11]

Further studies with different amounts and types of fiber are needed for a better understanding of the different effects on SCFA concentrations in the blood. These studies are useful for investigating the impact of acetate and propionate on glucose and lipid metabolism, which may explain some of the physiologic responses attributed to dietary fiber.

TABLE 1
Influence of Dietary Fiber on Fecal SCFA Excretion in Humans

Subjects	Experimental design	Fiber source	Control diet				Fiber diet				Ref.
			SCFA	C2	C3	C4	SCFA	C2	C3	C4	
Humans: 6 males, age 23–25	3 weeks control diet: 17 g fiber; Bran diet: 45 g	Bran	93 mmol/kg; 7 mmol/d				87 mmol/kg; 20 mmol/d				13
Humans: 5 males, age 21–32	9 d for each diet; Control diet; Cellulose; xylan and pectin: 0.5 g/kg of body w/d; Corn bran: 0.6 g/kg/d	Xylan; Pectin; Corn bran; Cellulose	**g/3 d** 1.47	0.57	0.39	0.22	**g/3 d** 1.16[a]; 2.02[a]; 2.68[a]; 1.8	0.38; 0.67; 1.29[a]; 0.61	0.37; 0.56[a]; 0.52[a]; 0.49[a]	0.20; 0.40[a]; 0.41[a]; 0.33[a]	14
			% of dry fecal solids 3.11	1.17	0.81	0.50	**% of dry fecal solids** 2.30[a]; 2.9[a]; 1.88[a]; 1.32[a]	0.97[a]; 0.99[a]; 0.95[a]; 0.46[a]	0.67; 0.79; 0.34[a]; 0.35[a]	0.33[a]; 0.56[a]; 0.27[a]; 0.24[a]	
Humans: 12 males, age 21–35	23 d; Beans: 100 g/d; Fiber: 41.7 ± 9.3; Control diet: fiber 33.3 ± 11.5	Beans	**mg/3 d** 3354	1428	688	900	**mg/3 d** 3887	1721	761	989	15
			mg/g feces 6.32	2.75	1.22	1.64	**mg/g feces** 6.87[a]	3.05	1.33	1.74	
Humans: 42 males/females, age 23–60	2 weeks: control diet; 3 weeks fiber; Cellulose (n 13 subjects):14 g/d; Pectin (n 12 subjects): 6 g/d; Placebo (n 14 subjects)	Cellulose; Pectin; Placebo	**g/7 d (week 2)** 3.0 ± 2.5; 2.7 ± 2.6; 3.6 ± 1.6	1.5 ± 1.1; 1.3 ± 1.1; 1.8 ± 0.7	0.5 ± 0.4; 0.5 ± 0.5; 0.7 ± 0.3	0.5 ± 0.6; 0.5 ± 0.6; 0.6 ± 0.3	**g/7 d (week 4)** 4.4 ± 2.7[a]; 3.8 ± 4.3[a]; 3.0 ± 1.4[a]; **g/7 d (week 5)** 4.3 ± 3.4[a]; 3.3 ± 3.7[a]; 2.4 ± 1.4[a]	2.1 ± 1.2; 2.0 ± 2.3; 1.4 ± 0.6; 2.1 ± 1.6; 1.7 ± 1.9; 1.2 ± 0.7	0.9 ± 0.5; 0.7 ± 0.8; 0.5 ± 0.3; 0.9 ± 0.7; 0.6 ± 0.7; 0.5 ± 0.3	0.8 ± 0.7; 0.7 ± 1.0; 0.5 ± 0.3; 0.7 ± 0.8; 0.6 ± 0.7; 0.4 ± 0.2	16

Subjects	Diet / duration & dose	Diet details		Mean of 5 d (24-28) of control diet	Diet	Mean of 5 d (24-28) of acarbose diet	Ref
Humans: 6 males, 5 female, age 23-31	4 weeks control diet; 4 weeks: acarbose; 150 mg: week 1, 300 mg: week 2, 600 mg: week 3-4	57.6 ± 4.1 7.6 ± 1.8 100.0		μmol/g (wet wt): 22.7 ± 1.9 16.0 ± 1.6 12.4 ± 1.3 mmol/d: 3.1 ± 0.9 2.0 ± 0.4 1.7 ± 0.5 %: 40.6 ± 1.6 26.8 ± 1.4 20.8 ± 1.4	supplemented with acarbose	μmol/g (wet wt): 65.8 ± 4.1 27.6 ± 1.7 14.0 ± 1.4 19.6 ± 2.6 mmol/d: 14.8 ± 2.5b 6.1 ± 1.0b 2.9 ± 0.5 4.8 ± 1.1b %: 100 41.8 ± 1.8 21.2 ± 2.1 29.9 ± 3.1a	17
Humans: 45 subjects with history of colonic adenomas (31 males, 14 females); 49 control subjects (26 males, 23 females)	2 weeks	Group 1: wheat bran: 24 g/d Group 2: oat bran: 25 g/d Control diet: 13-16 g fiber/d	mmol/l 71 ± 6 74 ± 8	14 ± 3 11 ± 3	Wheat bran Oat bran 62 ± 4 59 ± 6	mmol/l 15 ± 2 8 ± 2b	18
Humans: male adults	2 weeks	Control diet: 21.8 g/d fiber; Wheat fiber 53.2 g/d	mmol/l 57.7 ± 4.7	31.7 ± 2.8 13.2 ± 1.3 12.8 ± 1	Wheat fiber	mmol/l 66.1 ± 4.2 39.5 ± 2.5 12.3 ± 0.9 14.3 ± 1.2	19
Humans: 12 males	4 2-week followed by 1 3-week period with 26 g fibers of different sources	mmol/l 145 118 107a 166a			Coarse bran Fine bran Cellulose Cabbage	mmol/l 69 27 51 25 53 16 78 25	20

a Indicates a significant difference ($p < 0.05$).
b Indicates a significant difference ($p < 0.01$).

TABLE 2
Influence of Malabsorbed Carbohydrates on SCFA in Human Subjects

Subjects	Exp design	Fiber source	Specimen	Acetate (µmol/l)		Ref.
Humans: 8 subjects	Acute study during 6 h after lactulose ingestion	Lactulose, 10 g	Blood	Control, 44.3 ± 4.7	Lactulose, 114.4 ± 16.2[a]	24
Humans: 8 subjects — 7 males and 1 female, age 37.5 ± 3.0	2 weeks control diet 2 weeks lactulose	Lactulose, 18-25 g/d	Blood	At 13th d, 97 ± 10	At 13th d, 110 ± 12	25
Humans: 9 subjects — 6 males and 3 females, age 56.9 ± 16.8	Acute study during 180 min after ingestion	Lactulose, 20 g	Blood	110 ± 60	Lactulose, 230 ± 90[b]	26
Humans: 14 subjects — 12 males and 2 females, age 24.5	Acute study Lactulose: 360 min Pectin: 24 h	Lactulose, 5, 10, 20 g Pectin, 20 g	Blood	55.5 ± 11	Lactulose, 5 g: 98.6 ± 23.1, 10 g: 127.3 ± 18.2, 20 g: 181.3 ± 23.9 Pectin, 20 g: 95.8 ± 11.7	27
Humans: 8 subjects — 4 males and 4 females, age 27 ± 4	Acute study 14 h	Guar, 20 g Psyllium, 20 g	Blood	62 ± 4	Guar, 93 ± 6[b] Psyllium, 78 ± 6	28
Humans: 11 subjects — 8 males and 3 females, age 37	Acute study 14 h	Lactulose, 12.5 g Cold potatoes, 300 g Hot potatoes, 300 g Ripe banana, 300 g Unripe banana, 300 g	Blood taken (timed to coincide with the peak of breath hydrogen)	Fasting 56 ± 5 42 ± 11 38 ± 6 46 ± 8 54 ± 7	Lactulose, 66 ± 6 Hot potatoes, 45 ± 6 Cold potatoes, 48 ± 3 Ripe banana, 45 ± 9 Unripe, 104 ± 31[a]	29
Humans: 10 diabetics, 5 insulin- dependent — 6 males and 4 females, age 52.1 ± 11.8	Chronic study 6 weeks Mixed fiber	Control diet fiber/d 30 g High fiber diet, fiber/d 55 g High fiber + sucrose, fiber/d 55 g + 45 g sucrose	Blood	210 ± 60	High fiber diet, 280 ± 100[b] High fiber + sucrose 300 ± 100[b]	30
Humans: 7 short bowel 7 males, age 33.6 ± 2.4	Acute study 6 h	50 g carbohydrate as bread	Blood	144.9 ± 18.9	At 4 h, 167 ± 26.7[b]	31
	2 weeks	2600 kcal, 39.4 g mixed fiber Mixed diet in 3 meals (control) 17 meals: nibbling	Blood	Control, 104.2 ± 6.6	Nibbling mean over day 13, 79.2 ± 5.7	32

[a] Indicates a significant difference ($p < 0.01$).
[b] Indicates a significant difference ($p < 0.05$).

Acetate is potentially capable of interacting with glucose and lipid metabolism in a number of ways. In liver cells, acetate inhibits glycolysis[33,34] and stimulates gluconeogenesis.[33] In rats, acetate potentiates glucose-induced insulin secretion and improves glucose tolerance;[35] in isolated rat pancreatic islet cells, acetate increases the insulin response to glucose.[36] Acetate may reduce blood glucose in the long term by reducing serum free fatty acid concentrations,[37,38] but propionate may have the opposite effect, being a gluconeogenic substrate in ruminants and horses.[39,40] Intraportal propionate stimulates insulin secretion in sheep in the absence of raised blood glucose,[41] but long-term oral propionate in humans reduces the serum insulin response to oral glucose.[42]

Propionate may reduce serum cholesterol through inhibition of hydroxymethylglutaryl (HMG) CoA reductase, the rate-limiting step for cholesterol synthesis.[43] However, acetate is the primary substrate for such synthesis.[44]

Table 3 summarizes the studies on the effect of acetate and propionate on a number of metabolic parameters in humans. The data in Table 3 suggest that propionate is gluconeogenic and increases blood glucose in human subjects in the acute studies. In the long-term studies it appears to decrease fasting serum glucose. The blood glucose lowering effect of acetate seems to depend on the reduction of free fatty acids observed in most of the studies by suppression of tissue lipolysis. Carbohydrates, which are rapidly fermented in the colon, may raise serum lipid concentrations as a result of the increased production of acetate. Propionate alone seems not to change serum cholesterol levels, but when given together with acetate, it inhibits the acetate-induced rise in serum cholesterol.

The implication of these studies is that the effect of colonic fermentation on blood glucose and lipid concentrations may depend on the relative proportion of acetate and propionate produced by different substrates. The fall in free fatty acid levels associated with raised plasma acetate levels may contribute toward the beneficial effects on glycemic control observed with high-fiber diets.[47,48]

In conclusion, there is evidence that microbial fermentation of carbohydrates occurs in the large intestine of humans and that SCFA represent major end products. SCFA activities and the processes whereby they are produced and absorbed need to be more thoroughly investigated and understood. Changes in the diet seem to be the most likely way of modifying the extent to which SCFA are produced and absorbed by humans and their impact on glucose and lipid metabolism. Studies conducted with *in vitro* fermentation, analysis of SCFA in the feces, and detection of peripheral blood concentrations may together be useful methods to improve our knowledge of the effects of SCFA on glucose and lipid metabolism.

TABLE 3
Influence of SCFA on a Number of Metabolic Parameters in Humans (Concentrations in Serum or Plasma)

Subjects	Experimental design	Time (min)	Acetate (μmol/l)	Propionate (μmol/l)	Glucose (mmol/l)	Insulin (pmol/l)	C-peptide (pmol/l)	FFA (meq/l)	Glycerol (mmol/l)	Cholesterol (mmol/l)	Ref.
Humans: 3 females and 3 males, age 33 ± 3	Rectal infusion Isotonic saline control	0	93 ± 29		4.4 ± 0.1	37 ± 4	358 ± 32	0.407 ± 0.097		5.17 ± 0.41	38
		30	103 ± 23		4.4 ± 0.1	38 ± 5	349 ± 27			5.18 ± 0.41	
		60	83 ± 17		4.2 ± 0.1	37 ± 6	342 ± 27	0.379 ± 0.089		4.90 ± 0.35	
		90	106 ± 26		4.2 ± 0.1	37 ± 5	323 ± 38			4.99 ± 0.39	
		120	102 ± 19		4.2 ± 0.1	33 ± 4	304 ± 31	0.385 ± 0.089		5.01 ± 0.44	
	180 mM acetate 60 mM propionate	0	108 ± 19		4.5 ± 0.1	47 ± 9	443 ± 46[a]	0.385 ± 0.032		4.85 ± 0.26	
		30	356 ± 81[a]		4.5 ± 0.1	51 ± 7[a]	385 ± 66			4.77 ± 0.29	
		60	398 ± 66[a]		4.4 ± 0.1	54 ± 9[a]	450 ± 42[a]	0.242 ± 0.015		4.75 ± 0.31	
		90	372 ± 70[a]		4.4 ± 0.1[a]	44 ± 6	409 ± 41[a]			4.93 ± 0.27	
		120	286 ± 35[a]		4.4 ± 0.1	49 ± 10[a]	381 ± 44[a]	0.269 ± 0.047		4.87 ± 0.26	
Humans: 9 males, non-diabetic	Intravenous infusion Bicarbonate: (2.8 mM/min) (60 min)	Pre-infusion			4.8 ± 0.1			0.23 ± 0.12	0.086 ± 0.026		45
		Post-infusion			4.8 ± 0.1			0.38 ± 0.19[a]	0.050 ± 0.021		
	Acetate: 2.5 mM/min (60 min)	Pre-infusion			4.8 ± 0.3			0.31 ± 0.16	0.066 ± 0.030		
		Post-infusion			4.5 ± 0.2[a]			0.28 ± 0.09[a]	0.065 ± 0.026[a]		
6 male, NIDD	Bicarbonate	Pre-infusion			6.8 ± 1.9			0.54 ± 0.23	0.054 ± 0.022		
		Post-infusion			6.5 ± 1.8			0.63 ± 0.20[a]	0.055 ± 0.028		
	Acetate	Pre-infusion			7.1 ± 2.1			0.35 ± 0.07	0.068 ± 0.020		
		Post-infusion			6.8 ± 1.8			0.27 ± 0.09[a]	0.053 ± 0.018[a]		
Humans: 4 males and 2 females, age 29 ± 3	Rectal infusion: (120 min) Isotonic saline control	Mean of 30–120 min. increments compared to saline	μmol/l 0	0	0	0		% change 0		mmol/l	46
	Acetate, 180 mM		268 ± 66[a]	12 ± 21	−0.02 ± 0.06	9 ± 2[a]		−22 ± 6[a]		0.12 ± 0.05[a]	
	Propionate, 180 mM		35 ± 16	66 ± 27[a]	0.37 ± 0.13[a]	5 ± 3[a]		−16 ± 20[a]		0.03 ± 0.03[a]	

Humans: 7 males and 1 female, age 37.5 ± 3	Acetate 180 mM + Propionate, 60 mM	208 ± 53[a]	28 ± 29	0.14 ± 0.05[a]	-8 ± 7[a]	0.85 ± 0.10	-32 ± 5[a]	0.06 ± 0.04[a]
	2 weeks control diet	Control	97 ± 10		5.4 ± 0.1	162 ± 21		384 ± 46	4.27 ± 0.23
	2 weeks lactulose diet, 18-25 g/d	After lactulose	110 ± 12		5.4 ± 0.1	166 ± 25	0.83 ± 0.07	300 ± 29[a]	4.67 ± 0.29[b]

				mmol/l
5 NIDD, 5 IDD, 6 males and 4 females, age 52.1 ± 11.8	Control diet (31.2 ± 8.9 g fiber)	**All subjects (n = 10)**		
		Control	0.21 ± 0.06	10.8 ± 4.5	1.4 ± 0.70
	High fiber (HF) (6 weeks): 55.7 ± 2.8 g	HF	0.28 ± 0.10[a]	9.4 ± 3.2	1.03 ± 0.48
	High-fiber + 45 g sucrose (HFS) (6 weeks)	HFS	0.30 ± 0.10[a]	8.9 ± 4.8[a]	0.49 ± 0.22[a]
		NIDD (n = 5)		
		Control	0.22 ± 0.02	9.6 ± 2.9	1.49 ± 0.59
		HF	0.26 ± 0.07	9.9 ± 2.0	1.14 ± 0.59
		HFS	0.29 ± 0.02[a]	8.9 ± 2.6	0.48 ± 0.09[a]
		IDD (n=5)		
		Control	0.20 ± 0.11	12 ± 5	1.30 ± 0.81
		HF	0.30 ± 0.11[a]	8.9 ± 4.4[a]	0.96 ± 0.44
		HFS	0.31 ± 0.13[a]	8.8 ± 6.6	0.50 ± 0.29[a]

[a] Indicates a significant difference ($p < 0.05$).
[b] Indicates a significant difference ($p < 0.01$).

REFERENCES

1. **Cummings, J.H.,** Short chain fatty acids in the human colon, *Gut,* 22, 763, 1981.
2. **Schmitt, M.G., Soergel, K.H., and Wood, C.M.,** Absorption of short chain fatty acids from the human jejunum, *Gastroenterology,* 70, 211, 1976.
3. **Ruppin, H., Bar-Meir, S., Soergel, K.H., Wood, C.M., and Schmidt, M.G.,** Absorption of short-chain fatty acids by the colon, *Gastroenterology,* 78, 1500, 1980.
4. **McNeil, N.I., Cummings, J.H., and James, W.P.T.,** Short chain fatty acid absorption by the human large intestine, *Gut,* 19, 819, 1978.
5. **Cummings, J.H., Pomare, E.W., Branch, W.J., Naylor, C.P.E., and MacFarlane, G.T.,** Short chain fatty acids in human large intestine, portal, hepatic and venous blood, *Gut,* 28, 1221, 1987.
6. **Hoverstad, T., Fausa, O., Bjorneklett, A., and Bohmer, T.,** Short-chain fatty acids in the normal human feces, *Scand. J. Gastroenterol.,* 19, 375, 1984.
7. **Cummings, J.H. and Branch, W.J.,** Fermentation and the production of short-chain fatty acids in the human large intestine, in *Dietary Fiber: Basic and Clinical Aspects,* Vahouny, G.V. and Kritchevsky, D., Eds., New York, 1986, chap. 10.
8. **Lundquist, F., Sestoft, L., Damgaard, S.E., Clausen, J.P., and Trap-Jensen, J.,** Utilization of acetate in the human forearm during exercise after ethanol ingestion, *J. Clin. Invest.,* 52, 3231, 1973.
9. **Skutches, C.L., Holroyde, C.P., Myers, R.N., Paul, P., and Reichard, G.A.,** Plasma acetate turnover and oxidation, *J. Clin. Invest.,* 64, 708, 1979.
10. **Ballard, F.J.,** Supply and utilization of acetate in mammals, *Am. J. Clin. Nutr.,* 25, 773, 1972.
11. **Knowles, S.E., Jarret, I.G., Filsell, O.H., and Ballard, F.J.,** Production and utilization of acetate in mammals, *Biochem. J.,* 142, 401, 1974.
12. **Fleming, S.E.,** Influence of dietary fiber on the production, absorption, or excretion of short chain fatty acids in humans, in *Dietary Fiber in Human Nutrition,* Spiller, A.G., Ed., Plenum Press, London, 1993, chap. 6-8.
13. **Cummings, J.H., Hill, M.J., Jenkins, D.J.A., Pearson, J.R., and Wiggins, H.S.,** Changes in fecal composition and colonic function due to cereal fiber, *Am. J. Clin. Nutr.,* 29, 1468, 1976.
14. **Fleming, S.E. and Rodriguez, M.A.,** Influence of dietary fiber on fecal excretion of volatile fatty acids by human adults, *J. Nutr.,* 113, 1613, 1983.
15. **Fleming, S.E., O'Donnell, A.U., and Perman, J.A.,** Influence of frequent and long-term bean consumption on colonic function and fermentation, *Am. J. Clin. Nutr.,* 41, 909, 1985.
16. **Spiller, G.A., Chernoff, M.C., Hill, R.A., Gates, J.E., Nassar, J.J., and Shipley, E.A.,** Effect of purified cellulose, pectin, and a low-residue diet on fecal volatile fatty acids, transit time, and fecal weight in humans, *Am. J. Clin. Nutr.,* 33, 754, 1980.
17. **Scheppach, W., Fabian, C., Sachs, M., and Kasper, H.,** The effect of starch malabsorption on fecal short-chain fatty acid excretion in man, *Scand. J. Gastroenterol.,* 23, 755, 1988.
18. **Kashtan, H., Stern, H.S., Jenkins, D.J.A., Jenkins, A.L., Thompson, L.U., Hay, K., Marcon, N., Minkin, S., and Bruce, R.W.,** Colonic fermentation and markers of colorectal-cancer risk, *Am. J. Clin. Nutr.,* 55, 723, 1992.
19. **Cummings, J.H., Hill, M.J., Bone, E.S., Branch, W.J., and Jenkins, D.J.A.,** The effect of meat protein and dietary fiber on colonic function and metabolism. II. Bacterial metabolites in feces and urine, *Am. J. Clin. Nutr.,* 32, 2094, 1979.
20. **Ehle, F.R., Robertson, J.B., and Van Soest, P.J.,** Influence of dietary fibers on fermentation in the human large intestine, *J. Nutr.,* 112, 158, 1982.
21. **McBurney, M.I., Thompson, L.U., Cuff, D.J., and Jenkins, D.J.A.,** Comparison of ileal effluents, dietary fibers, and whole foods in predicting the physiological importance of colonic fermentation, *Am. J. Gastroenterol.,* 83, 536, 1988.
22. **Mortensen, P.B., Clausen, M.R., Bonnén, H., Hove, H., and Holtug, K.,** Colonic fermentation of ispaghula, wheat bran, glucose, and albumin to short-chain fatty acids and ammonia evaluated in vitro in 50 subjects, *Parent. Ent. Nutr.,* 16, 433, 1992.
23. **Mortensen, P.B., Holtug, K., and Rasmussen, H.S.,** Short-chain fatty acid production from mono- and disaccharides in a fecal incubation system: implications for colonic fermentation of dietary fiber in humans, *J. Nutr.,* 118, 321, 1988.
24. **Scheppach, W., Pomare, E.W., Elia, M., and Cummings, J.H.,** The contribution of the large intestine to blood acetate in man, *Clin. Sci.,* 80, 177, 1991.
25. **Jenkins, D.J.A., Wolever, T.M.S., Jenkins, A., Brighenti, F., Vuksan, V., Rao, A.V., Cunnane, S.C., Ocana, A., Corey, P., Vezina, C., Connelly, P., Buckley, G., and Patten, R.,** Specific types of colonic fermentation may raise low-density-lipoprotein-cholesterol concentration, *Am. J. Clin. Nutr.,* 54, 141, 1991.
26. **Akanji, A.O. and Hockaday, T.D.R.,** Breath hydrogen excretion or plasma acetate levels during the lactulose tolerance test?, *Afr. J. Med. Sci.,* 20, 101, 1991.

27. **Pomare, E.W., Branch, W.J., and Cummings, J.H.,** Carbohydrate fermentation in the human colon and its relation to acetate concentrations in venous blood, *J. Clin. Invest.,* 75(5), 1448, 1985.
28. **Wolever, T.M.S., Wal, P., Spadafora, P., and Robb, P.,** Guar, but not psyllium, increases breath methane and serum acetate concentrations in human subjects, *Am. J. Clin. Nutr.,* 55, 719, 1992.
29. **Cummings, J.H. and Englyst, H.N.,** Measurement of starch fermentation in the human large intestine, *Can. J. Physiol. Pharmacol.,* 69, 121, 1991.
30. **Akanji, A.O., Peterson, D.B., Humphreys, S., and Hockaday, T.D.R.,** Change in plasma acetate levels in diabetic subjects on mixed high fiber diets, *Am. J. Gastroenterol.,* 11, 1365, 1989.
31. **Royall, D., Wolever, T.M.S., and Jeejeebhoy, K.N.,** Evidence for colonic conservation of malabsorbed carbohydrate in short bowel syndrome, *Am. J. Gastroenterol.,* 87(6), 751, 1992.
32. **Brighenti, F., Ciappellano, S., Vuksan, V., Rao, A.V., Volever, T.M.S., Jenkins, A., Jenkins, D.J.A., and Testolin, G.,** Is colonic fermentation minimized by increasing meal frequency?, *Eur. J. Clin. Nutr.,* 45, 221, 1991.
33. **Anderson, J.W. and Bridges, S.R.,** Short-chain fatty acid fermentation products of plant fiber affect glucose metabolism of isolated rat hepatocytes (41958), *Proc. Soc. Exp. Biol. Med.,* 177, 372, 1984.
34. **Nomura, T., Iguchi, A., Sakamoto, N., and Harris, R.A.,** Effects of octanoate and acetate upon hepatic glycolysis and lipogenesis, *Biochim. Biophys. Acta,* 754, 315, 1983.
35. **Shah, J.H., Wongsurawat, N., and Aran, P.P.,** Effect of ethanol on stimulus-induced insulin secretion and glucose tolerance, *Diabetes,* 26, 271, 1977.
36. **Patel, D.G. and Singh, S.P.,** Effect of ethanol and its metabolites on glucose mediated insulin release from isolated islets of rats, *Metabolism,* 28, 85, 1979.
37. **Crouse, J.R., Gerson, C.D., DeCarli, L.M., and Lieber, C.S.,** Role of acetate in the reduction of plasma free fatty acids produced by ethanol in man, *J. Lipid Res.,* 9, 509, 1968.
38. **Wolever, T.M.S., Brighenti, F., Royall, D., Jenkins, A.L., and Jenkins, D.J.A.,** Effect of rectal infusion of short chain fatty acids in human subjects, *Am. J. Gastroenterol.,* 9, 1027, 1989.
39. **Judson, G.J., Anderson, E., Luick, J.R., and Leng, R.A.,** The contribution of propionate to glucose synthesis in sheep given diets of different grain content, *Br. J. Nutr.,* 22, 69, 1968.
40. **Ford, E.J.H. and Simmons, H.A.,** Gluconeogenesis from caecal propionate in the horse, *Br. J. Nutr.,* 53, 55, 1985.
41. **Brockman, R.P.,** Insulin and glucagon responses in plasma to intraportal infusions of propionate and butyrate in sheep *(Ovis aries), Comp. Biochem. Physiol.,* 73A, 237, 1982.
42. **Venter, C.S., Vorster, H.H., and Cummings, J.H.,** Effects of dietary propionate on carbohydrate and lipid metabolism in healthy volunteers, *Am. J. Gastroenterol.,* 85, 549, 1990.
43. **Chen, W.-J.L., Anderson, J.W., and Jennings, D.,** Propionate may mediate the hypocholesterolemic effects of certain soluble plant fibers in cholesterol fed rats, *Proc. Soc. Exp. Biol. Med.,* 175, 215, 1984.
44. **Bloch, K.,** The biological synthesis of cholesterol, *Science,* 150, 19, 1965.
45. **Akanji, A.O. and Hockaday, T.D.R.,** Acetate tolerance and the kinetics of acetate utilization in diabetic and nondiabetic subjects, *Am. J. Clin. Nutr.,* 51, 112, 1990.
46. **Wolever, T.M.S., Spadafora, P., and Eshuis, H.,** Interaction between colonic acetate and propionate in humans, *Am. J. Clin. Nutr.,* 53, 681, 1991.
47. **Jenkins, D.J.A., Wolever, T.M.S., and Nineham, R.,** Improved glucose tolerance four hours after taking guar with glucose, *Diabetologia,* 19, 21, 1980.
48. **Vorster, H.H., Venter, C.S., and Van Eeden, T.S.,** Benefits from supplementation of the current recommended diabetic diet with gel fibre, *Int. Clin. Nutr. Rev.,* 8, 140, 1988.

Chapter 3.6

EFFECT OF DIETARY FIBER ON PLASMA LIPOPROTEINS*

**David J. A. Jenkins, Peter J. Spadafora, Alexandra L. Jenkins, and
Cynthia G. Rainey-Macdonald**

INTRODUCTION

Until just over a decade ago, the dietary treatment for hyperlipidemia and hypercholesterolemia, in particular, included a reduction in saturated fat and cholesterol intakes, an increase in P:S ratio, elimination of alcohol consumption, and reduction of calories to achieve ideal body weight.

The major classes of drugs used were few and remain so. These include cholestyramine, clofibrate, nicotinic acid and their analogs, and HMG-CoA reductase inhibitors. However, more recently there has been an interest in the pharmacology of natural products as hypolipidemic agents. Foremost among these is dietary fiber. In addition, other food constituents, including the phytosterols (β-sitosterol), saponins, and plant proteins have attracted attention. Of these, only dietary fibers and vegetable proteins have been tested extensively in therapeutic diets for hyperlipidemia.

DIETARY FIBER AND POSSIBLE MECHANISMS OF ACTION

A number of possible mechanisms are likely to be involved in the hypocholesterolemic effect of dietary fiber, and different mechanisms are likely to predominate depending on the fiber.

The pioneer studies of Kritchevsky and colleagues demonstrated that a number of fiber sources were capable of binding bile acids *in vitro*,[1-3] and provided a rationale for the increased bile acid losses seen *in vivo*. Together with this mechanism of action, there are possibly three other broad reasons why fiber lowers serum cholesterol levels, none of which are mutually exclusive in providing an explanation of the mechanism for an individual food. Indeed, it is likely that for a given food more than one mechanism is operative.

INCREASED FECAL STEROL LOSSES

From the beginning, it was recognized that increased fecal sterol losses provided one explanation for the lipid lowering effect of fiber.[1] There is general agreement that purified viscous fiber[8] administration increases bile acid outputs by 20 to 80%,[79,89] but the effect of fiber in foods is less clear.[90,91] Studies have been limited in number due to the unsavory nature of this line of work, and further data are therefore urgently required.

INCREASED PROPIONATE GENERATION

Bacterial fermentation of fiber in the colon gives rise to short chain fatty acids (SCFA) which are absorbed. One of the SCFA, propionate, has been shown in pigs[92] and rats[93] to reduce serum cholesterol levels and to inhibit cholesterol synthesis in liver *in vitro*.[94] Propionate has also been demonstrated in man to inhibit, acutely, the acetate induced rise in serum cholesterol after rectal infusion.[95] However, human feeding studies of propionate[96,97] have not

---

* This chapter was previously published with the title "Fiber in the Treatment of Hyperlipidemia" in *CRC Handbook of Dietary Fiber in Human Nutrition, 2nd Edition*, Gene A. Spiller, ed., CRC Press, Boca Raton, FL, 1993.

demonstrated a clear effect in reducing LDL cholesterol levels. On the other hand, when colonic fermentation is increased using the nonabsorbable sugar, lactulose, LDL cholesterol levels appear to rise rather than fall.[105] The nature of the fermentation and the type of fiber may therefore be important in determining the final outcome. This area also requires further studies for its definition.

REDUCED INSULIN LEVELS

Increased insulin levels have been linked with CHD.[98-101] A common effect of the viscous fibers and high fiber foods which reduce serum cholesterol levels is that they produce relatively flat post prandial glucose and insulin responses.[102,103] Early studies demonstrated that hepatic cholesterol synthesis in the rat increased during periods of maximum insulin secretion.[104] The explanation was that insulin induced an increase in activity of hepatic HMG-CoA reductase, a rate-limiting step in cholesterol synthesis.[104] The cholesterol-lowering effect associated with reduced insulin levels has been confirmed using a model of altered food frequency ("Nibbling")[110] to mimic slow absorption and so this mechanism of action also deserves further investigation.

ALTERED LIPID ABSORPTION AND GENETIC FACTORS

Fiber delays the rate of nutrient absorption[103] and, in the longer term, may alter small intestinal morphology and lipid absorption.[106,107] Alteration in the rate and site of lipid absorption may alter the pattern of lipoprotein secretion[108] and catabolism. Vitamin A tolerance tests with added fiber indicate that some fibers appear to enhance chylomicronemia.[109] Added to this are the genetic differences which may make fiber more or less effective. Amongst genetic variants which influence serum lipids, differences in the apo E genotype have attracted much attention, and may influence the response to drugs such as gemfibrozil[80] and dietary change, including a prudent diet, dietary cholesterol, and vegetable protein.[81,82] In view of the association of E genotype with differences in remnant particle uptake, cholesterol absorption and bile acid excretion,[83-88] this genetic classification may be particularly useful in predicting responses to altered fat and fiber intakes. Other genetic markers have not, as yet, received this degree of scrutiny. At present, there are no data on the interaction of genetics and fiber, but these data are likely to appear in the near future. No detailed studies have been carried out at different levels of dietary fat to assess the effect on the different possible mechanisms of action of dietary fiber, i.e., whether some mechanisms are more or less effective at different levels of dietary fat intake. Hypothetically, fiber foods which induce a bile salt loss might be more effective if the bile salt pool is expanded through greater intakes of dietary fat. On the other hand, reduction in carbohydrate intake may minimize differences in glycemic response and hence mechanisms which relate to altered insulin secretion. In the absence of studies, however, these are simply speculations to be explored.

EXPERIENCE WITH SPECIFIC FIBERS

The early studies described the hypolipidemic effects of fibers in healthy volunteers before they were tested on patient groups. The literature relating to healthy volunteers has already been reviewed. The present discussion will therefore focus on the therapeutic use of fiber. In general, viscous fibers have proved useful in lowering serum lipids[5,16,21,25,53,64] while nonviscous fibers have for the most part been without effect.[10-12,52,61] There are exceptions to this generalization.[111]

LIGNIN AND CELLULOSE

Early on, the suggestion that lignin may be hypocholesterolemic by virtue of its bile acid binding ability resulted in two conflicting clinical studies.[9,10] The dosages were small (Table

1), and the work has not been repeated. Similarly, the particulate fiber, cellulose, was without effect on serum cholesterol or triglyceride levels.[11,12] More recent work has confirmed these findings (Table 1).[52,61]

PECTIN

The early studies of Palmer and Dixon[13] in normal volunteers were followed by those of Miettinen and colleagues[14] on hyperlipidemic patients who consumed relatively large doses of pectin (40 to 50 g/d) (Table 2). The observation that the resulting falls in serum cholesterol levels were associated with only modest increases in fecal sterol loss suggested that increased sterol excretion was likely to be only one of a number of mechanisms responsible. No changes were seen in serum triglycerides. Palmer and Dixon demonstrated that little cholesterol-lowering effect could be seen in healthy individuals taking 6 g or less of pectin daily.[13] Subsequent studies in hyperlipidemic patients confirmed this observation.[15] However, a substantial lowering of serum total and LDL cholesterol levels was observed even when as little as 12 g of pectin daily was taken.[16] A clear dose-response is not evident from these studies. The interest in pectin has continued, in general, to support the earlier work (Table 2).[52-54,76]

GUAR

Again following observations in healthy volunteers, studies testing the effects of guar were undertaken in hyperlipidemic patients.[5,17-23] The cholesterol-lowering results with guar were materially the same as those observed with pectin. The effect was predominantly reflected in the LDL-cholesterol fraction with much less or no change in the HDL-cholesterol fraction. Triglyceride levels were reduced, but the reduction was significant only when the guar was incorporated into very low fat, starchy carbohydrate foods, such as crisp bread or spaghetti.[5,18] The physicochemical nature of the guar and the formulation in which it is provided may be very important factors, since greatly differing effects were reported by different investigators when doses of guar of the order of 15 g/d were given.[5,17-23] Nevertheless, in a study in which the same guar was added in powder form to fruit juices and soup, baked into conventional breads, or incorporated into a dry crisp bread or melba toast-type formulation, all were equally effective.[18] These findings suggested that prehydration was not a prerequisite for the hypolipidemic action of this viscous fiber.

In 2-week studies in which the maximum acceptable dose of guar given in crisp bread form was compared with the maximum acceptable dose of cholestyramine in the same patients, the falls in total and LDL cholesterol were comparable.[18] This indicated that, in pharmacological terms, the effects of viscous fiber on lipid metabolism might have significant clinical utility.

In the majority of the guar and pectin studies, fiber was added to the patients' preexisting diet and/or drug therapy, which was then maintained constant. This included low fat, low cholesterol diets, with or without cholestyramine, clofibrate, or their analogs. It is not possible at present to say whether the mechanism or action of fiber overlap with those of the established hypolipidemic drugs and whether specific combinations might bestow an advantage. In view of the relatively small bile acid losses seen with pectin[4,14,24] and guar[4] compared with specific drugs, it is likely that the mechanisms of fiber will complement those of the bile salt binding (anion exchange) resins (e.g., cholestyramine). On the other hand, when the maximum effect has been achieved with clofibrate, it is possible that any further effect of the fiber may be reduced.[20] Recent work has contributed further support to previous evidence (Table 2).[55-61,67,78]

LOCUST BEAN GUM

This viscous fiber has also been used successfully in a range of hyperlipidemic patients to lower serum cholesterol[25] (Table 2). Its advantage has been claimed to be its superior taste (or lack of taste) by comparison with guar. However, the taste of guar depends on its purity and,

TABLE 1
Effects of Lignin and Cellulose on Serum Lipid Concentrations in Normal and Hyperlipidemic Subjects

Fiber type	Fiber form/dose	Study protocol	Background diet/drugs	Lipid disorder	Cholesterol initial (mg/dl), change %				TG initial (mg/dl), change % total	Comments	Ref.
					Total	LDL	VLDL	HDL			
Lignin	"Celluline" (99% lignin), 4 g/day	Cholestyramine 12 g/day for 2 to 5 months, then celluline 1.2 g/d for 2 to 5 months, then celluline 4 g/d for 2 to 3 months	Usual therapeutic diet	6 Type II	(416) -21%					Celluline duplicated or maintained effect of cholestyramine	9
	Capsules, 2 g/day	Control diet for 4 weeks, then fiber for 4 weeks, then control for 4 weeks		7 Type II	(304) +8%				(168) 0%	Total-C ↓ to nearly basal level during second control period	10
	High MW, 16.7% methoxyl content, mixed with food 12 g/day	4 Weeks, crossover with no fiber as control	Normal diet	10 Healthy	(228)* NC	—	—	(62)* 6.3 NS	(68)* -2.6 NS	NC HDL/Total-C ratio	52
Cellulose	60 g/day	Reducing		60 Normal, some HLP	(-) NC				(-) NC	Reducing diet ↓ total-C by 22% with or without cellulose	11

Fiber	Study design	Diet	Subjects	TG	LDL-C	HDL-C	Other	Ref.
Powdered, in bread	LL diet plus cellulose bread *for 3 months,* partially substituted with soy-hull cookies *for 3 months*	LL diet	14 Type IIa	(346) +8% NS	—	(99) +6% NS		12
Micro-crystalline cellulose, form not specified 15 g/day 99.9%	12 Weeks, crossover	Normal diet, 19 of 22 on oral hypoglycemic drugs	22 Diabetics type II	(236)* 1.6 NS	NC	NC	NC fasting blood glucose or HBA1c	61
Alpha cellulose fiber, mixed with food 15 g/day	4 Weeks, crossover with no fiber as control	Normal diet	10 Healthy	(209)* -5.6 NS	(66)* -11.8 NS	(74)* -10.7 NS	NC HDL/Total-C ratio	52

Note: TG, triglyceride; HLP, hyperlipoproteinemic; HC, hypercholesterolemic; LL, lipid-lowering; NC, no change; LDL-C, low-density lipoprotein cholesterol; Total-C, total cholesterol. Initial values are given in parentheses; values represent results for fiber treatment period or fiber-treated group (time period for fiber treatment is in italics). An asterisk indicates that values were converted from SI to traditional units, using factors C in mmol × 18.7-C in g/dl; TG in mmol × 88.5-TG in mg/dl.

TABLE 2
Effects of Soluble, Purified Fibers on Lipid Concentrations in Normal and Hyperlipidemic Subjects

Fiber type	Fiber form/dose	Study protocol	Background diet/drugs	Lipid disorder	Cholesterol initial (mg/dl), change %				TG initial (mg/dl), change % total	Comments	Ref.
					Total	LDL	VLDL	HDL			
Pectin	Powdered, in jam, 40-50 g/d	LL diet for 2 weeks, then pectin for 2 weeks	LL diet	4 Type IIa, 1 type IV, 2 secondary HLP, 2 healthy	(307)* -13%				(157) -5% NS	Fecal bile acids ↑57%; total fecal sterols ↑27%; increased plasma methyl sterols	14
	Lemon or apple, 6 g/d (10.5% methoxyl content)	LL diet for 6 weeks, then either (1) lemon pectin, (2) apple pectin, or (3) clofibrate	LL diet	18 Type IIa, 3 type IIb, 10 type III, 2 type IV, (or IIb)	(i)(301) +5% NS (ii)(321) -4% NS (iii)(306) -13% NS				(i)(142) +0.7% NS (ii)(109) +36% NS (iii)(139) -30%		15
	Apple, as granulate, 12 g/day (12% methoxyl content)	LL diet for 8 weeks then cholestyramine added for 10 weeks, then pectin added for 8 weeks	LL diet, cholestyramine, 16 g/d	6 Type IIa (male)	(377)*a -19%	(285) -22%	(-) NC	(45) -12% NS	(199) -0.4%	↓ Lipids evident at 4 weeks enhanced over 8 weeks; TG ↑ of 25% after LL diet + drug persisted with pectin	16
	Citrus pectin as gel, 15 g/d	4 Weeks, 1 week control, 3 weeks pectin	Normal Western diet	10 Healthy	(206) -18 p < 0.05	NC	—	NC	NC	Decline in anaerobe to aerobe ratio	53

Fiber, dose	Duration, design	Diet	Subjects						Comments	Ref.
Grapefruit pectin in capsule, 15 g/d	4 Weeks, crossover	Normal diet	27 HC	(275) -7.6 p ≤ 0.008	(195) -10.7 p ≤ 0.07	(38) 2.4 NS	(42) -1.2 NS	(188) 2.7 NS	Decline LDL:HDL 9.8% (p ≤ 0.03)	54
Breakfast cereal, 57 g/d	6 Weeks	AHA Step 1 diet	58 HC males	(220)* -2.1 NS	(152)* -3.9 NS	—	(46)* 2.5 NS	(111)* 3.7 NS	Body wt, serum glucose, and iron unchanged	76
Citrus pectin 9.3% methoxyl content mixed with food, 12 g/d	4 Weeks, crossover with no fiber as control	Normal diet	10 Healthy	(225)* NC	—	—	(62)* NC	(86)* 4.1 NS	NC HDL/Total-C ratio	52
Psyllium granule, 6-12 g/d	5 Weeks	Normal diet	9 Healthy males	(189) -14.4 NS	(LDL + VLDL 145) -25.6p < 0.01	—	(44) 22.1 NS	(161) -11.7 NS	Weight loss with psyllium	62
Metamucil, 10.6 g/d	8 Weeks, parallel design	Normal diet	14 HC males	(247) -14.8 p < 0.01	(162) -20.2 p < 0.01	—	(54) -6.5 NS	(147) -12.7 NS	LDL/HDL ratio decline 14.8% (p < 0.01)	63
Ground husk, 21 g/d	3 Weeks	Normal Western diet	7 Healthy	(187) -16 p < 0.002	(83)+ -18 p < 0.01	—	(49)+ -8.2 p < 0.03	—	Total fecal steroid excretion NC	64
Metamucil, 10.2 g/d	8 Weeks	LL Step 1 diet	40 HC	(228)* -4.2 p < 0.05	(156)* -7.7 p < 0.05	—	(53)* 1.9 NS	(101)* 2.6 NS	Apo B declined 8.8% with psyllium p < 0.01	65
Breakfast cereal, 57 g/d	6 Weeks	AHA Step 1 diet	58 HC males	(218)* -5.9 p = 0.001	(148)* -5.7 p = 0.0066	—	(47)* -1.6 NS	(106)* -10.9 NS	Body wt, serum glucose, and iron unchanged	76
Vi-Siblin, 30 g/d	11 Days	Low cholesterol, moderate	9 HLP type IIa, IIb	(321)* -6.0 p < 0.05	(255)* -9.1 p < 0.05	—	(39)* 20.0 p < 0.05	(133)* -6.0 NS	Increase in fecal bile acids and elimination of	67

TABLE 2
Effects of Soluble, Purified Fibers on Lipid Concentrations in Normal and Hyperlipidemic Subjects (continued)

Fiber type	Fiber form/dose	Study protocol	Background diet/drugs	Lipid disorder	Cholesterol initial (mg/dl), change % Total	LDL	VLDL	HDL	TG initial (mg/dl), change % total	Comments	Ref.
	Flavored Metamucil, 20.4 g/d	13 Weeks, after 7 weeks AHA phase 1	fiber intake AHA Phase I	27 HC	(266) -7.1 p < 0.01	(190) -8.6 p < 0.01	—	(44) 6.2 p < 0.05	(177) -10.9 NS	cholesterol LDL/HDL ratio declined 13.3% p < 0.05	77
Guar	Added to beverage or in soup, 15 g/d	Usual drugs for 2 years, LL diet for 3 months, then pectin for 2 weeks	LL diet, cholestyramine 12 to 16 g/d for 3 patients, clofibrate 1 g/d for 1 patient	10 Type II	(345) + -11%				(164) +13% NS		17
	In crispbread, 13 g/d	LL for 1 year, then crispbread exchanged for bread for 2 to 8 weeks	LL diet, 2 patients on cholestyramine, 2 on clofibrate	11 Type II or IV	(259)*b -14%	(259) -16%		(46) -2.6% NS	(186) -13% NS	Maximum ↓ in C evident in 2 to 3 weeks maintained at 8 weeks; patients on cholestyramine show similar responses	18
	Granules in food or drink (73% w/w guar, 15 g/d)	Drugs discontinued for at least 1 month, then patient randomized into guar, placebo, and control groups for 4 months	LL diet	32 Type II female	(320)* -3% NS			(65) 0%	(163) -3% NS	Body weight in guar-treated group, TG ↓ 29% in placebo group, seasonal variation of lipids in control group	19

Preparation	Diet	Protocol	Subjects						Comments	Ref
Granulate, 16 g/d	LL diet, Bezafibrate 600 g/d	Bezafibrate for 2 months, then guar added for either a second or third 2-month period	12 Type IIa	(351) -7%, additional to drug effect	(273) -14%	(-)NC	(55) -13% NS	(130) +28% NS	Apo-B lipoprotein ↓ 20% with nc Apo-A, ↑ total and LDL-C pretreatment levels on withdrawal of guar	20
Preparation with gel inhibitor, 18 g/d	LL diet	LL drugs discontinued for several months, LL diet for 1 year, then placebo for 4 weeks, then guar for 3 to 12 months	1 Type IIa, 16 type IIb	(302)* -13%	(232) -18%	(23) +17% NS	(45) -0.4% NS	(230) +8% NS	LL effects of guar evident in 3 months and sustained for 12 months; fecal acidic steroids ↑ 30%; % cholesterol absorption reduced in 4/5 normals	21
In pasta 20% w/w guar 14 to 19 g/d	LL diet, no LL drugs	(i) LL drugs discontinued, LL diet for 8 to 12 days, then guar pasta for 2 weeks	2 Type IIa, 2 type IIb, 4 type IV	(274)^b -32%				(510) -40%	↓ LDL-C calculated to be less than 5%	5
	LL diet, no LL drugs	(ii) LL diet for 3 months, then guar pasta for 8 weeks	4 Type IIb, 1 type IV	(290) -6%				(358) -27%	Total-C ↓ 11% and TG 40% in 3 patients followed for 20 weeks	
Granulate, 15 g/d	Normal diet, no LL drugs	Drugs discontinued for 6 months, then placebo for 4 weeks, then guar or placebo for 12 weeks (washout 4 weeks between periods)	14 HC, male	(329)* -15% @ 6 weeks, -8% NS @ 12 weeks	(231) -22% @ 6 weeks, -14% NS @ 12 weeks	(43) -21% NS	(46) -6% NS	(145) -4% NS	Guar effect negated between 6 and 12 weeks in 11/14 subjects	22
Powder, 16 g/d	Patients' own diets	60 Days guar on own diets, then 60 days control	Familial, 10 men, 2 women	(297) -10%	(226) -10%	(27) -23%	(54) 2%	(191) -22%	Very detailed study showing changes in apolipoproteins	23
Guargel in water, 16 g/d	Usual diet, sulphonylurea	6 Weeks	10 Diabetics type II	(190) -11 p < 0.05	(127) -18 p < 0.01	(17) 17.9 NS	(45) -1.8 NS	(119) -10.8 NS	Fasting BG declined 9.6% and insulin declined 23%	55

TABLE 2
Effects of Soluble, Purified Fibers on Lipid Concentrations in Normal and Hyperlipidemic Subjects (continued)

Fiber type	Fiber form/dose	Study protocol	Background diet/drugs	Lipid disorder	Cholesterol initial (mg/dl), change %				TG Initial (mg/dl), change % total	Comments	Ref.
					Total	LDL	VLDL	HDL			
	Guar bars 26.4–39.6 g/d	24 Weeks	Normal diet	8 Diabetics type II	(223) 3 NS	(141) -16 NS	—	(44) -9 NS	(186) 90 $p < 0.025$	($p < 0.05$) NC in apo B levels	56
	Baked into crispbread, 11.4 g/d	2 Weeks	Controlled mixed diet	6 Healthy males	(186) -16 p <0.05	—	—	—	(82.8) 1.8 NS	82-95% guar metabolized in bowel	57
	Guar pasta, 10g/d	4 Days	Normal weight maintenance	10 Obese women	(194) -11.9 $p < 0.05$	—	—	(46) 2.2 NS	(117) -6 NS	Significant decline blood glucose and insulin	58
	Granulated 20-30 g/d	50 Weeks	Normal diet	23 HC	(388)* -10 p <0.001	(312)* -14.9 $p < 0.01$	(46)* -23.5 NS	(45)* 17.9 NS	(215)* -9.9 NS	Apo AI/Apo B ratio increase 11.8% ($p < 0.05$)	59
	Cracker, liquid, 15 g/d	8 Weeks	Normal diet	32 HLP	(241) -5.8 NS	(163) -9.8 NS	—	(45) -2.2 NS	(157) 12 NS	Total- and LDL-C declined with high viscosity guar	60
	Form not specified 15 g/d	12 Weeks, crossover	Normal diet 19 of 22 on oral hypo-glycemic drugs	22 Diabetics type II	(255) -10.6 $p < 0.01$	—	—	(45) NC	NC	Trend towards lower glucose, HBA1c	61
	Gel, 40 g/d	11 Days	Low cholesterol, low fiber	8 Hospitalized patients with symptomatic diverticular disease	(201)* -9.6 p <0.05	(147)* -13.2 $p < 0.05$	(21)* -3.7 NS	(34)* -1.1 NS	(192)* -2.1 NS	Bile acid output doubled	67

Powder in fluid, 15 g/d	21 Days	Normal diet	13 Subjects moderately elevated lipids	(244) -10	(152) -17	(30) -3	(62) 1.6	(145) -1.4		78
Locust bean gum	18-30 g/d	LL diet, then 8 weeks of locust bean gum or 4 weeks locust, 4 weeks control, and 4 weeks locust	LL diet	18 familial HC, adults, and children, 10 normals	(260) -6 to -19%	(188) -10 to -19%	(22) -10 to -19%	(50) -0 to -17%	(95) -10%	25

Note: Abbreviations: TG, triglyceride, HLP, hyperlipoproteinemic, HC, hypercholesterolemic, LL, lipid-lowering, NC, no change, LDL-C, low-density lipoprotein cholesterol, Total-C, total cholesterol. Initial values are given in parentheses; values represent results for fiber treatment period or fiber-treated group (time period for fiber treatment is in italics). An asterisk indicates that values are converted from SI to traditional units, using factors: C in mmol × 38.7-C in g/dl; TG in mmol × 83.5, TG in mg/dl. A plus sign indicates that values are approximated from graphs.

[a] Initial values = @ 8 weeks diet + drug; % changes = @ 8 weeks diet + drug + pectin.
[b] 10 Patients also participated in earlier 2-week studies comparing effects of cholestyramine with guar in semihydrated (bread) or hydrated (soup) forms; values are for 7 patients given crispbread who were followed for 8 weeks.
[c] Values are for 14 patients who completed 12 months of treatment.

since both substances are galactomannans, direct comparative studies must be undertaken before any statement about their relative efficacy can be made.

WHEAT BRAN

Of the wheat brans, only hard red spring wheat bran has been convincingly shown to lower serum cholesterol levels in normal man.[26] This fiber source has not yet been tested in hyperlipidemic patients. As with normal volunteers, almost all studies which have used other wheat bran preparations have failed to show significant reductions in blood lipid levels of hyperlipidemic individuals[27,28] (Table 3), although there is one report of a significant rise in HDL cholesterol.[29] The lack of consistent effect of wheat bran on blood lipids is of interest from the standpoint of mechanisms, since the bile acid losses in the stool following bran consumption have been shown to be comparable to those following pectin,[4] which consistently lowers serum cholesterol.

OAT BRAN

Since the early studies of DeGroot,[30] it was realized that oat constituents may have hypocholesterolemic effects. Unlike wheat bran, oat fiber contains an appreciable proportion of viscous fiber (β-glucan), and it is likely that this constituent may be one of its active hypolipidemic ingredients. More recently, studies of Anderson and co-workers have demonstrated the lipid-lowering effect of oat bran given to hyperlipidemic patients (predominantly Types IIa and IIb) for 10 days to 2 years[7,31,32] (Table 3). Although there were highly significant falls in all cholesterol fractions together with serum triglyceride during the initial 3 weeks of fiber treatment, the HDL cholesterol level increased slowly, to almost approximately the starting value by 24 weeks. The other fractions remained low throughout the maintenance treatment period and for the four patients who were followed for 2 years.[32] Again, the increase in fecal acidic steroids was small and in proportion to the increase in fecal bulk, and unlikely to provide more than a small part of the explanation of the hypocholesterolemic action of oat bran. In this respect, it has been proposed that the volatile fatty acids from oat bran and other viscous fibers, which arise from colonic fermentation of fiber and are subsequently absorbed, may produce metabolic changes which favor reduced cholesterol synthesis.[7] Despite early enthusiasm and then apparent despair, the body of evidence supports that oat bran will have a significant, although small, beneficial effect on serum lipids (Table 3).[66,68,72,78]

DRIED LEGUMES

Cooked, dried legumes have been shown to lower serum cholesterol levels of middle-aged men,[33] although not of young student volunteers.[34] More recently, with interest in their effect of improving glucose tolerance[35] and other aspects of diabetic control,[36] high legume diets have been studied in types IIa and IV hyperlipidemic patients[7,32,37,39] (Table 3). All studies have shown falls in serum cholesterol irrespective of the class of hyperlipidemia studies. One investigator also recorded falls in serum triglyceride comparable to those seen with oat bran of equivalent soluble fiber content (20 g/day).[32] As with oat bran, the effects on blood lipids appear to be sustained for 4 months to 2 years.[32]

The reasons for the effects, however, are not clear. Increases in fecal output on 115 g of beans are small and not significant in the hyperlipidemic individuals.[7] In addition, where recorded, increases in fecal acidic steroid losses were not noted.

Nevertheless, the falls in blood lipids, especially triglycerides, may be related to the flatter postprandial glucose and insulin responses elicited by legumes.[35] These may result in a chronically reduced stimulus to hepatic triglyceride synthesis and hepatic lipid synthesis in general. Evidence for this hypothesis has been drawn together in the studies of Albrink et al.[6] in healthy volunteers and is supported by the observation, also in healthy volunteers, that 24-h urinary C-peptide outputs were reduced on high legumes diets.[40]

TABLE 3
Effects of Fiber-Rich Whole Foods and Supplements on Serum Lipids

Fiber type	Fiber form/dose	Study protocol	Background diet/drugs	Lipid disorder	Cholesterol initial (md/dl), change % Total	LDL	VLDL	HDL	TG initial (mg/dl), change % total	Comments	Ref.
Unprocessed wheat bran	50 g/d	Bran added to normal diet for 3 months	Normal diet	5 Type IV	(259)* +3% NS	(178) -2% NS	(39) +20% NS	(39) +30% NS	(283) 0%		27
	50 g/d	Bran added to normal diet for 2 months	Normal diet	8 Secondary HLP, male	(252)* -2% NS			NS	(195) +5% NS		28
Supplement wheat bran	"Fiberform" 10.5 g/d (8.2 g/d fiber)	Bran or placebo added to LL diet for 8 weeks (crossover)	LL diet	12 HC, male	(284)* +0.7% NS	(217) -2% NS	(24) -30% NS	(43) +23%	(166) -24%	↓ TG in lipoprotein fractions; NC during placebo period; seasonal variation may have masked C-lowering effect in bran-treated group	29
				14 Normal	(208)* +9%	(145) +8%		(47) -2%	(119) +9% NS		
Oat bran	In muffins, as cereal 100 g/d (26 g total,15 g soluble fiber) (23 g/d total fiber more than control diet)	No LL drugs for 3 months, then either/or bran-enriched diet for 10 days, sequence randomized	Normal control	6 Type II, 2	(269)	(184)		(49) -2% NS	(161) -9%	NC lipids during	31
			diet (identical except for bran content)	normal	-13%	-14%		-2% NS	NS	control period: fecal bile acids 54% higher and neutral sterols slightly lower on oat bran than on control diets	

TABLE 3
Effects of Fiber-Rich Whole Foods and Supplements on Serum Lipids (continued)

Fiber type	Fiber form/dose	Study protocol	Background diet/drugs	Lipid disorder	Cholesterol initial (md/dl), change %				TG initial (mg/dl), change % total	Comments	Ref.
					Total	LDL	VLDL	HDL			
	Baked in muffins, 17 g/d	28 Days	Normal diet	19 Healthy	(179) -5 NS	(108) -9 NS	—	(55) +2 NS	(77) -8 NS	—	69
	Flakes, Chex biscuits, 25 g/d	2 Weeks	Normal Western diet	12 HC males	(260)* -5.4 p <0.05	(180)* -8.5 p < 0.025	—	(38)* -3.3 NS	(202)* 8.7 NS	Decline in Apo B 9.8% NS	72
	Bread, muffins, 95 g/d	4 Weeks crossover	Low dietary fiber diet	24 HC males	(245)* -4.9	(176)* -6.8	—	(41)* 2.9	(62)* -3.1	NC blood glucose, insulin, blood pressure	66
	Powder in fluid, 77 g/d	21 Days	Normal diet	13 Adults moderately elevated lipids	(244) -3.3	(152) -5.9	(30) 3.3	(62) NC	(145) 5.5	200 calories extra added/day with oat bran	78
Mixed	52 fiber/2000 kcal (30 g/d more than regular diet) as part of modified U.K. LL diet	Habitual diet for 8 weeks, then test diet (iso- or hypo-caloric) for 5 to 11 months	Habitual diet	16 Type IIa, 11 type IIb, 1 type III, 3 type IV, 6 normal	(302)*,a -22%	(213) -25%	(48) -37% NS	(52) -4% NS	(235) -24%	NC HDL_2-C	50
Beans	50 g/d (100 g/2 d) dried	Beans added to usual diet 3 to	Usual Chinese diet, no LL	136 HLP	(282)[b] -17%				(-)[c] + 13 to 32% NS	NC for subjects given control diets; pro-	37

Food	Amount	Protocol / duration	Background diet	Subjects					Comments	Ref
Baked beans	Canned, 110 g/d (dry weight)	6 months; (18 responders continued up to 12 months); Clofibrate withdrawn for 1 year for Type IV patients, then 50% carbohydrate exchanged for beans for 2 weeks	drugs; LL diet, usual drugs (except for clofibrate)	106 Normal; 6 type IIa, 6 type IIb, 5 type IV	(205)[b] −9% (290)* −8%			(−)* + 13 to 32% NS (212) −25% NS	nounced LL effect was seen in 1 month	38
	Cooked or canned, 140 g/d (dry wt) = 15 g/d fiber	LL diet, then high legume diet for 4 months (increase of 15 g fiber per day for high legume diet)	LL diet	2 Type IIa, 1 type IIb, 4 type IV	(269) −7%	(189) −5% NS	(37) −15% NS	(238) −25%	Relatively high animal protein content (56%) of test diet could have negated effect of vegetable protein on cholesterol	39
	450 g/d in tomato sauce	2 Weeks	Normal diet	13 Healthy males	(192)* −12 p < 0.05	—	(52)* −14.8 p < 0.001	(94)* 7.5 NS	3.7% Decline in HDL: Total-C ratio NS	71
	120–162 g per day as pork and beans or in tomato sauce	3 Weeks, single or double dose	Normal Western diet	24 HLP type IIa or IIb	(295)* −10.4 p < 0.001	(201)* −8.4 p < 0.005	(42)* −6.9 p < 0.05	(256)* −10.8 p < 0.025	Decline in body weight with bean diet	73
Beans/ oat bran	(i) Oat bran, or beans (dry wt) 100 g/d (48 g total, 18 g soluble fiber per day)	No LL drugs for 3 months, control diet for 7 days, then randomized into oat bran or bean diets for 3 weeks	No LL drugs, LL diet	10 Type II, male	(294) −23%	(216) −23%	(37) −20%	(203) −21%		32
	(ii) 50 g day oat bran	Oat bran or bean supplement at	High carbohydrate, high	10 Type II, male	(226) −4% fall	(167) −2%,	(30) +17%	(161) −6%, fall	Reduced cholesterol and fat intake during	

TABLE 3
Effects of Fiber-Rich Whole Foods and Supplements on Serum Lipids (continued)

Fiber type	Fiber form/dose	Study protocol	Background diet/drugs	Lipid disorder	Cholesterol initial (md/dl), change %				TG initial (mg/dl), change % total	Comments	Ref.
					Total	LDL	VLDL	HDL			
	(dry wt) or 145 g beans (wet wt)	home for 24 weeks, 4 patients were followed for 99 weeks	fiber maintenance diet		maintained	fall maintained		NC	maintained	maintenance study may have contributed to results	
	100 g oat bran (dry wt)	Normal diet, oat bran for 3 weeks	Normal diet	10 Type IIa and type IIb	(280) -19%	(190) -23%	(—)	(31) -6% NS	(289) -18%	Modest ↑ in fecal weight and proportionate ↑ in bile acid losses	7
	115 g beans (dry wt)	Normal diet, beans for 3weeks	Normal diet	10 Type IIa and type IIb	(300) -10%	(221) -24%	(—)	(32) -12%	(233) -3% NS	No ↑ in fecal weight or bile acid losses	
Wheat bran	Bread, muffins, 35 g/d	4 Weeks crossover	Low dietary fiber diet	24 HC males	(245)* 0.8	(176)* 0.2	—	(41)* 3.8	(62)* 3.1	NC blood glucose, insulin, blood pressure	66
Oats	Instant oats, 56.7 g/d	8 Weeks, parallel with control	Normal diet	42 HC	(254)* -6.3	(177)* -9.2	(27)* 1.4	(50)* 0.8	(135)* 1.3	NC in blood pressure	68
Oatmeal	Breakfast cereal, muffins, 56 g/d	8 Weeks	AHA phase II	113 Healthy	(193) -3 NS	— NS	—	— NS	— NS	Total-C and LDL-C declined in subgroup ≥ 198 mg/dL (4 weeks)	70

Note: Abbreviations: TG, triglyceride; HLP, hyperlipoproteinemic; HC, hypercholesterolemic; LL, lipid lowering; NC, no change; LDL-C, low-density lipoprotein cholesterol; Total-C, total cholesterol. Initial values are given in parentheses; values represent results for fiber treatment period or fiber-treated group (time period for fiber treatment is in italics). An asterisk indicates that values are converted from SI to traditional units, using factors: C in mmol × 38.7-C in g/dl; TG in mmol × 83.5, TG in mg/dl. A plus sign indicates that values are approximated from graphs.

[a] Initial values = @ 8 weeks diet + drug; % changes = @ 8 weeks diet + drug + pectin.

[b] 10 Patients also participated in earlier 2-week studies comparing effects of cholestyramine with guar in semihydrated (bread) or hydrated (soup).

[c] Values are for 14 patients who completed 12 months treatment.

LEGUME PROTEIN

In addition to legume fibers (e.g., guar and locust bean gum), some legume proteins, notably soya protein, have been shown to reduce cholesterol levels of hyperlipidemic patients (total and LDL)[41-47] (Table 4). The effect does not appear to be related to associated fiber or saponins, since it is also found after administration of soya isolate. However, the capacity to lower blood lipids is not a universal property of all legume proteins, since fava protein isolate failed to achieve the desired effect.[46]

The mechanisms for the hypolipidemic action of soy protein remain obscure, but may relate to the amino acid profile[49] or the presence of specific pharmacologically active peptides liberated during digestion. With respect to the whole bean, although specific fiber and protein effects may be relevant to the action of some beans, other factors will have to be uncovered to explain the general hypolipidemic effect of legumes in hyperlipidemic individuals. Data continue to accumulate supporting the lipid lowering effect of legume protein (Table 4).[74,75]

EFFECTIVE FIBER DOSAGE

In general, the effective dose of the viscous fibers required to lower serum cholesterol levels has been of the order of 12 to 30 g/d. Interestingly, the levels of fiber in the oat bran- and bean-containing diets have also been of this order of magnitude, since soluble fiber comprises approximately 20% of the dry weight of both beans and oat bran. However, the effectiveness of the supplement in hyperlipidemia may be determined by its formulation, in addition to the background diet. This fact is well illustrated by the guar-enriched spaghetti studies of Gatti and colleagues, which showed the greatest lowering of cholesterol and triglyceride levels of all the studies to date.[5] Spaghetti is already recognized as a slowly digested carbohydrate form which causes an unexpectedly low rise in blood glucose.[50] This effect is likely to have been greatly enhanced by the addition of guar, resulting in the creation of a very effective sustained release carbohydrate source. The addition of guar would likely not only have enhanced the reduction of glycemic and insulinemic responses to the pasta, but may, in effect, have resulted in a proportion of the pasta starch being converted to "fiber", i.e., carbohydrate which is unavailable for small intestinal absorption, but which acts as an additional source for synthesis of SCFAs in the colon. The choice by Gatti and co-workers of spaghetti as the vehicle for delivery of the fiber may have been in large measure the reason for the success of their trial, since it encompassed many of the mechanisms responsible for reducing blood lipids. One of the important directions for future development in this field would, therefore, appear to lie in finding the most effective food vehicles in which to incorporate fiber.

CONCLUSION

Viscous fibers such as pectin, guar, and locust bean gum, and high fiber foods such as oat bran and dried beans, all providing 12 to 30 g fiber daily, have been shown to reduce serum total and LDL cholesterol levels by 10 to 20% and with a lesser fall in HDL cholesterol levels. When the fiber was provided in a starchy food such as pasta, crisp bread, oat bran, or beans, significant falls in serum triglyceride levels have also been observed.

The mechanisms of action of fiber are likely to be complex and possibly include increased bile salt loss, altered site and rate of absorption, reduced hepatic lipogenesis secondary to reduced postprandial glucose and insulin responses, and enhanced colonic synthesis and uptake of SCFAs.

Further developments to enhance the clinical utility of this approach should include not only a search for effective fiber types, but also the appropriate vehicles in which to deliver them.

TABLE 4

Effects of Legume Protein (± Saponins) on Serum Lipid Concentrations of Hyperlipidemic Subjects

Fiber type	Fiber form/dose	Study protocol	Background diet/drugs	Lipid disorder	Cholesterol initial (mg/dl) change %				TG initial (mg/dl), change %, total	Comments	Ref.
					Total	LDL	VLDL	HDL			
Textured soybean protein	Granules	A. LL diet for 3 months, then usual diet for 1 week, then soybean or LL for 3 weeks each (crossover)	No LL drugs Soy replaced ½ animal protein	9 Type IIa, 11 type IIb	(313) -19%	(220) -18%			(217) -17%	LL diet alone was ineffective; response varied linearly with initial cholesterol levels; for patients given soy diet first, cholesterol ↑ after withdrawal soy diet	41
		B. Same preparation as A, then soy diet for 6 weeks, 500 g cholesterol added (first or second 3-week period)	ibid	8 Type II	(315) -23%				(200) NC	Cholesterol intake did not influence LL effect of soy-protein diet	
	Granules, in liquid 60 to 100 g/d "cholsoy"	Standard LL diet for 6 months (then control LL diet) for 8 weeks, then soy diet replacing all animal protein for 8 weeks, then control LL diet for 6 weeks		127 Type II	(335)+b -20%		(—) NC		(152) -11% NS	LL effect slightly greater in females than in males, familial cases less sensitive to soy diets, but show greater rise after withdrawal soy than nonfamilial cases	42
	"Cholsoy", 60 to 100 g/d	LL animal protein diet for 4 weeks, then LL soy diet		27 Type II	(340)+b -26% @ 1st soy	(260) -33% during first soy peroid,		(44) +15% NS @			51

					period, -5% NS more @ 18 months	-4% NS more during 18 months	18 mos				Ref.
Cholsoy-P (texturized, lecithinated) 60 to 100 g/d	for 4 weeks, then LL animal protein diet for 6 weeks, then LL soy diet, 6 meals per week for 18 months. LL animal protein. for 4 weeks, then LL soy diet for 4 weeks	19 Type IIa	(330)+b -21%	(255) -26%			(40) +13%			Reduced cholesterol associated with ↑ serum arginine concentrations	45
In mixed foods	LL animal protein for 2 weeks, then test diet in which ½ animal protein exchanged for soy for 2 weeks	1 Type IIa, 4 type IIb, 1 type III	(275)* -10%	(185) -9%	(70) -17% NS		(36) 0%	(230) 11% NS		LL animal protein diet reduced total-C by 25%	44
In mixed meals	Beef-protein LL diet for 3 weeks, then soy replacing all beef protein for 4 weeks	12 Type II, 9 type IV, 1 normal	(241) -3% NS	(163) -0.8% NS	(29) +3% NS		(49) -12% NS	(104) +3% NS	Beef-protein LL diets ↓ total-C by 18% and LDL-C by 19%; NC lipid levels in a second study using same diets and crossover design	45

TABLE 4

Effects of Legume Protein (± Saponins) on Serum Lipid Concentrations of Hyperlipidemic Subjects (continued)

Fiber type	Fiber form/dose	Study protocol	Background diet/drugs	Lipid disorder	Cholesterol initial (mg/dl) change % Total	LDL	VLDL	HDL	TG initial (mg/dl), change %, total	Comments	Ref.
	Granules, "Cholsoy" L 70-80 g/d	4 Weeks LL diet followed by 4 weeks on soy protein replaced diet followed by 4 weeks on LL diet	LL diet	21 Familial HC unrelated	(392)* -20.8 p < 0.01	(313)* -25.8 p < 0.01	(19)* 32.6 NS	(43)* -7.2 NS	(143)* -69 NS	Apo B declined 14% p < 0.05 and Apo E declined 17% p < 0.05	74
	Flavored soy protein beverage, 250 mls	4 Weeks test followed by 4 weeks washout then 4 weeks test, crossover	LL diet	9 Familial HC children 6-12 years	(307)* -0.5 NS	(245)* -0.2 NS	(17)* -20 p < 0.05	(45)* 4 p < 0.04	(88)* -19 p < 0.05	NC in apo A-I or apo B levels	75
Soybean flour	Defatted, in biscuits, 50 g/d (35% of protein)	Saponin-rich (22 g/kg) or saponin-depleted (4 g/kg) soyflour diets for 4 weeks each (crossover)	Usual Australian, no saponin-rich foods, no drugs	10 HC, free-living males	(264*)[a] +1% NS	(—) NC	(—) NC	(—) NC	(192) -4% NS	Neither diet affected lipid levels or distribution, or bile acid or neutral sterol excretion	47
Favabean	Fava bean protein concentrate, 11% dietary calories	2 Consecutive 18-d periods of fat reduced diet, (i) fava bean protein, (ii) egg white protein for 18 days	Fat reduced from 32 to 26%, carbohydrate increased from 48 to 59%	6 Type IIa, 1 type IIb, 1 normal	(i) (305) -13%; (ii)(320) -13% NS	(235) -17%; (240) -8% NS	(18) -5% NS; (19) -5% NS	(40) -7%; (40) -5% NS	(—) NC; (—) NC	No significant differences between fava and egg white diets apart from lower blood glucose on fava	46

Note: Abbreviations: TG, triglyceride; HLP, hyperlipoproteinemic; HC, hypercholesterolemic; LL, lipid-lowerings; NC, no change; LDL-C, low-density lipoprotein cholesterol; Total-C, total cholesterol. Initial values are given in brackets; values represent results for fiber treatment period or fiber-treated group (time period for fiber treatment is in italics). An asterisk indicates values are converted from SI to traditional units, using factors: C ion mmol × 38.7-C in g/dl; TG in mmol × 83.5, TG in mg/dl. A plus sign indicates that values are approximated from graphs.

[a] Values are means for 2, 3, and 4 weeks on diet, for saponin-rich diets; saponin-depleted diets showed similar results (final values only given for lipoprotein fractions).

[b] Values are averaged for men and women.

REFERENCES

1. **Kritchevsky, D. and Story, J.A.,** Binding of bile salts in vitro by nonnutritive fiber, *J. Nutr.,* 104, 458, 1974.
2. **Kritchevsky, D. and Story, J.A.,** *Am. J. Clin. Nutr.,* 28, 305, 1975.
3. **Vahouny, G.V., Tombes, R., Cassidy, M.M., Kritchevsky, D., and Gallo, L.L.,** Dietary fibers. V. Binding of bile salts, phospholipids and cholesterol from mixed micelles by bile sequestrants and dietary fibers, *Lipids,* 15, 1012, 1980.
4. **Jenkins, D.J.A.,** Dietary fibre, diabetes and hyperlipidemia, *Lancet,* ii, 1287, 1979.
5. **Gatti, E., Catenazzo, G., Camisasca, E., Torri, A., Denegri, E., and Sirtori, C.R.,** Effects of guar-enriched pasta in the treatment of diabetes and hyperlipidemia, *Ann. Nutr. Metab.,* 28, 1, 1984.
6. **Albrink, M.J., Newman, T., and Davidson, P.C.,** Effect of high- and low-fiber diets on plasma lipids and insulin, *Am. J. Clin. Nutr.,* 32, 1486, 1979.
7. **Anderson, J.W., Story, L., Sieling, B., Chen, W.-J.L., Petro, M.S., and Story, J.,** Hypocholesterolemic effects of oat-bran or bean intake for hypercholesterolemic men, *Am. J. Clin. Nutr.,* 40(b), 1146, 1984.
8. **Jenkins, D.J.A., Wolever, T.M.S., Leeds, A.R., Gassull, M.A., Haisman, P., Dilawari, J., Goff, D.V., Metz, G.L., and Alberti, K.G.M.M.,** Dietary fibres, fibre analogues and glucose tolerance: importance of viscosity, *Br. Med. J.,* 1, 1372, 1978.
9. **Thiffault, C., Belanger, M., and Pouliot, M.,** Traitement de l'hyperlipoprotéinémie essentielle de type II par un nouvel agent pièrapeutique, la celluline, *Can. Med. Assoc. J.,* 103, 165, 1970.
10. **Lindner, P. and Moller, B.,** Lignin: a cholesterol-lowering agent?, *Lancet,* 2, 1259, 1973.
11. **Huth, K. and Fettel, M.,** Bran and blood lipids, *Lancet,* 2, 456, 1975.
12. **Palumbo, P.J., Esperanza, R., Briones, M.S., and Nelson, R.A.,** High fiber diet in hyperlipidemia. Comparison with cholestyramine treatment in Type IIa hyperlipoproteinemia, *JAMA,* 2403, 223, 1978.
13. **Palmer, G.H. and Dixon, D.G.,** Effect of pectin dose on serum-cholesterol levels, *Am. J. Clin. Nutr.,* 18, 437, 1966.
14. **Miettinen, T.A. and Tarpila, S.,** Effect of pectin on serum cholesterol, fecal bile acids and biliary lipids in normolipidemic and hyperlipidemic individuals, *Clin. Chim. Acta,* 79, 471, 1977.
15. **Delbarré, F., Rondier, J., and de Géry, A.,** Lack of effect of two pectins in ideopathic or gout-associated hyperdyslipidemia hypercholesterolemia, *Am. J. Clin. Nutr.,* 30, 463, 1977.
16. **Schwandt, P., Richter, W.O., Weisweiler, P., and Neureuther, G.,** Cholestyramine plus pectin in treatment of patients with familial hypercholesterolemia, *Atherosclerosis,* 44, 379, 1982.
17. **Jenkins, D.J.A., Leeds, A.R., Slavin, B., Mann, J., and Jepson, E.M.,** Dietary fiber and blood lipids: reduction of serum cholesterol in type II hyperlipidemia by guar gum, *Am. J. Clin. Nutr.,* 32, 16, 1979.
18. **Jenkins, D.J.A., Reynolds, D., Slavin, B., Leeds, A.R., Jenkins, A.L., and Jepson, E.M.,** Dietary fiber and blood lipids: treatment of hypercholesterolemia with guar crispbread, *Am. J. Clin. Nutr.,* 33, 575, 1980.
19. **Tuomilehto, J., Voutilainen, E., Huttunen, J., Vinni, S., and Homan, K.,** Effect of guar gum on body weight and serum lipids in hypercholesterolemic females, *Acta Med. Scand.,* 208, 45, 1980.
20. **Wirth, A., Middlehoff, G., Brauning, C.H., and Schlierf, G.,** Treatment of familial hypercholesterolemia with a combination of bezafibrate and guar, *Atherosclerosis,* 45, 291, 1982.
21. **Simons, L.A., Gayst, S., Balasubramaniam, S., and Ruys, J.,** Long-term treatment of hypercholesterolaemia with a new palatable formulation of guar gum, *Atherosclerosis,* 45, 101, 1982.
22. **Aro, A., Uusitupa, M., Voutilainen, E.V., and Korhonen, T.,** Effects of guar gum in male subjects with hypercholesterolemia, *Am. J. Clin. Nutr.,* 39, 911, 1984.
23. **Bosello, O., Cominacini, L., Zocca, I., Garbin, U., Ferrari, F., Davoli, A.,** Effects of guar gum on plasma lipoproteins and apolypoproteins C-II and C-III in patients affected by familial combined hyperlypoproteinemia, *Am. J. Clin. Nutr.,* 40(b), 1165, 1984.
24. **Kay, R.M.,** Dietary fiber, *J. Lipid Res.,* 23, 221, 1982.
25. **Zavoral, J.H., Hannan, P., Fields, D.J., Hanson, M.N., Frantz, I.D., Kuba, K., Elmer, P., and Jacobs, D.R.,** The hypocholesterolemic effect of locust bean food products in hypercholesterolemic adults and children, *Am. J. Clin. Nutr.,* 38, 285, 1983.
26. **Munoz, J.M., Sandstead, H.H., Jacob, R.A., Logan, G.M., Jr., Reck, S.J., Klevay, L.M., Dintzis, F.R., Inglett, G.E., and Shuey, W.C.,** Effects of some cereal brans and textured vegetable protein on plasma lipids, *Am. J. Clin. Nutr.,* 35, 580, 1979.
27. **Bremner, W.F., Brooks, P.M., Third, J.L.H.C., and Lawrie, T.D.V.,** Bran in hypertriglyceridaemia: a failure of response, *Br. Med. J.,* 3, 574, 1975.
28. **Brooks, P.M., Bremner, W.F., and Third, J.L.H.C.,** Bran, hypertriglyceridaemia and urate clearance, *Med. J. Aust.,* 2, 753, 1976.
29. **Lindegarde, F. and Larsson, L.,** Effects of a concentrated bran fibre preparation on HDL-cholesterol in hypercholesterolaemic men, *Hum. Nutr. Clin. Nutr.,* 38C, 39, 1984.
30. **DeGroot, A.P., Luyken, R., and Pikaar, N.A.,** Cholesterol-lowering effect of rolled oats, *Lancet,* 2, 303, 1963.

31. **Kirby, R.W., Anderson, J.W., Sieling, B., Rees, E.D., Chen, W.L., Miller, R.E., and Kay, R.M.,** Oat-bran intake selectively lowers serum low-density lipoprotein cholesterol concentrations of hypercholesterolemic men, *Am. J. Clin. Nutr.*, 34, 824, 1981.
32. **Anderson, J.W., Story, L., Sieling, B., and Chen, W.L.,** Hypocholesterolemic effects of high-fibre diets rich in water-soluble plant fibres, *J. Can. Dietet. Assoc.*, 45, 2, 140, 1984.
33. **Grande, F., Anderson, J.T., and Keys, A.,** Effect of carbohydrates and leguminal seeds, wheat and potatoes on serum cholesterol concentration in man, *J. Nutr.*, 86, 313, 1965.
34. **Grande, F., Anderson, J.T., and Keys, A.,** Sucrose and various carbohydrate containing foods and serum lipids in man, *Am. J. Clin. Nutr.*, 27, 1043, 1974.
35. **Jenkins, D.J.A., Wolever, T.M.S., Jenkins, A.L., Thorne, M.J., Lee, R., Kalmusky, J., Reichert, R., and Wong, G.S.,** The glycemic index of foods tested in diabetic patients: a new basis for carbohydrate exchange favouring the use of legumes, *Diabetologia*, 24, 257, 1983.
36. **Simpson, H.C.R., Simpson, R.W., Lonsley, S., Carter, R.D., Geekie, M., Hockaday, T.D.R., and Mann, J.I.,** A high carbohydrate leguminous fibre diet improves all aspects of diabetic control, *Lancet*, 1, 1, 1981.
37. **Bingwen Liu, Zhaofeng Wu, Wanzhen Liu, and Rongjue Zhang,** Effects of bean meal on serum cholesterol and triglycerides, *Chinese Med. J.*, 94, 7, 455, 1981.
38. **Jenkins, D.J.A. and Jepson, E.M.,** Leguminous seeds and their constituents in the treatment of hyperlipidemia and diabetes, in *Lipoproteins and Coronary Atherosclerosis*, Noseda, G., Fragiacomo, C., Fumagalli, R., and Paoletti, R., Eds., Elsevier, Amsterdam, 1982, 247.
39. **Jenkins, D.J.A., Wong, G.S., Patten, R.P., Bird, J., Hall, M., Buckley, G.C., McGuire, V., Reichert, R., and Little, J.A.,** Leguminous seeds in the dietary management of hyperlipidemia, *Am. J. Clin. Nutr.*, 38, 567, 1983.
40. **Burke, B.J., Hartog, M., Heaton, K.W., and Hooper, S.,** Assessment of the metabolic effects of dietary carbohydrate and fibre by measuring urinary excretion of C-peptide, *Hum. Nutr. Clin. Nutr.*, 36C, 373, 1982.
41. **Sirtori, C.R., Agradi, E., Conti, F., Mantero, O., and Gatti, E.,** Soybean-protein diet in the treatment of Type II hyperlipoproteinemia, *Lancet*, i, 275, 1977.
42. **Descovich, G.C., Ceredi, C., Gaddi, A., et al.,** Multicentre study of soybean protein diet for out-patient hypercholesterolemic patients, *Lancet*, ii, 709, 1980.
43. **Descovich, G.C., Benassi, M.S., Cappelli, M., Gaddi, A., Grossi, G., Piazzi, S., Songiorgi, Z., Mannino, G., and Lenzi, S.,** Metabolic effects of lecithinated and non-lecithinated textured soy protein treatment in hypercholesterolemia, in *Liproproteins and Coronary Atherosclerosis*, Noseda, G., Fragiacomo, C., Fumagalli, R., and Paoletti, R., Eds., Elsevier, Amsterdam, 1982, 279.
44. **Vessby, B., Karlstrom, B., Lithell, H., Gustafsson, I.B., and Werner, I.,** The effects on lipid and carbohydrate metabolism of replacing some animal protein by soy-protein in a lipid-lowering diet for hypercholesterolemic patients, *Hum. Nutr. Appl. Nutr.*, 36Z, 179, 1982.
45. **Holmes, W.L., Rubel, G.B., and Hood, S.S.,** Comparison of the effect of dietary meat versus dietary soybean protein on plasma lipids of hyperlipidemic individuals, *Atherosclerosis*, 36, 379, 1980.
46. **Contaldo, F., DiBiase, G., Giacco, A., Pacioni, D., Moro, L.O., Grasso, L., Mancini, M., and Fidanza, F.,** Evaluation of the hypocholesterolemic effect of vegetable proteins, *Prevent. Med.*, 12, 138, 1983.
47. **Calvert, G.D. and Blight, L.,** A trial of the effects of soya-bean flour and soya-bean saponins on plasma lipids, fecal bile acids and neutral sterols in hypercholesterolemic men, *Br. J. Nutr.*, 45, 277, 1981.
48. **Wolfe, B.M., Giovanetti, P.M., Cheng, D.C.H., Roberts, D.C.K., and Carroll, K.K.,** Hypolipidemic effects of substituting soybean protein isolate for all meat and dairy protein in diets of hypercholesterolemic men, *Nutr. Rep. Int.*, 24, 1187, 1981.
49. **Kritchevsky, D., Tepper, S.A., and Story, J.A.,** Influences of soy protein and casein on atherosclerosis in rabbits, *Fed. Proc. Fed. Am. Soc. Exp. Biol.*, 37, 747, 1978.
50. **Jenkins, D.J.A., Wolever, T.M.S., Jenkins, A.L., Lee, R., Wong, G.S., and Josse, R.,** Glycemic response to wheat product: reduced response to pasta but no effect of fiber, *Diabetes Care*, 6, 155, 1983.
51. **Choudhury, S., Jackson, P., Katon, M.B., Marenah, C.B., Cortese, C., Miller, N.E., and Lewis, B.,** A multifactorial diet in the management of hyperlipidemia, *Atherosclerosis*, 50, 93, 1984.
52. **Hillman, L.C., Peters, S.G., Fisher, C.A., and Pomare, E.W.,** The effects of the fiber components pectin, cellulose and lignin on serum cholesterol levels, *Am. J. Clin. Nutr.*, 42, 207, 1985.
53. **Vargo, D., Doyle, R., and Floch, M.H.,** Colonic bacterial flora and serum cholesterol: alterations induced by dietary citrus pectin, *Am. J. Gastroenterol.*, 80(5), 361, 1985.
54. **Cerda, J.J., Robbins, F.L., Burgin, C.W., Baumgartner, T.G., and Rice, R.W.,** The effects of grapefruit pectin on patients at risk for coronary heart disease without altering diet or lifestyle, *Clin. Cardiol.*, 11, 589, 1988.
55. **Tagliaferro, V., Cassader, M., Bozzo, C., Pisu, E., Bruno, A., Marena, S., Cavallo-Perin, P., Cravero, L., and Pagano, G.,** Moderate guar-gum addition to usual diet improves peripheral sensitivity to insulin and lipemic profile in NIDDM, *Diab. Metabol.*, 11, 380, 1985.
56. **McIvor, M.E., Cummings, C.C., Van Duyn, M.A., Leo, T.A., Margolis, S., Behall, K.M., Michnowski, J.E., and Mendeloff, A.I.,** Long-term effects of guar gum on blood lipids, *Atherosclerosis*, 60, 7, 1986.

57. **Penagini, R., Velio, P., Bozzani, A., Castagnone, D., Ranzi, T., Bianchi, P.A., and Vigorelli, R.,** The effect of dietary guar on serum cholesterol, intestinal transit and fecal output in man, *Am. J. Gastroenterol.,* 81(2), 123, 1986.

58. **Tognarelli, M., Miccoli, R., Giampietro, O., Cerri, M., and Navalesi, R.,** Guar-pasta: a new diet for obese subjects, *Acta Diabetol.,* 23, 77, 1986.

59. **Tuomilehto, J., Silvasti, M., Aro, A., Koistinen, A., Karttunen, P., Gref, C-G., Ehnholm, C., and Uusitupa, M.,** Long term treatment of severe hypercholesterolemia with guar gum, *Atherosclerosis,* 72, 157, 1988.

60. **Superko, H.R., Haskell, W.L., Sawrey-Kubicek, L., and Farquhar, J.W.,** Effects of solid and liquid guar gum on plasma cholesterol and triglyceride concentrations in moderate hypercholesterolemia, *Am. J. Cardiol.,* 62, 51, 1988.

61. **Niemi, M.K., Keinanen-Kiukaanniemi, S.M., and Salmela, P.I.,** Long-term effects of guar gum and microcrystalline cellulose on glycemic control and serum lipids in type II diabetes, *Eur. J. Clin. Pharmacol.,* 34, 427, 1988.

62. **Nakamura, H., Ishikawa, T., Tada, N., Kagami, A., Kondo, K., Miyazima, E., and Takeyama, S.,** Effect of several kinds of dietary fibers on serum and lipoprotein lipids, *Nutr. Rep. Int.,* 26(2), 215, 1982.

63. **Anderson, J.W., Zettwoch, N., Feldman, T., Tietyen-Clark, J., Oeltgen, P., and Bishop, C.W.,** Cholesterol lowering effects of psyllium hydrophilic mucilloid for hypercholesterolemic men, *Arch. Int. Med.,* 148, 292, 1988.

64. **Abraham, Z.D. and Mehta, T.,** Three-week psyllium-husk supplementation: effect on plasma cholesterol concentrations fecal steroid excretion, and carbohydrate absorption in men, *Am. J. Clin. Nutr.,* 47, 67, 1988.

65. **Bell, L.P., Hectorne, K., Reynolds, H., Balm, T.K., and Huninghake, D.B.,** Cholesterol-lowering effects of psyllium hydrophilic mucilloid, *JAMA,* 261(23), 3419, 1989.

66. **Kestin, M., Moss, R., Clifton, P.M., and Nestel, P.J.,** Comparative effects of three cereal brans on plasma lipids, blood pressure and glucose metabolism in mildly hypercholesterolemic men, *Am. J. Clin. Nutr.,* 52, 661, 1990.

67. **Miettinen, T.A. and Tarpila, S.,** Serum lipids and cholesterol metabolism during guar gum, plantago ovata and high fiber treatments, *Clinica Chim Acta,* 183, 253, 1989.

68. **Van Horn, L., Moag-Stahlberg, A., Liu, K., Ballew, C., Ruth, K., Hughes, R., and Stamler, J.,** Effects on serum lipids of adding instant oats to usual American diets, *Am. J. Public Health,* 81(2), 183, 1991.

69. **Gold, K.V. and Davidson, D.M.,** Oat bran as a cholesterol-reducing dietary adjunct in a young, healthy population, *West J. Med.,* 148, 299, 1988.

70. **Van Horn, L., Emidy, L.A., Liu, K., Liao, Y., Ballew, C., King, J., and Stamler, J.,** Serum lipid response to a fat-modified oatmeal-enhanced diet, *Prev. Med.,* 17, 377, 1988.

71. **Shutler, S.M., Bircher, G.M., Tredger, J.A., Morgan, L.M., Walker, A.F., and Low, A.G.,** The effect of daily baked bean (Phaseolus vulgaris) consumption on the plasma lipid levels of young, normo-cholesterolemic men, *Br. J. Nutr.,* 61, 257, 1988.

72. **Anderson, J.W., Spencer, D.B., Hamilton, C.C., Smith, S.F., Tietyen, J., Bryant, C.A., and Oeltgen, P.,** Oat-bran cereal lowers serum total and LDL cholesterol in hypercholesterolemic men, *Am. J. Clin. Nutr.,* 52, 495, 1990.

73. **Anderson, J.W., Gustafson, N.J., Spencer, D.B., Tietyen, J., and Bryant, C.A.,** Serum lipid response of hypercholesterolemic men to single and divided doses of canned beans, *Am. J. Clin. Nutr.,* 51, 1013, 1990.

74. **Gaddi, A., Ciarrocchi, A., Matteucci, A., Rimondi, S., Ravaglia, G., Descovich, G.C., and Sirtori, C.R.,** Dietary treatment for familial hypercholesterolemia — differential effects of dietary soy protein according to the apolipoprotein E phenotypes, *Am. J. Clin. Nutr.,* 53, 1191, 1991.

75. **Laurin, D., Jacques, H., Moorjani, S., Steinke, F.H., Gagne, C., Brun, D., and Lupien, P-J.,** Effects of a soy-protein beverage on plasma lipoproteins in children with familial hypercholesterolemia, *Am. J. Clin. Nutr.,* 54, 98, 1991.

76. **Bell, L.P., Hectorn, K.J., Reynolds, H., and Hunninghake, D.B.,** Cholesterol lowering effects of soluble fiber cereals as part of a prudent diet for patients with mild to moderate hypercholesterolemia, *Am. J. Clin. Nutr.,* 52, 1020, 1990.

77. **Neal, G.W. and Balm, T.K.,** Synergistic effects of psyllium in the dietary treatment of hypercholesterolemia, *South Med. J.,* 83(10), 1131, 1990.

78. **Spiller, G.A., Farquhar, J.W., Gates, J.E., and Nichols, S.F.,** Guar gum and plasma cholesterol. Effect of guar gum and oat fiber source on plasma lipoproteins and cholesterol in hypercholesterolemic adults, *Arteriosclerosis and Thrombosis,* 11, 1204, 1991.

79. **Kay, R.M. and Truswell, A.S.,** Effect of citrus pectin on blood lipids and fecal steroid excretion in man, *Am. J. Clin. Nutr.,* 30, 171, 1977.

80. **Manttani, M., Koskinen, P., Ehnholm, C., Hutlunen, J.K., and Manninen, V.,** Apolipoprotein E polymorphism influences the serum cholesterol response to dietary intervention, *Metabolism,* 40, 217, 1991.

81. **Kesaniemi, Y.A., Ehnholm, C., and Miettinen, T.A.,** Intestinal cholesterol absorption efficiency in man is related to Apoprotein E phenotype, *J. Clin. Invest.,* 80, 578, 1987.

82. **Gatti, A., Clanocchi, A., Matteucci, A., Rimondi, S., Ravalia, G., Descovich, G.C., and Sirtori, C.R.,** Dietary treatment of familial hypercholesterolemia — differential effects of soy protein according to the Apolipoprotein E phenotypes, *Am. J. Clin. Nutr.,* 53, 1191, 1991.

83. **Gregg, R.E., Zech, L.A., Schaefer, E.J., and Brewer, H.B.,** Apolipoprotein E metabolism in normolipoproteinemic human subject, *J. Lipid Res.,* 25, 1167, 1984.

84. **Gregg, R.E., Zech, L.A., Cabelli, C., and Brewer, H.B.,** Apo E modulates the metabolism of apo B containing lipoproteins by multiple mechanisms, in *Cholesterol Transport Systems and Their Relation to Atherosclerosis,* Steimetz, A., Kalparik, H., and Schneider, J., Eds., Springer Verlag, Berlin, 1989, 11.

85. **Breslow, J.L.,** Genetic basis of lipoprotein disorders, *J. Clin. Invest.,* 84, 373, 1989.

86. **Gotto, A.M.,** Cholesterol intake and serum cholesterol level, *New Engl. J. Med.,* 324, 912, 1991.

87. **Kern, F.,** Normal plasma cholesterol in an 88-year old man who eats 25 eggs a day. Mechanisms of adaptation, *New Engl. J. Med.,* 324, 896, 1991.

88. **Miettinen, T.A. and Kesaniemi, Y.A.,** Cholesterol absorption: regulation of cholesterol synthesis and elimination and within-population variations of serum cholesterol levels, *Am. J. Clin. Nutr.,* 49, 629, 1989.

89. **Jenkins, D.J.A., Leeds, A.R., Gassull, M.A., Houston, H., Goff, D.V., and Hill, M.J.,** The cholesterol lowering properties of guar and pectin, *Clin. Sci. Mol. Med.,* 51, 8, 1976.

90. **Raymond, T.L., Connor, W.E., Lin, D.S., Warner, S., Fry, M.M., and Connor, S.L.,** The interaction of dietary fibers and cholesterol upon the plasma lipids and lipoproteins, sterol balance and bowel function in human subjects, *J. Clin. Invest.,* 60, 1429, 1977.

91. **Anderson, J.W., Story, L., Sieling, B., Chen, W.J.L., Petro, M.S., and Story, J.,** Hypocholesterolemic effects of oat-bran or bean intake for hypercholesterolemic men, *Am. J. Clin. Nutr.,* 40(b), 1146, 1984.

92. **Thacker, P.A., Salomons, M.O., Aherne, F.X. et al.,** Influence of propionic acid on the cholesterol metabolism of pigs fed hypercholesterolemic diets, *Can. J. Anim. Sci.,* 61, 969, 1981.

93. **Chen, W.L., Anderson, J.W., and Jennings, D.,** Propionate may mediate the hypocholesterolemic effects of certain soluble plant fibers in cholesterol-fed rats (41791), *Proc. Soc. Exp. Biol: Med.,* 175, 215, 1984.

94. **Chen, W.J.L. and Anderson, J.W.,** Hypercholesterolemic effects of soluble fibers, in *Dietary Fiber: Basic and Clinical Aspects,* Kritchevsky, D. and Yahouny, G. V., Eds., Plenum Press, New York, 1986, 275.

95. **Wolever, T.M.S., Brighenti, F., and Jenkins, D.J.A.,** Serum short chain fatty acids after rectal infusion of acetate and propionate in man, *J. Clin. Nutr. Gastroenterol.,* 3, 42, 1988.

96. **Venter, C.S., Vorster, H.H., and Cummings, J.H.,** Effects of dietary propionate on carbohydrate and lipid metabolism in healthy volunteers, *Am. J. Gastroenterol.,* 85(5), 549, 1990.

97. **Todesco, T., Rao, A.V., Bosello, O., and Jenkins, D.J.A.,** Propionate lowers blood glucose and alters lipid metabolism in healthy subjects, *Am. J. Clin. Nutr.,* 54, 1991.

98. **Reaven, G.M.,** Banting Lecture 1988. Role of insulin resistance in human disease, *Diabetes,* 37, 1595, 1988.

99. **Ducimetiere, P., Eschwege, E., Papoz, L., Richard, J.L., Claude, J.R., and Rosselin, G.,** Relationship of plasma insulin levels to the incidence of myocardial infarction and coronary heart disease mortality in a middle-aged population, *Diabetologia,* 19, 205, 1980.

100. **Welborn, T.A. and Wearne, K.,** Coronary heart disease incidence and cardiovascular mortality in Busselton with reference to glucose and insulin concentrations, *Diabetes Care,* 2, 154, 1979.

101. **Pyorala, K.,** Relationship of glucose tolerance and plasma insulin to incidence of coronary heart disease: results from two population studies in Finland, *Diabetes Care,* 2, 131, 1979.

102. **Jenkins, D.J.A., Leeds, A.R., Gassull, M.A., Cochet, B., and Alberti, K.G.M.M.,** Decrease in postprandial insulin and glucose concentrations by guar and pectin, *Ann. Int. Med.,* 86, 20, 1977.

103. **Jenkins, D.J.A., Wolever, T.M.S., Leeds, A.R., Gassull, M.A., Dilawari, J.B., Goff, D.V., Metz, G.L., and Alberti, K.G.M.M.,** Dietary fibers, fiber analogues and glucose tolerance: importance of viscosity, *Br. Med. J.,* 1, 1392, 1978.

104. **Lakeshmanan, M.R., Nepokroeff, C.M., Ness, G.C., Dugan, R.E., and Porter, J.W.,** Stimulation by insulin of rat liver Beta-hydroxy-Beta-methylglutaryl coenzyme A reductase and cholesterol-synthesizing activities, *Biochem. Biophys. Res. Commun.,* 50, 704, 1973.

105. **Jenkins, D.J.A., Wolever, T.M.S., Jenkins, A.L., Brighenti, F., Vuksan, V., Rao, A.V., Cunnane, S., Ocana, A.M., Corey, P., Vezina, C., Connelly, P., Buckley, G., and Patten, R.,** Specific types of colonic fermentation may raise low-density-lipoprotein-cholesterol concentrations, *Am. J. Clin. Nutr.,* 54, 141, 1991.

106. **Vahouny, G.V. and Cassidy, M.M.,** Dietary fiber and intestinal adaptation, in *Dietary Fiber: Basic and Clinical Aspects,* Vahouny, G. V. and Kritchevsky, D., Eds., Plenum Press, New York, 1986, 181.

107. **Story, J.A.,** Modification of steroid excretion in response to dietary fiber, in *Dietary Fiber: Basic and Clinical Aspects,* Vahouny, G. V. and Kritchevsky, D., Eds., Plenum Press, New York, 1986, 253.

108. **Schneeman, B.O., Cimmarusti, J., Cohen, W., Downes, L., and Lefevre, M.,** Composition of high density lipoproteins in rats, *J. Nutr.,* 45, 564, 1976.

109. **Kasper, H., Rabast, U., Fassl, H., and Fehle, F.,** The effect of dietary fiber on the postprandial serum vitamin A concentration in man, *Am. J. Clin. Nutr.,* 32, 1847, 1979.
110. **Jenkins, D.J.A., Wolever, T.M.S., Ocana, A.M., Vuksan, V. et al.,** Metabolic effects of reducing rate of glucose ingestion by single bolus versus continuous sipping, *Diabetes,* 39, 775, 1990.
111. **Spiller, G., Gates, J., Nichols, S., Jensen, C., and Whittam, J.,** The relationship of water soluble dietary fiber (WSDF) structure to plasma cholesterol-lowering efficacy in humans. (Abstract) *FASEB J.,* 6, A1654, 1992.

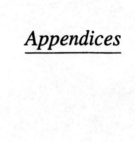

Appendices

Appendix I

POLYUNSATURATED, MONOUNSATURATED, AND SATURATED FATS IN SOME COMMON FOODS

Thomas J. Hudson

Fatty Acid Content in Selected Fats and Foods

	g/100 g			
	Polyunsaturated	Monounsaturated	Saturated	Total fat
Fats and Oils				
Almond	27	66	7	100
Apricot kernal	33	62	5	100
Avocado	15	69	16	100
Babasso	2	15	83	100
Butterfat	3	31	66	100
Castor	3	96	1	100
Chinese tallow	0	27	73	100
Chicken fat	20	49	31	100
Cod liver	35	50	15	100
Cocoa butter	2	36	62	100
Coconut	1	7	92	100
Cohune	1	10	89	100
Corn	59	35	16	100
Cottonseed	83	9	8	100
Date pit	8	44	48	100
Flaxseed	72	18	10	100
Groundnut	30	50	20	100
Herring	58	27	15	100
Kapok seed	34	46	20	100
Lard	11	46	43	100
Linseed	68	21	11	100
Macadamia	4	84	12	100
Mustardseed	32	62	6	100
Murumuru tallow	1	9	90	100
Oat	44	36	20	100
Oiticica	84	6	10	100
Olive	7	83	10	100
Ouri-curi	2	13	85	100
Palm	8	37	55	100
Palm kernel	2	14	84	100
Papaya seed	8	72	20	100
Peanut	32	49	19	100
Perillo	84	8	8	100
Pilchard	59	18	23	100
Poppyseed	62	30	8	100
Rapeseed	23	71	6	100
Rapeseed-low erucic	28	66	6	100
Rice bran	34	46	20	100
Safflower	78	13	9	100
Safflower-high oleic	12	80	8	100
Salmon	51	28	21	100
Sardine	56	22	22	100

Fatty Acid Content in Selected Fats and Foods (continued)

| | g/100 g | | |
	Polyunsaturated	Monounsaturated	Saturated	Total fat
Sesame seed	43	42	15	100
Shark liver	44	39	17	100
Shea butter	4	49	47	100
Soybean	59	25	16	100
Sperm—body[a]	1	86	13	100
Sperm—head[a]	1	55	44	100
Sunflower seed	70	18	12	100
Tallow—beef	4	43	53	100
Tallow—mutton	5	43	52	100
Teaseed	8	83	9	100
Tucum	3	13	84	100
Tung	79	15	6	100
Ucuhuba tallow	3	7	90	100
Whale	18	55	27	100
Wheat germ	66	18	16	100

| | g/100 g edible portion | | |
	Unsaturated	Saturated	Total fat

Raw Whole Foods

Animal			
Fish			
Catfish	3	1	4
Eel	14	4	18
Herring	13	3	16
Mackerel	9	4	13
Salmon	7	2	9
Trout	8	3	11
Tuna (albacore)	5	3	8
Meats (lean, raw)			
Beef	13	12	25
Mutton	9	6	15
Pork	12	19	31
Rabbit	5	3	8
Venison	1	3	4
Whale	7	1	8
Poultry/eggs			
Chicken w/skin	13	7	20
Chicken—lean	10	4	14
Egg	8	4	12
Turkey w/skin	11	4	15
Turkey—lean	5	2	7
Milk			
Buffalo	4	5	9
Cow	2	2	4
Butter	44	46	81
Goat	2	2	4
Human	2	2	4

Fatty Acid Content in Selected Fats and Foods
(continued)

	g/100 g edible portion		
	Unsaturated	Saturated	Total fat
Plant			
Cereal and grains			
Maize	3	1	4
Millet	2	1	3
Oats	6	1	7
Rice	1	1	2
Rye	1	1	2
Sorghum	3	1	4
Wheat	1	2	3
Nuts and seeds			
Almond	50	4	54
Beechnut	46	4	50
Brazilnut	54	13	67
Cashew	38	8	46
Coconut	5	31	36
Filbert	60	5	65
Groundnut	40	10	50
Hickorynut	63	6	69
Peanut	38	10	48
Pecan	45	5	71
Pilinut	38	25	63
Pinenut	49	6	55
Pistachio	49	5	54
Pumpkin	39	8	47
Safflower	55	5	60
Soybean	15	3	18
Sunflower	41	6	47
Walnut	66	7	63

[a] Fatty acid averages of glyceride content: 34% — sperm body oil; 26% — sperm head oil. Balance is wax esters.

Food and Agriculture Organization of the United Nations, Dietary Fats and Oils in Human Nutrition, report of an expert consultation, Rome, Italy, 1977.

USDA, *Composition of Foods: Raw, Processed, Prepared,* Agriculture Handbook No. 8, 1963, U.S. Department of Agriculture, Washington, D.C., 1975.

Erasmus, U., *Fats and Oils. The Complete Guide to Fats and Oils in Health and Nutrition,* Alive Books, Vancouver, B.C., 1986.

Swern, D., Ed., *Bailey's Industrial Oil and Fat Products,* Vol. 1 and 2, 4th ed., John Wiley & Sons, New York, 1979.

Appendix II

PHYSICAL PROPERTIES OF FATTY ACIDS

Gene A. Spiller

Fatty acid	Molecular weight	Melting point	Boiling point
16:0	256.4	62.8	268
18:0	284.5	69.6	291
18:1ω9	282.4	13	286
18:2ω6	280.4	-12	202
18:3ω3	278.4	-14.5	230
20:4ω6	304.5	-49.5	169
20:5ω3	302.5	-54.4	
22:5ω3	330.5	-78	
22:6ω3	328.5	-44.4	

Data from Salem, N., Jr., Omega-3 fatty acids: molecular and biochemical aspects, in *New Protective Roles For Selected Nutrients*, Spiller, G. A. and Scala, J., Eds., Alan R. Liss, New York, 1989, 114.

0-8493-4248-1/96/$0.00+$.50
© 1996 by CRC Press Inc.

Appendix III

CHOLESTEROL IN SOME COMMON ANIMAL FOODS (mg/100 g)

Gene A. Spiller

All values have been rounded. Values are expressed in increments of 5 mg, as intermediate values show a nonexisting precision. For values lower than 10 mg/100 ml, increments of 1 mg/100 g have been used. Published data sometimes use intermediate values (e.g., 72 mg/100 g), but this precision can be misleading and of little consequence in clinical and nutritional calculations. For land animals, factors that can affect values include growing or living conditions (free range vs. restricted) and type of feed; for fishes, temperature (tropical vs. nordic climates) and location (such as Atlantic vs. Pacific Oceans).

Food	Cholesterol
Milk Products and Eggs	
Cow, whole milk	15
Cow, 2% fat	7
Cow, 1% fat	4-5
Goats milk, whole	10
Butter	230
Cheese	
Camembert	70
Cheddar	70
Cottage cheese	15
Danish blue	80
Edam	70
Parmesan	90
Stilton	120
Yogurt, low fat, natural	5
Eggs, whole, raw or boiled	250-400
Egg yolk, raw or boiled	1000-1200
Egg white	0
Meat	
Bacon	
Raw, lean and fat	60
Fried, lean and fat	80
Grilled, lean and fat	75
Beef, raw	
Lean and fat	65
Lean only	60
Beef, cooked	
Lean and fat	80
Lean only	80
Pork, raw	
Lean and fat	70
Lean only	70
Pork, cooked	
Lean and fat	110
Lean only	110
Chicken	
Raw, light meat	60-70
Raw, dark meat	100-110

Food	Cholesterol
Boiled, light meat	70-80
Boiled, dark meat	110
Roasted, light meat	75
Roasted, dark meat	120
Turkey	
Raw, light meat	50
Raw, dark meat	80-85
Roasted, light meat	60-70
Roasted, dark meat	100
Liver (calf), raw	370
Sausage (beef)	
Raw	40
Fried	40
Sausage (pork)	
Raw	45
Fried	50
Salami	80

Fish

Lean fish	
Cod	
Raw	50
Baked or steamed	60
Haddock	60
Raw	
Steamed	75
Halibut	
Raw	50
Steamed	60
Sole	60
Fatty fish	
Herring	70
Mackerel	80
Salmon	40-70
Trout	80
Tuna, canned in oil	65
Crustacea	
Crab	100
Lobster	150
Prawns	200
Shrimps	200
Mollusks	
Mussels	100
Oysters	50
Scallops	40

Data derived from the following publications:

Paul, A. A. and Southgate, D. A. T., *The Composition of Foods,* Crown, London, 1978.

Health Research and Studies Center, analytical data.

Leveille, G. A., Zabik, M. E., and Morgan, K. J., *Nutrients in Foods,* The Nutrition Guild, Cambridge, 1983.

U.S. Department of Agriculture, miscellaneous publications.

Appendix IV

STEROLS IN PLANT FOODS

Gene A. Spiller

The data in these tables are from Weihrauch, J. L. and Gardner, J. M., Sterol content of foods of plant origin, J. Am. Diet. Assoc., 73, 39, 1978.

TABLE 1
Sterol Content of Plant Oils (mg/100 g Oil)

Oil	Total sterol[a]	Beta-sitosterol	Campesterol	Stigmasterol	Δ^5-Avenasterol	Δ^7-Stigmasterol	Δ^7-Avenasterol	Brassicasterol	Alpha-spinasterol	24-Methylcholest-7-enol	Cholesterol
Alfalfa seed	2080	—[b]	—	38	—	766	96	—	880	134	—
Almond nut	266	214	10	5	27	5	5	—	—	—	tr[c]
Avocado	404	366	24	4	—	—	—	10	—	—	—
Cashew nut	348	268	26	tr	29	tr	tr	—	—	—	3
Castor	282	133	27	59	42	3	2	—	—	—	tr
Chestnut	5350	4420	535	396	—	—	—	—	—	—	—
Cocoa butter	201	122	18	49	7	2	—	tr	—	—	3
Coconut	133	77	8	21	18	7	—	tr	—	—	2
Refined	91	53	6	14	13	5	—	tr	—	—	1
Refined and hydrogenated	64	37	4	10	9	3	—	—	—	—	1
Coffee	2390	1199	336	459	211	35	tr	tr	—	—	tr
Corn	1390	989	259	98	36	11	tr	tr	—	—	—
Refined	952	690	158	76	22	6	tr	—	—	—	tr
Cottonseed	431	400	26	tr	5	tr	tr	tr	—	—	—
Refined	327	303	20	tr	4	tr	tr	tr	—	—	—
Refined and hydrogenated	241	223	14	tr	3	tr	tr	tr	—	—	tr
Grapeseed	130	98	13	15	2	2	tr	tr	—	—	tr
Hazelnut	120	111	6	1	2	2	—	—	—	—	—
Hemp seed	372	165	64	55	9	9	3	tr	—	—	tr

Illipe butter	550	323	74	32	28	5	tr	tr	—	—	tr
Linseed	412	219	122	35	28	—	—	4	—	—	4
Refined	338	180	100	29	23	—	—	3	—	—	3
Mustard seed	624	342	180	tr	6	6	tr	66	—	—	tr
Olive	232	202	7	3	20	tr	tr	—	—	—	—
Refined	176	153	6	2	15	tr	tr	—	—	—	—
Palm	117	72	23	14	2	—	—	tr	—	—	5
Refined	49	30	10	6	1	—	—	tr	—	—	2
Palm kernel	140	97	12	19	7	2	tr	tr	—	—	3
Refined	95	66	8	13	5	1	tr	tr	—	—	2
Peanut	337	217	49	36	26	6	2	tr	—	—	—
Refined	206	131	31	19	20	4	1	tr	—	—	—
Pecan nut	152	123	6	3	18	1	1	—	—	—	—
Pine nut	276	168	28	tr	77	3	tr	—	—	—	—
Pistachio nut	201	154	11	4	15	2	tr	tr	—	—	tr
Poppy seed	276	188	61	8	10	10	—	tr	—	—	tr
Pumpkin seed	523	—	25	8	—	118	50	—	164	13	tr
Rapeseed	513	284	156	2	13	—	—	55	—	—	3
Refined	250	138	76	1	6	—	—	27	—	—	2
Rice bran	3225	1745	658	252	355	71	142	—	—	—	—
Refined	1055	571	215	82	116	23	46	—	—	—	—
Rye germ	2425	—	—	—	—	—	—	—	—	—	—
Safflower	494	257	55	45	6	106	25	—	—	—	—
Refined	444	231	49	40	5	95	22	—	—	—	—
Sesame seed	2950	1735	661	245	265	41	—	—	—	—	6
Refined	865	509	193	72	79	11	28	—	—	—	2
Shea butter	357	—	—	—	—	94	137	tr	—	—	—
Spinach seed	1827	—	34	64	—	770	2	—	110	15	tr
Soybean	327	183	68	64	5	5	tr	2	599	137	—
Refined	221	123	47	47	1	1	—	tr	—	—	—
Refined and hydrogenated	132	76	26	30	—	—	—	—	—	—	—

TABLE 1
Sterol Content of Plant Oils (mg/100 g Oil) (continued)

Oil	Total sterol[a]	Beta-sito-sterol	Campe-sterol	Stigma-sterol	Δ⁵-Avena-sterol	Δ⁷-Stigma-sterol	Δ⁷-Avena-sterol	Brassi-casterol	Alpha-spina-sterol	24-Methyl-cholest-7-enol	Choles-terol
Sunflower	725	465	69	75	28	60	22	—	—	—	tr
Tea seed	102	—	—	—	—	34	2	tr	60	4	—
Tobacco seed	150	90	11	20	—	—	—	—	—	—	24
Walnut	176	155	10	tr	10	tr	tr	tr	—	—	—
Wheat germ	1970	1320	433	tr	118	59	39	tr	—	—	—
Refined	553	370	122	tr	33	17	11	—	—	—	tr

a Total sterol may differ from the sum of individual sterols by an amount equivalent to sterols not reported elsewhere in the table.
b Dashes (—) = not detected.
c tr = trace; denotes an amount <0.5 mg.

TABLE 2
Sterol Content of Plant Foods (mg/100 g Edible Portion[a])

Food	Total sterol	Beta-sito-sterol	Campe-sterol	Stigma-sterol	Other[b]
			Vegetables		
Artichoke, Jerusalem- interior dry	23				
Asparagus greens	24	14	3	5	2
Bamboo shoot	19	15	3	2	—[c]
Barley seedlings, dry	234	98	33	101	1
Beans					
Common seedlings	121				
Kidney with pods, immature	14	7	1	5	1
Mung, sprouts	15	5	1	9	—
Beet					
Greens	21	13	tr[d]	6	3
Root	25				
Brussel sprouts	24	17	6	—	1
Cabbage	11	7	2	—	2
Carrot	12	7	1	3	tr
Cauliflower	18	12	3	2	1
Celery	6	2	tr	3	tr
Chives	9	7	1	tr	—
Corn					
Immature, high-oil strain	70				
Immature, low-oil strain	60				
Cucumber	14	14	—	—	—
Eggplant	7	3	tr	2	1
Garlic, Chinese	1	1	tr	—	tr
Ginger root	15	10	1	4	—
Gourd					
Bottle	9	5	—	—	4
Wax	7	5	—	—	2
White, dry	119	89	2	—	28
Lettuce					
Asparagus	11	6	1	4	—
Garden	38	21	2	11	4
Head	10	5	1	4	—
Melon, Oriental pickling	28	16	—	—	12
Mustard greens					
Brown	6	5	1	—	tr
Chinese	2	2	tr	—	tr
Potherb	4	3	1	—	1
Okra	24	15	3	6	tr
Onion	15	12	1	—	2

TABLE 2
Sterol Content of Plant Foods (mg/100 g Edible Portion[a])
(continued)

Food	Total sterol	Beta-sito-sterol	Campe-sterol	Stigma-sterol	Other[b]
Parsley	5	2	tr	2	tr
Pea seedlings	108				
Pepper					
Red	13	7	3	2	1
Sweet	9	6	2	1	tr
Potato					
Sweet	12	8	3	1	1
White	5	3	tr	1	1
Pumpkin	12	12	—	—	—
Radish					
Greens	34	22	6	—	7
Root	11	6	5	—	tr
Shallot	5	4	tr	tr	tr
Soybean, immature	50	30	9	11	—
Spinach	9	—	tr	—	9[c]
Taro	19	11	3	6	
Tomato	7	3	1	3	tr
Turnip					
Greens	12	9	2	—	2
Root	7	5	1	—	—
Vetch, dry	52				
Yam	10	7	2	1	tr

Fruits

Food	Total sterol	Beta-sito-sterol	Campe-sterol	Stigma-sterol	Other[b]
Apple	12	11	1	—	—
Apricot	18	16	1	—	1
Banana	16	11	2	3	1
Cherry	12	12	tr	—	tr
Fig	31	27	1	3	1
Grape	4	3	tr	tr	tr
Grapefruit	17	13	2	2	1
Lemon					
Whole	12	8	2	1	1
Peelings	35	22	4	7	1
Loquat	2	2	tr	—	tr
Muskmelon	10	8	—	—	2
Orange					
Juice, bitter	2	1	tr	tr	tr
Navel	24	17	4	2	tr
Peelings, navel	34	26	4	3	1
Peach	10	6	1	3	tr
Pear	8	7	—	—	tr
Persimmon	4	3	—	1	—
Pineapple	6	4	1	—	tr
Plum	7	6	tr	tr	tr
Pomegranate	17	16	tr	—	1
Strawberry	12	10	tr	tr	1
Watermelon	2	1	tr	tr	tr

Seeds and Nuts

Food	Total sterol	Beta-sito-sterol	Campe-sterol	Stigma-sterol	Other[b]
Almond	143	122	5	3	13

TABLE 2
Sterol Content of Plant Foods (mg/100 g Edible Portion[a])
(continued)

Food	Total sterol	Beta-sito-sterol	Campe-sterol	Stigma-sterol	Other[b]
Cashew	158	130	13	tr	16
Chestnut	22	18	2	2	—
Coconut					
Meat	47	27	3	7	10
Milk	1	1	tr	tr	tr
Pecan	108	88	4	2	14
Pine nut	141	84	14	tr	41
Pistachio	108	90	6	2	10
Sesame seed	714	443	91	78	102
Shea nut	29	—	—	1	28
Sunflower seed	534	349	61	75	49
Walnut	108	87	6	—	15

Spices, dry

Food	Total sterol	Beta-sito-sterol	Campe-sterol	Stigma-sterol	Other[b]
Allspice	61	57	1	3	—
Basil	106	79	9	18	—
Caraway	76	42	3	31	—
Cardamom	46	36	4	6	—
Celery	60	21	5	30	3
Cinnamon	26	21	3	2	—
Clove	256	242	—	14	—
Coriander	46	22	3	18	2
Cumin	68	30	5	31	2
Dill	124	49	9	42	25
Fennel	66	31	1	31	—
Fenugreek	140	100	18	7	15
Garlic	8	7	tr	1	tr
Ginger	83	56	10	18	—
Horseradish	91	71	20	—	—
Laurel	151	145	6	—	—
Mace	73	56	6	11	—
Marjoram	60	52	3	5	—
Mustard, white	118	84	34	—	—
Nutmeg	62	46	3	13	—
Onion	87	51	5	4	27
Oregano	203	177	12	15	—
Paprika	175	119	29	18	9
Pepper					
Black	92	65	6	21	—
Japanese	65	46	9	10	—
Red	83	53	15	12	3
White	55	38	3	14	—
Poppy seed	89	62	20	7	—
Rosemary	58	52	1	5	—
Sage	244	209	12	23	—
Savory	31	28	1	1	—
Star anise	114	93	13	8	—
Tarragon	81	41	3	33	5
Thyme	163	152	3	8	—
Turmeric	82	51	9	22	—

TABLE 2
Sterol Content of Plant Foods (mg/100 g Edible Portion[a])
(continued)

Food	Total sterol	Beta-sito-sterol	Campe-sterol	Stigma-sterol	Other[b]
Cereals					
Buckwheat	198	164	20	8	5
Corn	178	120	32	21	4
Oats, dry	58				
Rice bran	1325	735	257	289	44
Sorghum	178	97	35	36	10
Wheat					
Bran (HRS wheat)	154				
Bran (soft wheat)	89				
Flour	60				
Whole	69	40	27	—	2
Legumes					
Beans					
Azuki	76	37	1	36	3
Broad	124	95	8	9	12
Common	76				
Kidney	127	91	3	31	3
Mung	23	13	2	8	1
Sasage	99	43	6	42	9
Peanuts	220	142	24	23	31
Peas					
Chickpea	35				
Common	135	106	10	10	9
Soybeans	161	90	23	40	7
Miscellaneous Foods					
Coffee	92	53	11	24	3
Raw beans					
Roasted beans	94	52	12	27	4
Powdered extract	15	9	2	4	1
Cocoa, powdered	59	33	5	18	3
Mayonnaise	238	105	51	31	51[f]
Peanut butter	102	66	15	9	12
Tea					
Black	260	213	—	—	47
Coarse	1140	1085	—	—	54
Green	345	223	—	—	122

a   Edible portion is the fresh weight in all cases unless specifically indicated.

b   Amount of "other" sterols, including the minor Δ^5- and Δ^7-sterols, may be underestimated, since, in many cases, high resolution analyses with the GLC stationary phase of OV-17 were not performed.

c   Dashes (—) = not detected.

d   tr = trace amounts (<0.5 mg sterol).

e   Includes 8 mg alpha-spinosterol.

f   Includes 51 mg cholesterol.

Appendix V

DEFINITIONS AND COMPOSITION OF VARIOUS OLIVE OILS*

Gene A. Spiller

Olive oil is the oil obtained solely from the fruit of the olive tree (*Olea europaea sativa Hoffm. et Link*), to the exclusion of oils obtained using solvents or reesterification processes and of any mixture with oils of other kinds. It is marketed in accordance with the following designations and definitions:

Virgin olive oil is the oil obtained from the fruit of the olive tree solely by mechanical or other physical means under conditions, particularly thermal conditions, that do not lead to alterations in the oil, and which has not undergone any treatment other than washing, decantation, centrifugation, and filtration.

Virgin olive oil fit for consumption as it is includes:

> *Extra virgin olive oil* — Virgin olive oil that has an organoleptic rating of 6.5 or more and a free acidity, expressed as oleic acid, of not more than 1 g/100 g, with due regard for the other criteria laid down in this standard.

> *Virgin olive oil* — (the qualifier "fine" may be used at the production and wholesale stage) Virgin olive oil that has an organoleptic rating of 5.5 or more and a free acidity, expressed as oleic acid, of not more than 2 g/100 g, with due regard for the other criteria laid down in this standard.

> *Ordinary virgin olive oil* — Virgin olive oil that has an organoleptic rating of 3.5 or more and a free acidity, expressed as oleic acid, of not more than 3.3 g/100 g, with due regard for the other criteria laid down in this standard.

Virgin olive oil not fit for consumption as it is, designated *lampante virgin olive oil,* is virgin olive oil that has an organoleptic rating of less than 3.5 and/or a free acidity, expressed as oleic acid, of more than 3.3 g/100 g, with due regard for the other criteria laid down in this standard. It is intended for refining or for technical purposes.

Refined olive oil is the olive oil obtained from virgin olive oils by refining methods which do not lead to alterations in the initial glyceridic structure.

Olive oil is the oil consisting of a blend of refined olive oil and virgin olive oil fit for consumption as it is.

Olive-pomace oil is the oil obtained by treating olive pomace with solvents, to the exclusion of oils obtained by reesterification processes and of any mixture with oils of other kinds. It is marketed in accordance with the following designations and definitions:

Crude olive-pomace oil is olive-pomace oil intended for refining with a view to its use in food for human consumption, or intended for technical purposes.

Refined olive-pomace oil is the oil obtained from crude olive-pomace oil by refining methods which do not lead to alterations in the initial glyceridic structure.

Olive-pomace oil is the oil comprising the blend of refined olive-pomace oil and virgin olive oil fit for consumption as it is. In no case shall this blend be called "olive oil".

* Data and official classifications and definitions supplied by International Olive Oil Council, Madrid, Spain.

TABLE 1
Sterol Composition (% of Total Sterols)

	Olive oils and olive-pomace oils
Brassicasterol	≤ 0.1
Campesterol	≤ 4.0
Stigmasterol	< Campesterol in edible olive oils
Delta-7-stigmastenol	≤ 0.5
Beta-sitosterol + delta-5-avenasterol + delta-5–23-stigmastadienol + clerosterol + sitostanol + delta 5–24-stigmastadienol	≥ 93.0

TABLE 2
Total Sterol Content

	(mg/100 g)
Virgin olive oils, refined olive oil, olive oil	≥ 100
Crude olive-pomace oil	≥ 250
Refined olive-pomace oil	≥ 180
Olive-pomace oil	≥ 160

TABLE 3
Fatty Acid Composition Using Gas-Liquid Chromatography

	% m/m of methyl esters
Myristic acid	≤ 0.05
Palmitic acid	7.5–20.0
Palmitoleic acid	0.3–3.5
Heptadecanoic acid	≤ 0.3
Heptadecenoic acid	≤ 0.3
Stearic acid	0.5–5.0
Oleic acid	55.0–83.0
Linoleic acid	3.5–21.0
Linolenic acid	≤ 0.9
Arachidic acid	≤ 0.6
Gadoleic acid (eicosanoic)	≤ 0.4
Behenic acid	≤ 0.2
Lignoceric acid	≤ 0.2

TABLE 4
Saturated Fatty Acid Content in the 2-Position in the Triglycerides: The Maximum Acceptable Level Is the Sum of the Palmitic and Stearic Acids

Virgin olive oil	≤ 1.5%
Refined olive oil	≤ 1.8%
Olive oil	≤ 1.8%
Crude olive-pomace oil	≤ 2.2%
Refined olive-pomace oil	≤ 2.2%

TABLE 5
Trans Fatty Acid Content

	C18:1 (%)	C18:2 + C18:3 (%)
Edible virgin olive oils	≤ 0.03	≤ 0.03
Lampante virgin olive oil	≤ 0.10	≤ 0.10
Refined olive oil	≤ 0.20	≤ 0.30
Olive oil	≤ 0.20	≤ 0.10
Crude olive-pomace oil	≤ 0.40	≤ 0.35
Olive-pomace oil	≤ 0.40	≤ 0.35

Appendix VI

A NOTE ON PROPOSED TECHNOLOGIES FOR EXTRACTING CHOLESTEROL FROM FOOD

David Oakenfull

INTRODUCTION

Recently there has been intensive research activity worldwide into the development of technologies for extracting or removing the cholesterol from food, particularly egg and dairy products. As a result, numerous patents have been applied for, covering a number of very different approaches.

OUTLINE OF PROPOSED TECHNOLOGIES

CONVERSION OF CHOLESTEROL INTO INNOCUOUS COMPOUNDS

A number of microorganisms, such as *Arthrobacter, Bacillus, Brevibacterium, Corynebacterium, Mycobacterium, Norcardia, Serratia,* and *Streptomyces,* have been found to convert cholesterol into completely harmless compounds (such as carbon dioxide and water).[1] These reactions have been exploited in a number of patented processes for removing cholesterol from dairy products.[2,3] Also, enzymes produced by Eubacteria, normally present in the large bowel, can be isolated and used to convert cholesterol into coprostanol which is not absorbed by the gut.[2]

EXTRACTION WITH SUPERCRITICAL CARBON DIOXIDE

At high pressure (greater than 74 bar) and above 31°C, carbon dioxide becomes a supercritical fluid. This supercritical fluid can be used to extract cholesterol from anhydrous butter fat or dry egg yolk powder.[4,5] Removal of up to 90% of the cholesterol in butterfat has been claimed[6] and this fat can be reemulsified with skim or used for the manufacture of other dairy products. Other recent experiments have shown that supercritical carbon dioxide (at 310 bar pressure and 45°C) can be used to remove 65% of the cholesterol and 33% of the lipid from egg powder without significantly affecting its functional properties (although some yellow pigments are also lost in the extraction process).[5]

Because of the high pressures involved, the process is necessarily expensive and is currently in use only for production of flavor extracts and high value food products such as decaffeinated coffee. Another major disadvantage for cholesterol extraction is that the extraction process can only be carried out on isolated anhydrous fat; it cannot be used directly on milk, cream, or liquid egg.

EXTRACTION BY STEAM STRIPPING

The OmegaSource Corporation (Burnsville, Minneapolis) has patented a process called "steam stripping" which they claim removes 93% of the cholesterol in butter oil.[7] The liquid fat is deaerated under vacuum (1 to 7 mmHg) and mixed with steam to raise its temperature to between 220 and 290°C. The mixture is then fed into a flash chamber where it is flash vaporized, leaving a "par-treated" oil phase. The par-treated oil is subsequently cascaded down an alternating series of jackets and disks heated against a steam countercurrent to vaporize the cholesterol-rich fractions.

The stripped fat can then be recombined with skim and used to manufacture dairy products. As was the case with extraction with supercritical carbon dioxide, this process is limited by

0-8493-4248-1/96/$0.00+$.50
© 1996 by CRC Press Inc.

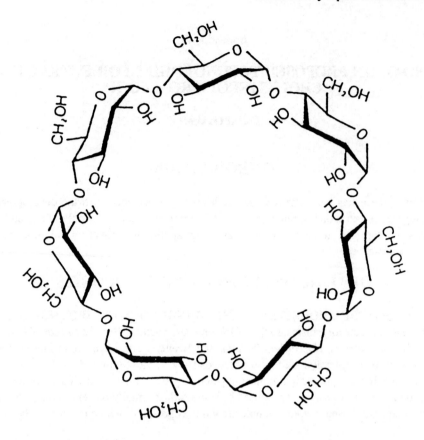

FIGURE 1. Chemical structure of β-cyclodextrin. (From Oakenfull, D. G., Pearce, R. J., and Sidhu, G. S., *Aust. J. Dairy Technol.*, 46, 110, 1991. With permission.)

the fact that it can only be used on the isolated fat and there is also the requirement for elaborate, specialized equipment.

ADSORPTION OF CHOLESTEROL ON ACTIVATED CHARCOAL

In this process, patented by the New Zealand Dairy Institute,[6] anhydrous butter fat is treated with activated charcoal. They claim that 95% of the cholesterol can be removed, but many other components are removed at the same time — including the yellow pigments that have to be replaced with β-carotene.

ADSORPTION BY SAPONINS

The ability of cholesterol to form an insoluble complex with digitonin is well known.[8] Certain food grade saponins apparently also have this ability and can be used to extract cholesterol from anhydrous fat or fat droplets in an oil-in-water emulsion.[6] No details are yet available.

COMPLEXATION WITH β-CYCLODEXTRIN

β-Cyclodextrin is a cyclic oligosaccharide of seven glucose units (Figure 1). The molecule is doughnut shaped, as shown in Figure 2. The central cavity is hydrophobic, giving the molecule its affinity for nonpolar molecules such as cholesterol. The radius of the cavity is such as to accommodate a cholesterol molecule almost exactly, which means that β-cyclodextrin forms a particularly strong inclusion complex with cholesterol. This reaction forms the basis of several patented processes for extracting cholesterol from dairy fat[9-14] and egg.[12-15]

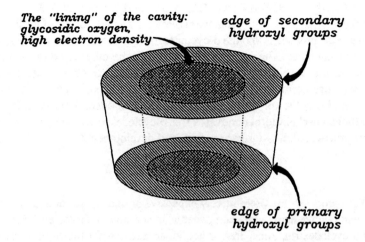

FIGURE 2. Schematic diagram of the β-cyclodextrin molecule showing the hydrophobic cavity. (From Oakenfull, D. G., Pearce, R. J., and Sidhu, G. S., *Aust. J. Dairy Technol.*, 46, 110, 1991. With permission.)

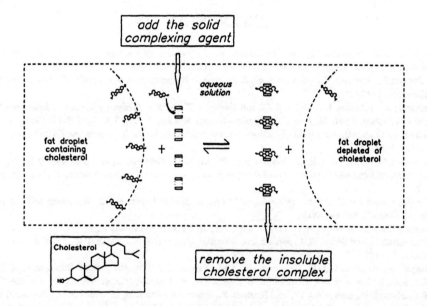

FIGURE 3. Schematic diagram showing the extraction of cholesterol from a fat droplet by formation of an inclusion complex with β-cyclodextrin. (From Oakenfull, D. G., Pearce, R. J., and Sidhu, G. S., *Aust. J. Dairy Technol.*, 46, 110, 1991. With permission.)

The SIDOAK™* Process is a particularly efficient form of this technology which exploits the fact that the fat in milk and egg yolk is dispersed in an oil-in-water emulsion. Because cholesterol is slightly surface-active,[16] the cholesterol molecules in these foods are concentrated at the oil-water interface and are accessible to β-cyclodextrin in the aqueous phase (Figure 3). The reaction takes place in the cold and the cholesterol complex that is precipitated can be easily removed by centrifugation.[12,13] Unlike other processes, the fat does not have to be first isolated and later recombined. Also, because the product remains cold throughout, there is minimal risk of microbiological spoilage or flavor deterioration.

* Registered trademark of CSIRO, Limestone Avenue, Campbell, ACT, Australia.

USE OF β-CYCLODEXTRIN ATTACHED TO A SOLID SUPPORT

When free β-cyclodextrin is used to remove cholesterol, a small residue of the β-cyclodextrin always remains in the product. Although β-cyclodextrin is nontoxic, its use in food is severely restricted in some countries. The problem of residues can be minimized or avoided entirely if the β-cyclodextrin is attached to a solid support, such as silica or polystyrene beads. The chemistry of the attachment must be such that the β-cyclodextrin cavity is accessible to cholesterol molecules at the oil-water interface. Solid-state adsorbents have been reported which are as efficient in cholesterol removal as free β-cyclodextrin on the laboratory scale, but scaling up this process requires further research and development.[17]

CONCLUSIONS

Many of these proposals are prohibitively expensive or totally impractical in that they are non-selective and remove many other components, in addition to cholesterol. Methods based on the use of β-cyclodextrin, either free or bound, appear to offer the highest probability of success and are already being used commercially.[18]

REFERENCES

1. **Aihara, H., Watanabe, K., and Nakamura, R.**, Degradation of cholesterol in hen's egg yolk and its lipoprotein by extracellular enzymes of *Rhodococcus equi* No 23, *Lebensm. Wiss. Technol.*, 21, 342, 1988.
2. **Chosan, P., Deshayes, C., and Franknet, J.**, Procède de Biodegradation des Sterols, FP-A-0278793, Brevet European, 1988.
3. **Behal, S.S., Johnson, J.A., Beitz, D.C., and Young, J.W.**, Alfalfa (*Medicago sativa*): A Source of Cholesterol Reductase, Abstr. 72, American Chemical Society Meeting, Dallas, TX, April 9-14, 1989.
4. **Zadow, J.G.**, Supercritical fluid extraction — a new technology for the food industry, *CSIRO Food Res. Q.*, 48, 25, 1988.
5. **Forning, G.W., Wehling, R.L., Cuppett, S.L., Pierce, M.M., Neiman, L., and Siekman, D.K.**, Extraction of cholesterol and other lipids from dried egg yolk using supercritical carbon dioxide, *J. Food Sci.*, 55, 95, 1990.
6. **Anon.**, Cholesterol Reduction Technologies Overview, Special Report No. 2, Wisconsin Milk Marketing Board Research Review, 1989.
7. **Anon.**, Cholesterol-reduced fats: they're here!, *Prepared Foods*, p. 99, July 1989.
8. **Oakenfull, D. and Sidhu, G.S.**, Saponins, in *Toxicants of Plant Origin*, Cheeke, P.R., Ed., CRC Press, Boca Raton, FL, 1989, 97.
9. **Bayol, A., Maffrand, J.-P., Gonzalez, B., and Frankinet, J.**, Process for the Elimination of Steroid Compounds Contained in a Substance of Biological Origin, Australian Patent Application No. 28449/89.
10. **Roderbourg, H., Dalemans, D., and Bouhon, R.**, Process for Reducing the Content of Cholesterol and of Free Fatty Acids in an Animal Fat Material and Fat Material So Obtained, Australian Patent Application No. 51259/90.
11. **Graille, J., Pioch, D., Serpelloni, M., and Mentink, L.**, Process for Preparing Dairy Products with a Low Content of Sterols by Use of Cyclodextrin, Australian Patent Application No. 58601/90.
12. **Oakenfull, D.G., Sidhu, G.S., and Rooney, M.L.**, Cholesterol Removal, Australian Patent Application No. 54768/90.
13. **Oakenfull, D.G., Sidhu, G.S., and Rooney, M.L.**, Cholesterol Reduction, Australian Patent Application No. 55112/90.
14. **Oakenfull, D.G., Sidhu, G.S., and Rooney, M.L.**, Modified Cyclodextrins, Australian Patent Application No. PK8222/91.
15. **Cully, J., Vollbrecht, H.R., and Weismaller, J.**, Process for the Removal of Cholesterol or Cholesterol Esters from Egg Yolk, Australian Patent Application No. 60836/90.
16. **Gilbert, D.B., Tanford, C., and Reynolds, J.A.**, Cholesterol in aqueous solution: hydrophobicity and self-association, *Biochemistry*, 14, 444, 1975.
17. **Sidhu, G.S. and Oakenfull, D.G.**, Cyclodextrin-Cholesterol Complexation and Technology for Removing Cholesterol from Egg and Dairy Products, Abstr. L47, 6th Int. Cyclodextrin Symp., Chicago, IL, April 21-24, 1992.
18. **Anon.**, Michael Foods introduces whole eggs with 80% less cholesterol, *Milling and Baking News*, p. 40, March 10, 1992.

Index

INDEX

Lightning Source UK Ltd.
Milton Keynes UK
UKOW01n1825280314

229036UK00001B/3/P